Hydrolases

Edited by Sajjad Haider, Adnan Haider and Angel Catalá

Published in London, United Kingdom

Hydrolases
http://dx.doi.org/10.5772/intechopen.95149
Edited by Sajjad Haider, Adnan Haider and Angel Catalá

Contributors
Toshiyuki Habu, Jiyeong Kim, Julius T. Tieroyaare Dongdem, Cletus Adiyaga Wezena, Michael Anthony Massiah, Erin Meghan Gladu, Iman Sayed, Katsumi Kitagawa, Yohei Niikura, Jose Edgar Edgar Zapata Montoya, Angie Franco Sanchez, Yaremis Beatriz Meriño-Cabrera, Maria Goreti de Almeida Oliveira, Mesut Işık, Nagendra Kumar Chandrawanshi, Deepali Koreti, Anjali Kosre, Ashish Kumar, Ekta Shukla, Ameya D. Bendre, Sushama M. Gaikwad

First published in London, United Kingdom, 2022 by IntechOpen
IntechOpen is the global imprint of INTECHOPEN LIMITED, registered in England and Wales, registration number: 11086078, 5 Princes Gate Court, London, SW7 2QJ, United Kingdom
Printed in Croatia

British Library Cataloguing-in-Publication Data
A catalogue record for this book is available from the British Library

Additional hard and PDF copies can be obtained from orders@intechopen.com

Hydrolases
Edited by Sajjad Haider, Adnan Haider and Angel Catalá
p. cm.

This title is part of the Biochemistry Book Series, Volume 29
Topic: Chemical Biology
Series Editor: Miroslav Blumenberg
Topic Editors: Şükrü Beydemir and Deniz Ekinci

Print ISBN 978-1-80355-162-3
Online ISBN 978-1-80355-163-0
eBook (PDF) ISBN 978-1-80355-164-7
ISSN 2632-0983

For EU product safety concerns:
IN TECH d.o.o., Prolaz Marije Krucifikse Kozulić 3, 51000 Rijeka, Croatia,
info@intechopen.com or visit our website at intechopen.com.

IntechOpen Book Series
Biochemistry
Volume 29

Aims and Scope of the Series

Biochemistry, the study of chemical transformations occurring within living organisms, impacts all of the life sciences, from molecular crystallography and genetics, to ecology, medicine and population biology. Biochemistry studies macromolecules - proteins, nucleic acids, carbohydrates and lipids –their building blocks, structures, functions and interactions. Much of biochemistry is devoted to enzymes, proteins that catalyze chemical reactions, enzyme structures, mechanisms of action and their roles within cells. Biochemistry also studies small signaling molecules, coenzymes, inhibitors, vitamins and hormones, which play roles in the life process. Biochemical experimentation, besides coopting the methods of classical chemistry, e.g., chromatography, adopted new techniques, e.g., X-ray diffraction, electron microscopy, NMR, radioisotopes, and developed sophisticated microbial genetic tools, e.g., auxotroph mutants and their revertants, fermentation, etc. More recently, biochemistry embraced the 'big data' omics systems. Initial biochemical studies have been exclusively analytic: dissecting, purifying and examining individual components of a biological system; in exemplary words of Efraim Racker, (1913 –1991) "Don't waste clean thinking on dirty enzymes." Today, however, biochemistry is becoming more agglomerative and comprehensive, setting out to integrate and describe fully a particular biological system. The 'big data' metabolomics can define the complement of small molecules, e.g., in a soil or biofilm sample; proteomics can distinguish all the proteins comprising e.g., serum; metagenomics can identify all the genes in a complex environment e.g., the bovine rumen.
This Biochemistry Series will address both the current research on biomolecules, and the emerging trends with great promise.

Meet the Series Editor

Miroslav Blumenberg, Ph.D., was born in Subotica and received his BSc in Belgrade, Yugoslavia. He completed his Ph.D. at MIT in Organic Chemistry; he followed up his Ph.D. with two postdoctoral study periods at Stanford University. Since 1983, he has been a faculty member of the RO Perelman Department of Dermatology, NYU School of Medicine, where he is codirector of a training grant in cutaneous biology. Dr. Blumenberg's research is focused on the epidermis, expression of keratin genes, transcription profiling, keratinocyte differentiation, inflammatory diseases and cancers, and most recently the effects of the microbiome on the skin. He has published more than 100 peer-reviewed research articles and graduated numerous Ph.D. and postdoctoral students.

Meet the Volume Editors

Dr. Sajjad Haider is an Associate Professor in the Department of Chemical Engineering, King Saud University, Riyadh, Saudi Arabia. He received his MSc in 1999 and M Phil in 2004 from the Institute of Chemical Sciences, University of Peshawar, KPK, Pakistan, and his Ph.D. degree in 2009 from the Department of Polymer Science and Engineering, Kyungpook National University, Taegu, South Korea. His research focuses on electrospun nanofibers, biopolymer composite, metal oxide/polymer composites, and polymer hydrogels to develop scaffolds for tissue engineering, drug delivery, and water treatment applications.

Dr. Adnan Haider is an Assistant Professor in the Department of Biological Sciences, National University of Medical Sciences (NUMS), Pakistan. He holds an MSc from Kohat University of Science and Technology, Pakistan, and an MS leading to a Ph.D. from Kyungpook National University, Taegu, South Korea. Dr. Haider has a post-doctorate from Yeungnam University, South Korea. His research focuses on electrospun nanofibers, biopolymer composites, and polymer hydrogels to develop scaffolds for tissue regeneration, drug delivery, and water treatment applications.

Angel Catalá (National University of La Plata) is a member of the National Research Council (CONICET), Argentina, and the Argentine Society for Biochemistry and Molecular Biology (SAIB). His laboratory has been focused on lipid peroxidation of biological membranes from various tissues and different species. Dr. Catalá has directed twelve doctoral theses, published more than 100 papers in peer-reviewed journals, written several chapters in books, and edited twelve books. He received numerous awards for his work, and he belongs to the editorial board of several journals.

Contents

Preface

This book provides an up-to-date overview of the relationships between the structure and function of hydrolases and ligases and strategies for better utilization of these important enzymes. The book is divided into two sections: "Cleavage" and "Ligation". Hydrolases are the largest and most diverse family of enzymes and provide an opportunity to study the structural diversity underlying their various biological functions. In light of increasing scientific developments, there is a desire to reevaluate and update our understanding of the functional and structural modifications of these enzymes.

Hydrolases are well-known proteolytic enzymes belonging to the class of complex enzymes known as proteases. Microbial proteolytic enzymes are preferred because they have a fast generation time, genetic manipulation of microbes is easy, and there are many different species in nature. Macrofungi, such as fungi, are extremely important and play an important role in the degradation of lignocellulosic materials. They effectively degrade cellulose and produce extracellular enzymes such as xylanases, cellulases, and ligninolytic enzymes.

Serine proteases are the other major class of protein-digesting enzymes found in the midgut of numerous lepidopteran species and are the subject of this study, with trypsin and chymotrypsin being the best studied enzymes. They are involved in a variety of physiological processes involving not only digestion but also activation of specific proteins (e.g., in coagulation cascades) in the insect and plant immune systems, in the development and production of biologically active peptides, in signal transduction, hormone activation, and development.

ChE enzymes damage the cholinergic system by hydrolyzing the neurotransmitter acetylcholine (ACh). ChE inhibitors, which play an important role in the cholinergic system, are used in the treatment of Alzheimer's disease because they maintain ACh levels in brain regions and prevent Aβ accumulation by inhibiting ChE. In this context, it is crucial to develop a large number of synthetic and naturally occurring ChE inhibitors for the treatment of cholinergic system disorders and diseases with neuropsychiatric symptoms.

Ubiquitination by ligases is critical for protein function and regulation. E3 ubiquitin ligases, of which there are > 600 putative in humans, form a family of highly heterogeneous proteins and protein complexes. E3 ubiquitin ligases play a critical role in subcellular signaling cascades in eukaryotes. Dysfunctional E3 ubiquitin ligases therefore often have dramatic effects on human health and can lead to the development of various diseases such as Parkinson's disease, amyotrophic lateral sclerosis, Alzheimer's disease, cancer and so on.

CENP-A is a centromere-specific histone H3 variation that is essential for normal chromosome segregation. Its function is highly conserved in a variety of organisms, including the yeast *Saccharomyces cerevisiae*. The ability of E3 ligase to ubiquitinate/ sumoylate and regulate CENP-A protein has recently emerged as an essential regulatory paradigm in several species. Protein ubiquitination is a post-translational modification that regulates protein concentration, function, and localization to govern key biological processes. RING E3 ligases are an important part of a three-enzyme cascade that enables protein ubiquitination. RING-type E3 ligases are a form of E3 ligase that binds the substrate protein and ubiquitin-conjugating enzymes (E2s).

This book is useful for students and teachers alike for its thoughtful account of core concepts, as well as a source of interpretations and references to additional knowledge sources.

Sajjad Haider
Department of Chemical Engineering,
College of Engineering,
King Saud University,
Riyadh, Saud Arabia

Adnan Haider
Department of Biological Sciences,
National University of Medical Sciences,
Rawalpindi, Punjab, Pakistan

Angel Catalá
National University of La Plata,
La Plata, Argentina

Section 1

Cleavage

Chapter 1

Hydrolases: The Most Diverse Class of Enzymes

Ekta Shukla, Ameya D. Bendre
and Sushama M. Gaikwad

Abstract

Being the largest and most diverse class of enzymes, hydrolases offer an opportunity to explore the conformational diversity which forms the basis of their differential biological functions. In recent times, there is an urge to re-evaluate and update our existing knowledge on functional and conformational transitions of these enzymes, in the context of emerging scientific trends. In this chapter, we discuss hydrolases in terms of their diversity, classification, and different nomenclature styles that exist. Further, the concepts of protein stability and significance of studying the structure–function relationship of hydrolases are mentioned in detail taking serine protease as an example. The chapter talks about multiple ways by which an enzyme's structure and function can be explored. The available information and literature survey on hydrolases have been systematically summarized for an easy understanding. Various experimental methods and techniques involving artificial intelligence are introduced in the later sections. The knowledge obtained by these strategies contributes to our current knowledge of the interplay between the stability, structure, and function of these enzymes. This, in turn, can help in designing and engineering these proteins with improved functional and structural features toward the goal of increasing their applicability in biotechnology.

Keywords: hydrolysis, catalysis, nomenclature, structure–function relationship, protein stability

1. Introduction

Hydrolase is a class of hydrolytic enzymes that are commonly used as biochemical catalysts which utilize water as a hydroxyl group donor during the substrate breakdown. In simple words, a hydrolase is an enzyme that catalyzes the hydrolysis of a chemical bond in biomolecules. This, in turn, divides a large molecule into two smaller ones. Hydrolases are hence important for the environment since they digest large molecules into small fragments for the synthesis of biopolymers as well as for the degradation of toxins. In biochemistry,

IntechOpen

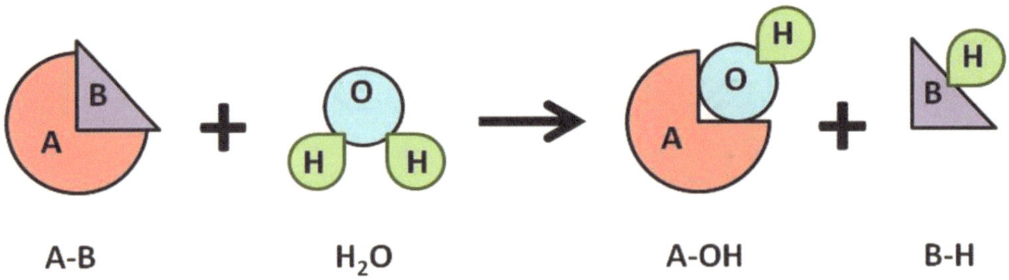

Hydrolases is the largest and most diverse class of enzymes with more than 200 enzymes that catalyze the hydrolysis of several types of compounds. They catalyze the hydrolytic cleavage of carbon–oxygen (C–O), carbon–nitrogen (C–N), carbon–carbon (C–C), phosphorus–nitrogen (P–N) bonds, etc. Systematic names of hydrolases are formed as "substrate hydrolase." However, common names are typically in the form '*substrate*ase', such as nuclease refers to an enzyme that hydrolyses nucleic acids. Examples of some common hydrolases include esterases, proteases, glycosidases, and lipases.

2. Applications/significance of hydrolases

Enzymes of this class carry out important degradative reactions in the body. Hydrolases cleave large molecules into smaller fragments used for synthesis, excretion of waste materials, or as sources of carbon for the production of energy. These are involved in digestion, transport, excretion, regulation and signalling processes, etc.; for example, digestive enzymes like cholinesterase, carboxylesterase, lysosomal hydrolases, etc. To be specific, hydrolase expressed by *Lactobacillus spp.* in the human gut could stimulate the liver to secrete bile salts which facilitate the digestion of food [1].

Hydrolytic enzymes are not only physiologically important, playing role in various cellular processes, but also have myriad commercial applications too. The industrial importance of hydrolases exceeds that of other classes of enzymes holding the highest share of enzymes used for industrial purposes. Almost 75% of all industrial enzymes are hydrolytic enzymes. Carbohydrases, proteases, and lipases dominate the enzyme market, accounting for more than 70% of all enzyme sales. Many industrial sectors, such as the detergent, leather, textiles, pulp and paper, foods and feeds, dairy, biofuels, and waste treatment industries, depend on hydrolases. Proteases remain the dominant enzyme type, because of their extensive use in the detergent and dairy industries. Various carbohydrases (glycosidases), primarily amylases and cellulases, used in industries, such as the starch, textile, detergent, and baking industries, represent the second largest group [2–4].

3. Classification of hydrolases

Apart from the common names given to certain hydrolases, there exist systematic nomenclature systems to name these enzymes.

3.1 Based on enzyme commission (EC) numbers

Hydrolases belong to enzyme class 3 (EC 3) and are further categorized based on the type of bond they cleave [5]. The four-digit code includes the nature of the *bond* hydrolyzed, then the nature of the *substrate*, and lastly the *enzyme* (**Table 1**).

3.2 Based on active site

The active site geometry of different hydrolases is different, in spite of the same catalytic method, i.e., hydrolysis. Thus, a **H**ierarchical classification of hydrolases **C**atalytic **S**ites (**HCS**) has been proposed which is based on the amino acids involved in catalysis [6, 7]. The relation between a class and its subclass in the hierarchy is that the catalytic site of the subclass refines the catalytic site of the base class. Serine hydrolase like esterase, with unusual catalytic dyad Ser-His, belongs to class **S.01** (serine hydrolases with Ser-His dyad) while hydrolases, such as trypsin or subtilisin, are further categorized into subclass **S.01.01** (hydrolases with Ser-His-Asp/Glu triad), i.e. subclass contains all residues of the basic class and some additional ones. Currently, only hydrolases are included in such a classification since they are the most studied and most abundant enzymes (**Table 2**).

Let us understand these nomenclatures by taking an example of a single hydrolase, say a serine protease.

Subclass (hydrolase acting upon)	Sub-subclass example	Enzyme example
3.1 Ester bonds (esterases)	3.1.1 Lipases	3.1.1.3 Triacylglycerol lipase
3.2 Sugars	3.2.1 Glycosidase	3.2.1.1 α-amylase
3.3 Ether bonds	3.3.2 Ether hydrolase	3.3.2.6 Leukotriene-A4 hydrolase
3.4 Peptide bonds (peptidases)	3.4.21 Serine endopeptidase	3.4.21.1 Chymotrypsin
3.5 C-N bonds (other than peptide bonds)	3.5.1 In linear amides	3.5.1.1 Asparaginase
3.6 Acid anhydrides	3.6.1 In P-containing anhydrides	3.6.1.1 Inorganic diphosphatase
3.7 C-C bonds	3.7.1 In ketonic substances	3.7.1.1 Oxaloacetase
3.8 Halide bonds	3.8.1 In C-X compounds	3.8.1.1 Alkylhalidase
3.9 P-N bonds	3.9.1 On P-N bonds	3.9.1.1 Phosphoamidase
3.10 S-N bonds	3.10.1 On S-N bonds	3.10.1.1 N-Sulfoglucosamine sulfohydrolase
3.11 C-P bonds	3.11.1 On C-P bonds	3.11.1.1 Phosphonoacetaldehyde hydrolase
3.12 S-S bonds	3.12.1 On S-S bonds	3.12.1.1 Trithionate hydrolase
3.13 C-S bonds	3.13.1 On C-S bonds	3.13.1.1 UDP-Sulfoquinovose synthase

Adapted from ExplorEnz database: http://www.enzyme-database.org/downloads/ec3.pdf

Table 1.
Classification of hydrolase based on EC numbers.

Sr. No.	Base class	Subclass
1	**(A)** Carboxyl (aspartyl and glutamyl) hydrolases	**(A.01)** Pepsin-like proteases **(A.02)** Glycosidases **(A.03)** Hydrolases with covalent aspartyl-substrate intermediate **(A.04)** Epoxide hydrolase-like
2	**(C)** Cysteine hydrolases	**(C.01)** Cys hydrolases with Cys-His dyad **(C.02)** N-terminal cysteine hydrolases **(C.03)** Cys hydrolases with His as a proton donor **(C.04)** Tyrosine phosphatase-like **(C.05)** Cys hydrolases with carboxyl group as the proton acceptor
3	**(H)** Histidine hydrolase	**(H.01)** Ribonuclease-like **(H.02)** Hydrolases with covalent His-substrate intermediate **(H.03)** Hydrolases of carbon–carbon bond
4	**(M)** Metal-dependent hydrolases	**(M.01)** Zinc-dependent hydrolases **(M.02)** Magnesium-dependent hydrolases **(M.03)** Calcium-dependent hydrolases **(M.04)** Bimetallic (Zn and Mg-dependent) hydrolases **(M.05)** Iron-dependent hydrolases **(M.06)** Manganese-dependent hydrolases **(M.X)** Hydrolases without specific metal ion requirements
5	**(P)** N-terminal proline hydrolases	
6	**(S)** Serine hydrolases	**(S.01)** Ser hydrolases with Ser-His dyad **(S.02)** Ser hydrolases with the amino group as the proton acceptor **(S.03)** Ser hydrolases with carboxyl group as the proton acceptor
7	**(T)** Threonine hydrolases	**(T.01)** Asparaginase-like **(T.02)** N-terminal threonine hydrolases
8	**(U)** Unclassified hydrolases	**(U.01)** Proteins without hydrolase activity **(U.02)** Hydrolases without known catalytic domain structure
9	**(Y)** Tyrosine hydrolases	**(Y.01)** Sialidases
10	**(Z)** Substrate-assisted or cofactor-dependent hydrolases	**(Z.01)** Phosphatases with substrate's phosphate as the catalytic base **(Z.02)** NAD(+)-dependent deacetylases **(Z.03)** Hydrolases with oxidation/reduction steps

Adapted from http://www.enzyme.chem.msu.ru/hcs/classes.html

Table 2.
Classification of hydrolase based on active site residues.

4. Classification of serine protease

As per the earlier discussed classification systems for hydrolases, the categorization of serine proteases can be viewed in the schematic, as shown in **Figure 1**.

4.1 Based on site of cleavage

Proteases are further subdivided into exopeptidases and endopeptidases depending on the site of enzyme action. Exopeptidases catalyze the hydrolysis of the peptide bonds near the N- or C-terminal ends of the substrate and can be classified into aminopeptidases and carboxypeptidases. Endopeptidases cleave peptide bonds within and distant from the ends of a polypeptide chain [8, 9]. Serine proteases are also divided into endo- and exo- serine peptidases.

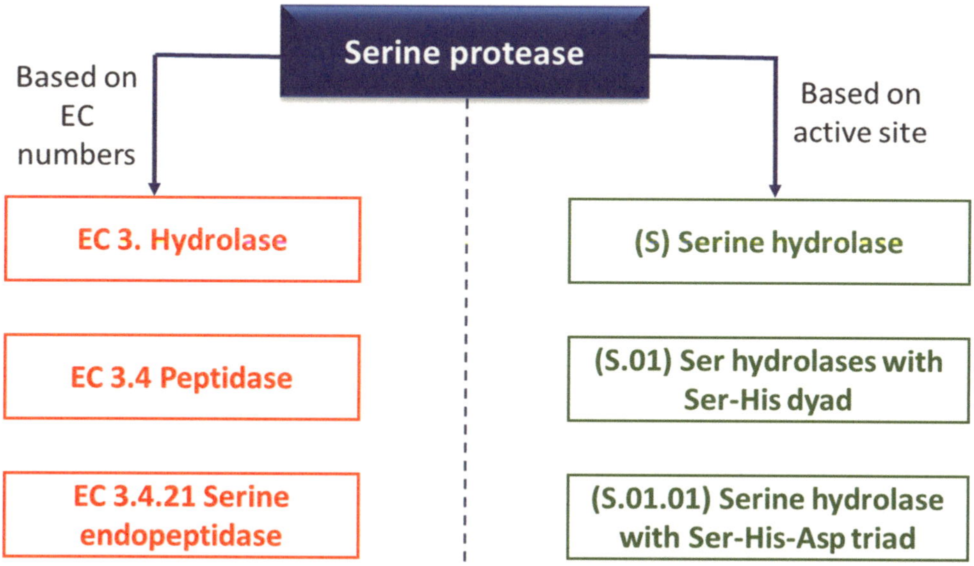

Figure 1.
Classification scheme of serine protease based on EC numbers and active site residues.

4.2 MEROPS classification system

According to MEROPS database version 9.9 (https://merops.sanger.ac.uk) over 183,000 serine proteases are known with >250 structure depositions in PDB (Protein Data Bank).

Clan	Families	Representative members	Fold	Catalytic residues	PDB
PA	12	Trypsin	Greek-key β-barrels	His, Asp, Ser	1DPO
PB	1	Protease from *Thermoplasma acidophilum*	α/β/β/α	His, Glu, Ser	1PMA
PC	1	Aspartyl dipeptidase	α/β/α	Ser, His	1FYE
SB	2	Subtilisin, sedolisin	3-layer sandwich	Asp, His, Ser	1SCN
SC	2	Prolyl oligopeptidase	α/β hydrolase	Ser, Asp, His	1QFS
SE	6	D-Ala–D-Ala carboxypeptidase	α-helical bundle	Ser, Lys	3PTE
SF	3	LexA peptidase	all β	Ser, Lys/His	1JHH
SH	2	Cytomegalovirus assemblin	α/β Barrel	His, Ser, His	1LAY
SJ	1	Lon peptidase	α + β	Ser, Lys	1RR9
SK	2	Clp peptidase	αβ	Ser, His, Asp	1TYF
SP	3	Nucleoporin	all β	His, Ser	1KO6
SQ	1	Aminopeptidase DmpA	4-layer sandwich	Ser	1B65
SR	1	Lactoferrin	3-layer sandwich	Lys, Ser	1LCT
SS	14	L,D-Carboxypeptidase	β-sheet+ β-barrel	Lys, Ser	1ZRS
ST	5	Rhomboid	α-barrel	His, Ser	2IC8

Adapted from reference Rawlings et al. [10].

Table 3.
Known diversity of serine peptidase structure and catalytic mechanism.

This classification system divides peptidases into clans based on catalytic mechanisms and families on the basis of common ancestry. The serine peptidases have been classified into 15 clans comprising numerous families. A summary of catalytic units in all serine peptidase families and their characteristic folds is provided in **Table 3**.

4.2.1 PA clan of serine peptidases

The PA clan (**P**roteases of the mixed nucleophile, superfamily **A**) of endopeptidases is the most abundant, and over two-thirds of this clan are comprised of the S1 family of serine proteases, which bear the archetypal trypsin fold and have a catalytic triad in the order histidine, aspartate, serine. Members have a trypsin/chymotrypsin-like fold

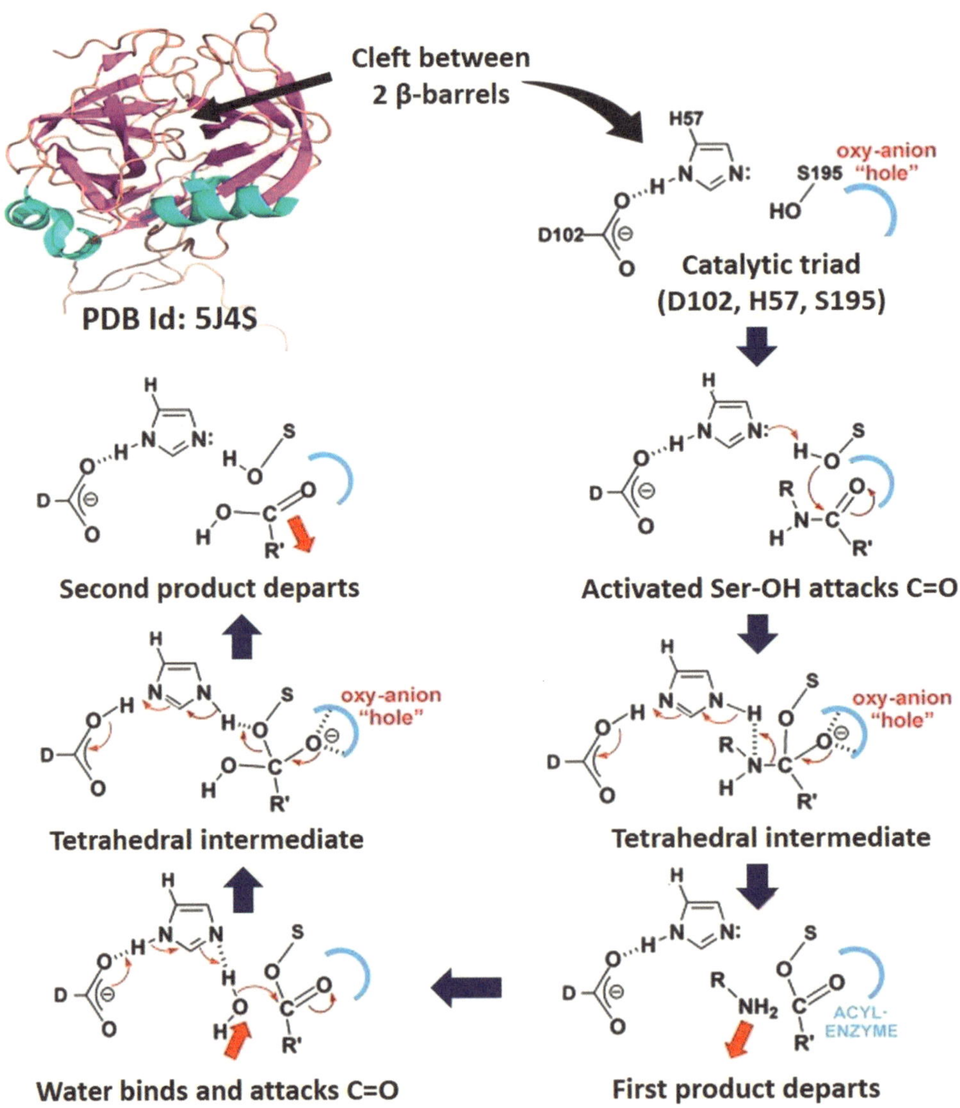

Figure 2.
Schematic illustration of the general catalytic mechanism for serine proteases with chymotrypsin as an example.

and similar proteolysis mechanisms but sequence identity of <10%. PA clan proteases share a core motif of two β-barrels arranged perpendicularly with the covalent catalysis occurring at the interface of both barrels. **Figure 2** shows the crystal structure of bovine chymotrypsin, deposited in PDB (PDB Id: 5J4S). The core structure and active site geometry of these proteases are of interest for many applications [10, 11].

5. Catalytic mechanism of hydrolases

To understand the mechanism in a simple way, let us again take the example of a serine protease. Serine proteases are widely distributed in nature and found in all kingdoms of cellular life as well as many viral genomes. Over one-third of all known proteolytic enzymes are serine peptidases [9]. All of the serine proteases contain three residues at their active site—a serine, a histidine, and an aspartate, comprising the characteristic 'catalytic triad'. Some serine proteases are synthesized as larger, inactive, precursors. As an example, chymotrypsinogen is converted to chymotrypsin by the excision of two dipeptides, 14–15 and 147–148 [12]. Interestingly, the structures of chymotrypsinogen and chymotrypsin are almost superimposable, i.e., the conformational change involved in the conversion process appears to be fairly small. The implication is that even relatively small structural changes can result in dramatic changes in activity. The serine proteases also differ in their sequence and substrate specificity. For instance, the bacterial protease subtilisin will cleave essentially any substrate, while other enzymes, Factor Xa (involved in blood clotting) requires a specific residue recognition sequence, Ile-Glu-Gly-Arg, to uniquely hydrolyze its polypeptide substrate after the arginine. Similarly, trypsin is specific for cleavage after Lys and Arg residues.

Almost all clan PA peptidases utilize the canonical catalytic triad of Ser195, Asp-102, and His-57 (chymotrypsin numbering). Catalysis proceeds through the formation of an H-bond between Asp-102 and His-57, which facilitates the abstraction of the proton from Ser195 and generates a potent nucleophile [10]. The catalytic triad is stabilized through a network of additional H-bonds formed by conserved amino acid residues surrounding the triad, which are Thr54, Ala56, and Ser214. The reaction pathway involves two tetrahedral intermediates. Initially, the hydroxyl O atom of Ser195 attacks the carbonyl of the peptide substrate as a result of His57 in the catalytic triad acting as a base. The backbone N atoms of Gly193 and Ser195 stabilize the tetrahedral intermediate and generate a positively charged pocket within the active site known as the oxyanion hole. The tetrahedral intermediate collapse results in the formation of an acyl-enzyme intermediate. In the second half of the mechanism, a water molecule displaces the free polypeptide fragment and attacks the acyl-enzyme intermediate. Again, the oxyanion hole stabilizes the second tetrahedral intermediate of the pathway and the collapse of this intermediate liberates a new C terminus in the substrate.

In the next section, we discuss how an enzyme's or protein's structure is maintained. There are various kinds of inter- and intra-molecular forces that are involved in preserving the native functional structure of a protein.

6. Forces involved in enzyme's (protein's) stability

Protein stability is predominantly dictated by forces that help in maintaining the native structure of a protein which include covalent interactions, such as disulfide bonds, and weak (non-covalent) interactions, such as hydrogen bonds, hydrophobic and ionic

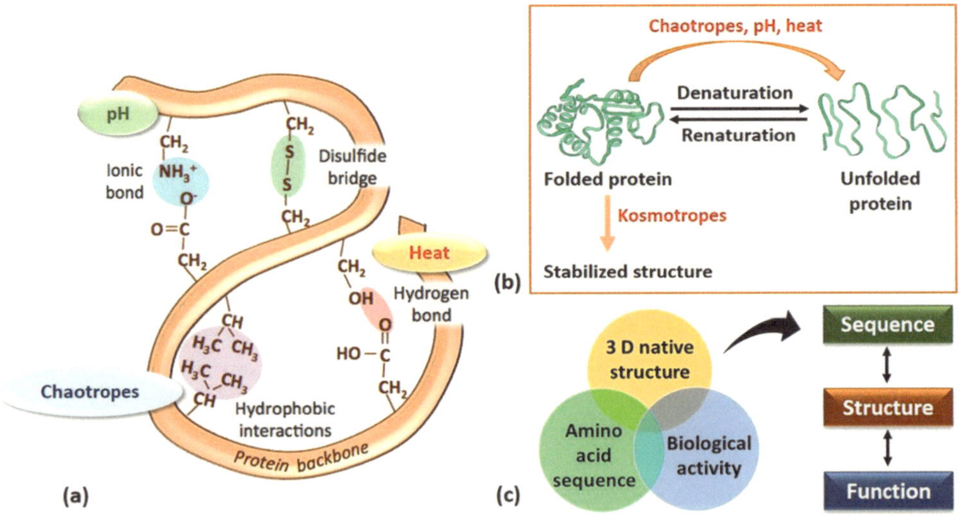

Figure 3.
(a) Molecular interactions which stabilize the protein structure. (b) Effect of various physicochemical conditions on protein structure. (c) Sequence-structure–function triad important for proteins.

interactions (**Figure 3a**). For instance, the primary structure is associated with the covalent bonds (peptide bonds) between the amino acid residues, making up the protein backbone. The secondary structures involve primarily hydrogen bonding between the atoms, thereby creating stable local conformations and structures. Sometimes, it also involves disulfide linkages between two cysteine residues of the same or different chains in a protein [13].

The ultimate three-dimensionally folded tertiary structure of a whole globular protein is formed and maintained by various weak ionic and hydrophobic interactions. Hydrophobic interactions play an important role in stabilizing a protein conformation where the interior of a protein generally consists of a densely packed core of hydrophobic amino acid side chains. Though covalent bonds (such as disulfide bonds) are much stronger than individual weak interactions, i.e., approximately 200–460 kJ/mol, are required to break a single covalent bond, whereas weak interactions can be disrupted by a mere 4–30 kJ/mol. Yet, due to their sheer number, the weak interactions predominate as the stabilizing force in protein structure [14]. The protein folding code is, thus, written in the side chains and not in the backbone hydrogen bonding, because it is through the side chains that one protein differs from another. The number of protein conformational diseases that are now recognized is an indication of the importance of proteins achieving and maintaining their correct fold.

7. Structure–function relationship of proteins

For a protein to be functional, it needs to fold into a specific three-dimensional native structure. The sequence of amino acids determines the structure of a protein which ultimately governs its function. Protein folding is an extremely active field of research, which requires converging expertise from biology, chemistry, computer science, and physics. Understanding the relationship between protein structure and function is a complicated puzzle and remains a primary focus in structural biology. Protein molecules display a remarkable relationship between their amino acid sequence, their three-dimensional structure, and their function at the molecular level (**Figure 3c**) [15]. A polypeptide can also adopt a less rigid or more flexible conformation, different from its functional native form, responding to changes in the environment. To understand this structure–function paradigm, one of the widely used approaches is to subject the native and active conformation of a protein to various physicochemical stress conditions and monitor the changes occurring in its conformation and function at each step.

7.1 Studying folding/unfolding transitions of proteins

So, we understood that any disturbance in the delicate balance of these interactions could lead to loss of the native structure of the protein. Therefore, understanding the protein folding and the unfolding mechanism is equally essential as learning its function. A protein exists in equilibrium with unfolded conformational states in solution with the folded ensemble being favored at ambient conditions. This equilibrium between the folded and the unfolded states can be perturbed by changing the thermodynamic state of the system (temperature, pressure, and pH) or by changing the composition by the addition of co-solvents to the solution [16]. Interestingly, the effect of co-solvents on the protein can alter this equilibrium in any direction. For example, urea and guanidium hydrochloride (GdnHCl) induce disorder and favor the unfolded state of proteins, and are, therefore, known as denaturants/chaotropes. On the other hand, protective osmolytes/kosmotropes, such as trimethylamine N-oxide (TMAO), dimethyl sulphoxide (DMSO), glycine, betaine, glycerol, and sugars, induce stabilization of the folded proteins (**Figure 3b**). Studies on solvent-mediated structural and conformational transitions of proteins can provide insight into their stability, folding pathways, and intermolecular aggregation behavior [16, 17].

Native proteins are generally marginally stable, i.e. free energy gap separating the folded and unfolded states in typical proteins under physiological conditions is quite small (20 to 65 kJ/mol). Therefore, when the delicate balance between the interactions involved in stabilizing or destabilizing a particular structure is disturbed by harsh environments, such as extreme temperature, pH, and chaotropes; it may lead to structural and functional alterations in protein [14]. A loss of the three-dimensional structure of a protein, sufficient to cause loss of function is called denaturation. The denatured state does not always equate with the complete unfolding of the protein (**Figure 3b**). Denaturation can be either partial or complete and it can also be reversible or irreversible. Under most conditions, denatured proteins exist in a set of partially folded states that are poorly understood [18]. In some cases, the structure of an enzyme remains stable, but the labile active site tends to lose its geometry and hence the activity. Contrary to this, the active site may get unusually stabilized and highly active. A polypeptide can also adopt a less rigid or more flexible conformation different from its functional native form, responding to changes in the environment [17, 19, 20].

Exploring structure–function relationships of proteins/enzymes can help in establishing the factors responsible for their stability. Furthermore, knowledge of the overall stability of protein molecules is important, especially when the protein in question is useful in industrial-scale biotechnology, where they may be subject to conditions, such as high temperature, low pH, and presence of co-solvents [21]. The optimization of biological stability is also an important criterion while considering the application of biomolecules (such as proteins/enzymes) as therapeutic agents [22]. Interestingly, novel proteins are now designed as variants of existing proteins or from non-natural amino acids or *de novo*. Moreover, new polymeric materials called **foldamers** are finding applications in biomedicine as antimicrobials, lung surfactant replacements, etc. (reviewed in [23]).

7.2 Tools for probing protein structure and conformational transitions/dynamics

The protein folding problem fascinates the scientist, the educated layman, and the entrepreneur. The full understanding of a molecular system comes from careful examination of the sequence-structure–function triad. Over the last 30 years, detailed experimental and theoretical studies of a number of proteins have advanced our understanding of protein folding and dynamics.

Technique	**Structural parameter probed**
Fluorescence	
Intrinsic	Environment of Trp and Tyr
ANS binding	Exposure of hydrophobic surface area
Substrate binding	Formation of the active site
FRET	Inter-residue distances
Anisotropy	Depolarization of the fluorescence emission
Fluorescence Correlation Spectroscopy	Autocorrelation analysis of fluctuations in fluorescence emission due to internal dynamics
2-D fluorescence lifetime correlation spectroscopy	Correlation of the fluorescence photon pairs with respect to the excitation–emission delay times
Single-molecule spectroscopy (Sm-FRET and sm-PET)	Distance between fluorophores dynamics
Red Edge Excitation Shift	Rate of solvent relaxation around an excited state fluorophore in a protein
Circular dichroism	
Far UV	Secondary-structure information
Near UV	Tertiary-structure information
Protein engineering	Role of individual residues in stabilizing intermediates and transition states
Small-angle X-ray scattering (SAXS)	Dimension and shape of a polypeptide
Absorbance (near UV)	Environment of aromatic residues or co-factors
FTIR	Secondary-structure information
NMR	
Real time	Environment of individual residues
Dynamic NMR	Lineshape analysis provides folding–unfolding rates close to equilibrium
Native state HX	Global stability and metastable states
Pulsed HX ESI MS	Folding populations
Force spectroscopy (AFM/optical tweezers)	Unfolding forces and unfolding-rate constants of single molecules
Differential scanning calorimetry (DSC)	Energetics
Differential scanning fluorimetry (DSF)	Environment of fluorescent dye and intrinsic fluorescence
Differential static light scattering (DSLS)	Temperature of aggregation of a protein

Table modified from references [24, 25].

Table 4.
Experimental techniques used to measure folding/unfolding and dynamics.

7.2.1 Experimental approach

The experimental techniques for studying protein structural transitions monitor the gradual folding/unfolding of proteins and observe conformational changes under various conditions. **Table 4** summarizes a few of the standard biophysical techniques based on fluorescence, absorbance and circular dichroism, etc. which are often used to probe such transitions in protein structure.

Figure 4 shows conformational and functional transitions in a serine protease isolated from *Conidiobolus brefeldianus* upon thermal denaturation. The changes in the protein structure are quite clear upon increasing temperature which is in corroboration with the loss in activity of the enzyme. These conformational changes were probed with fluorescence spectroscopy and Far-UV circular dichroism spectroscopy while functional activity assays were carried out using casein proteolysis. The enzyme apparently begins to lose its structure and function above 55°C. Thus, it becomes clear that a change in the native conformation of a protein affects its function.

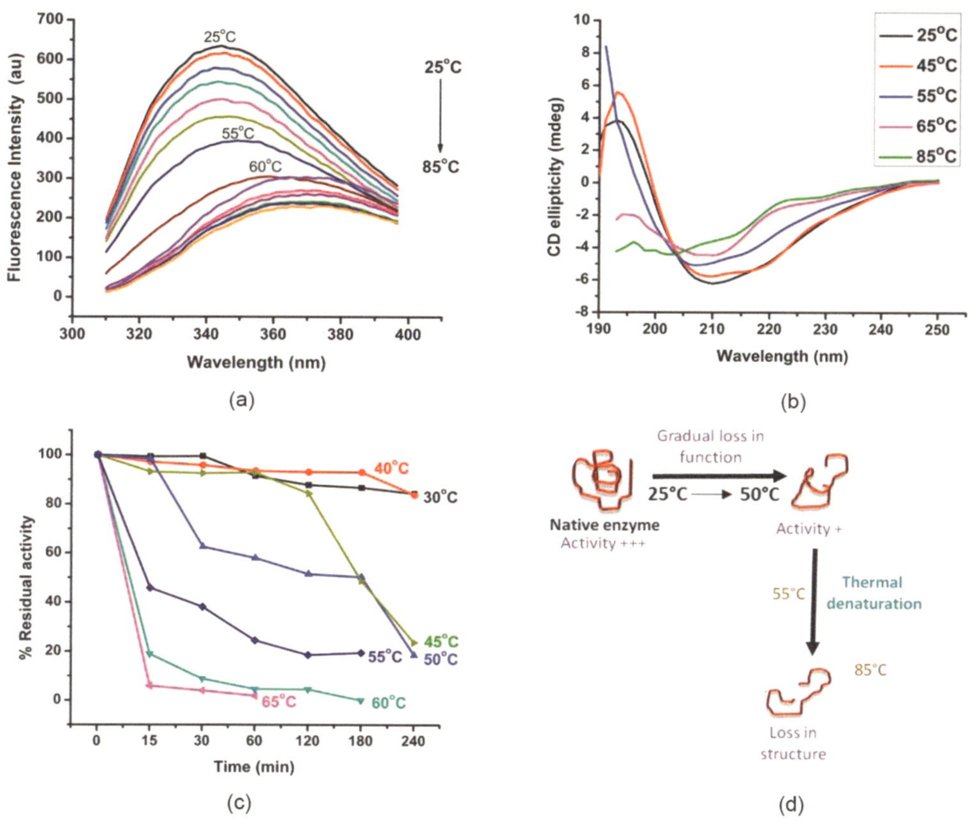

Figure 4.
Thermal denaturation of serine protease from Conidiobolus brefeldianus: (a) fluorescence spectra at different temperatures showing the redshift in λmax of intrinsic fluorescence at increasing temperatures. (b) Far-UV CD spectra at different temperatures clearly mark the change in protein conformation above 55°C. (c) Activity profile over various time points at different temperatures. This is in accordance with the loss of native structure above 55°C which hampers the catalytic activity of the enzyme correspondingly. (d) Cartoon representation of the changes occurring in structure and function of the enzyme at higher temperatures. (Figure credit: Shukla et al. [26]).

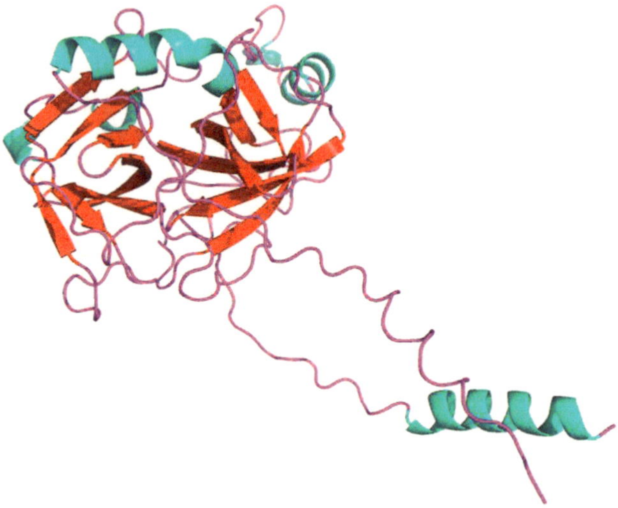

Figure 5.
Cartoon representation of homology model of a serine protease from rat (Prss30): It selectively cleaves synthetic peptide substrates of trypsin and activates the epithelial sodium channel. The model is derived from the AlphaFold database (https://alphafold.ebi.ac.uk). The protein structure for this physiologically and commercially significant serine protease has not been solved. Thus, the predicted model provides us with valuable information about the protein folds, secondary structure arrangements, and catalytic triad of this important enzyme. β-sheets are shown in red color, cyan color marks the α-helices, and magenta color highlights the loops.

7.2.2 Theoretical approach

New theoretical and computational approaches have emerged, including various bioinformatics tools, artificial intelligence (AI) based methods, deep evolutionary analysis, structure-prediction web servers, physics-based force fields, etc. These techniques are employed to complement the experiments in providing an overall picture of the protein structure. The computer-based protein-structure prediction has been advanced by Molt and colleagues, in an event initiated in 1994 called CASP: Critical Assessment of protein Structure Prediction [27]. Currently, all successful structure-prediction algorithms are based on the assumption that similar sequences lead to similar structures. These methods depend heavily on the PDB for template sequences. There are several computational methods for protein structure determination, including homology modeling, fold recognition via threading and *ab initio* methods [15, 28]. In recent times, computational methods that can predict protein structures with atomic accuracy, even in cases where no similar structure is available, have been developed. Recently developed programs and databases, such as AlphaFold and RosettaFold which are neural-based networks, are a great success in this field (**Figure 5**) [29, 30].

The tremendous increase in the amount of sequence and structural data of proteins, together with the advances in the experimental and bioinformatics methods are improving our knowledge about the relationship between the protein sequence, structure, dynamics, and function [31]. This knowledge, in turn, helps us to understand how proteins interact with their substrate and other molecules, such as small molecules or ligands, which can become a drug candidate [32]. Predicting the binding modes and affinities of different compounds upon interaction with the protein binding sites is the main goal of 'structure-based drug design' and is achieved by the 'docking' approach. There are a number of programs written to carry out such analysis. In general, a large number of conformations are generated for the small molecule (substrate or ligand), either prior to docking or during the docking routine. Each conformation is positioned in the active site in a variety of orientations, the combination of conformation and orientation being known as a 'pose'. Further, many such poses are selected and ranked by a scoring function to determine the overall best pose [15, 33] and the binding energy and affinity being calculated. The new frontiers now lie in physics-based modeling and AI to predict conformational changes, understand protein dynamics, design synthetic proteins, and improve protein modeling based on the laws of physics.

8. Conclusions

Hydrolases could participate in a variety of biological processes due to their diversification. Being the largest and most diverse class of enzymes, hydrolase offers an opportunity to explore the conformational/topological diversity which forms the basis of their differential biological activity. Thus, there is an urge to re-evaluate our existing knowledge on the functional and conformational transitions of these enzymes, in the context of emerging scientific trends. In this chapter, we discuss hydrolases in terms of their diversity, classification, the importance of the structure–function relationship of hydrolases taking serine protease as an example. The ongoing pandemic (SARS CoV-2 infections) further illustrates the importance to study hydrolases from the therapeutic point of view. To let the virus enter the host cell, viral spike protein plays a very important role and it is further activated by the serine protease 2 (TMPRSS2). The host proteases thus are involved in an intricate play in SARS-CoV-2 infection along with other viral infections and in designing antiviral therapeutic strategies [34].

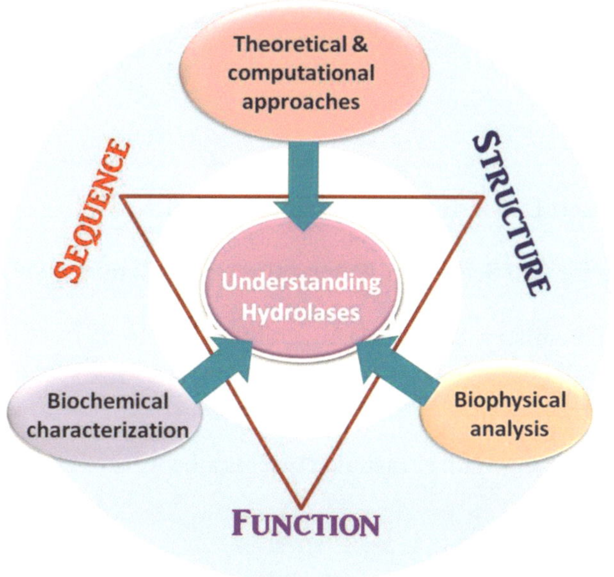

Figure 6.
The sequence-structure–function triad necessary for understanding enzymes. In this chapter, it has been discussed wrt hydrolases.

Furthermore, the available information and literature survey on selected hydrolases have been systematically summarized for easy understanding. Knowledge of the relationships between protein structure and function at the molecular level remains a primary focus in structural biology. So, various experimental and *in silico* methods and techniques have been mentioned in the chapter which contributes to our knowledge of the interplay among the stability, structure, and function of these enzymes (**Figure 6**). It can serve as a structural toolbox to improve their efficiency in the future by helping in engineering these proteins with improved functional and structural features.

Acknowledgements

Authors thank Dr. Radha Chauhan, Scientist, NCCS, Pune, for her kind support.

Author details

Ekta Shukla[1]*, Ameya D. Bendre[2] and Sushama M. Gaikwad[3]

1 National Centre for Cell Science, S.P. Pune University Campus, Pune, India

2 Department of Chemistry, S.P. Pune University, Pune, India

3 CSIR-National Chemical Laboratory, Pune, India

*Address all correspondence to: ektashukla19@gmail.com

References

[1] Chand D, Avinash VC, Yadav Y, Pundle AV, Suresh CG, Ramasamy S. Molecular features of bile salt hydrolases and relevance in human health. Biochimica et Biophysica Acta - General Subjects. 2017;**1861**:2981-2991

[2] Li S, Yang X, Yang S, Zhu M, Wang X. Technology prospecting on enzymes: Application, marketing and engineering. Computational and Structural Biotechnology Journal. 2012;**2**:3

[3] Shimizu K, Morse DE. Silicatein: A unique silica-synthesizing catalytic triad hydrolase from marine sponge skeletons and its multiple applications. In: Moore BS, editor. Methods in Enzymology. Vol. 605. Massachusetts, US: Academic Press; 2018. pp. 429-455

[4] Mustafa M, Ali L, et al. Heterologous expression and characterization of glycoside hydrolase with its potential applications in hyperthermic environment. Saudi Journal of Biological Sciences. 2021. DOI: 10.1016/j.sjbs.2021.09.076. (in press)

[5] Bairoch A. The ENZYME database in 2000. Nucleic Acids Research. 2000;**28**(1):304-305

[6] Gariev A, Varfolomeev SD. Hierarchical classification of hydrolases catalytic sites. Bioinformatics. 2006;**22**(20):2574-2576

[7] McDonald AG, Boyce S, Tipton KF. ExplorEnz: The primary source of the IUBMB enzyme list. Nucleic Acids Research. 2009;**37**:D593-D597

[8] Berg JM, Tymoczko JL, Stryer L. Biochemistry. In: Section 10.5, Many Enzymes Are Activated by Specific Proteolytic Cleavage. 5th ed. New York: W H Freeman; 2002 Available from: https://www.ncbi.nlm.nih.gov/books/NBK22589/

[9] Rao MB, Tanksale AM, Ghatge MS, Deshpande VV. Molecular and biotechnological aspects of microbial proteases. Microbiology and Molecular Biology Reviews. 1998;**62**(3):597-635

[10] Rawlings ND, Barrett AJ, Bateman A. MEROPS: The database of proteolytic enzymes, their substrates and inhibitors. Nucleic Acids Research. 2012;**40**(D1):D343-D350

[11] Page MJ, Di Cera E. Serine peptidases: Classification, structure and function. Cellular and Molecular Life Sciences. 2008;**65**(7-8):1220-1236

[12] Gurung N, Ray S, Bose S, Rai V. A Broader View: Microbial Enzymes and their Relevance in Industries, Medicine, and beyond. London, UK: Hindawi Publishing Corporation, BioMed Research International; 2013 Article ID 329121

[13] Berg JM, Tymoczko JL, Stryer L. Chapter 2: Protein composition and structure. In: Biochemistry. 7th ed. New York: W.H. Freeman; 2007

[14] Nelson D, Cox M. Chapter 4: Three dimensional structure of proteins. In: Lehninger Principles of Biochemistry. 4th ed. New York: W.H. Freeman and Company; 2005

[15] Breda A, Valadares NF, Norberto de Souza O, et al. Protein structure, modelling and applications [updated 2007 Sep 14]. In: Gruber A, Durham AM, Huynh C, et al., editors. Bioinformatics in Tropical Disease Research: A Practical and Case-Study Approach. Bethesda (MD): National Center for Biotechnology Information (US); 2006 2008. Chapter A06

[16] Varshney N, Rabbani G, Badr G, Khan RH. Cosolvents induced unfolding

and aggregation of keyhole limpet Hemocyanin. Cell Biochemistry and Biophysics. 2014;**69**(1):103-113

[17] Suresh Kumar R, Suresh CG, Brannigan JA, Dodson GG, Gaikwad SM. Bile salt hydrolase, the member of ntn-hydrolase family: Differential modes of structural and functional transitions during denaturation. IUBMB Life. 2007;**59**(2):118-125

[18] Bischof JC, He X. Thermal stability of proteins. Annals. New York Academy of Sciences. 2005;**1066**:12-33

[19] Rohamare SB, Dixit VS, Nareddy PK, Sivaramakrishna D, Swamy MJ, Gaikwad SM. Polyproline fold—in imparting kinetic stability to an alkaline serine endopeptidase. Biochimica et Biophysica Acta. 2013;**1834**:708-716

[20] Dalal S, More SV, Shankar S, Laxman RS, Gaikwad SM. Subtilase from Beauveria sp.: Conformational and functional investigation of unusual stability. European Biophysics Journal. 2014;**43**:393-403

[21] Eijsink VG, Bjørk A, Gåseidnes S, Sirevåg R, Synstad B, van den Burg B, et al. Rational engineering of enzyme stability. Journal of biotechnology. 2004;**113**(1):105-120

[22] Gecse KB, Khanna R, van den Brink G, Ponsioen CY, Löwenberg M, Jairath V, et al. Biosimilars in IBD: Hope or expectation? Gut. 2013;**62**(6):803-807

[23] Dill K, MacCallum J. The protein-folding problem, 50 years on. Science. 2012;**338**:1042-1046

[24] Radford SE. Protein folding: Progress made and promises ahead. Trends in Biochemical Sciences. 2000;**25**(12):611-618

[25] Senisterra GA, Finerty PJ Jr. High throughput methods of assessing protein stability and aggregation. Molecular BioSystems. 2009;**5**(3):217-223

[26] Shukla E, Agrawal SB, Gaikwad SM. Conformational and functional transitions and *in silico* analysis of a serine protease from *Conidiobolus brefeldianus* (MTCC 5185). International Journal of Biological Macromolecules. 2017;**98**:387-397

[27] Brockwell DJ, Smith DA, Radford SE. Protein folding mechanisms: New methods and emerging ideas. Current Opinion in Structural Biology. 2000;**10**:16-25

[28] Laskar A, Rodger EJ, Chatterjee A, Mandal C. Modeling and structural analysis of PA clan serine proteases. BMC Research Notes. 2012;**5**:256

[29] Jumper J, Evans R, Pritzel A, Green T, Figurnov M, et al. Highly accurate protein structure prediction with AlphaFold. Nature. 2021;**596**:583-592

[30] Baek M et al. Accurate prediction of protein structures and interactions using a three-track neural network. Science. 2021;**373**(6557):871-876

[31] John M, Krzysztof F, Andriy K, Torsten S, Anna T. Critical assessment of methods of protein structure prediction (CASP)-round x. Proteins. 2014;**82**(0 2):1-6

[32] Dodson EJ. Computational biology: Protein predictions. Nature. 2007;**450**:176-177

[33] Alonso H, Bliznyuk AA, Gready JE. Combining docking and molecular dynamic simulations in drug design. Medicinal Research Reviews. 2006;**26**:531-568

[34] Saadat YR, Khatibi SMH, Vahed SZ, Ardalan M. Host serine proteases: A potential targeted therapy for COVID-19 and influenza. Frontiers in Molecular Biosciences. 2021;**8**:725528

Chapter 2

Proteolytic Enzymes Derived from a Macro Fungus and Their Industrial Application

Nagendra Kumar Chandrawanshi, Deepali Koreti, Anjali Kosre and Ashish Kumar

Abstract

Proteolytic enzymes are well known for catalyzing hydrolytic reactions. These enzymes fall under the group of large and complex, also known as proteases. Proteolytic enzymes mainly derived from microbial origin are favored because they have a short generation time, ease of genetic manipulation of microorganisms, and the availability of diverse species in nature. Macro fungi are significant and played an excellent role in degrading lignocellulosic compounds, such as mushrooms. They efficiently degrade cellulose and produce extracellular enzymes such as xylanases, cellulases, and ligninolytic enzymes. Furthermore, proteases play a significant role in fungi physiology, such as metalloproteinase, subtilases, aspartate, etc. Many worldwide researchers have reported the mycelial secretion of proteases from basidiomycetes. Thus, many protease extraction methods have been developed from the various categories of mushroom species, i.e., Pleurotusostreatus, Phanerochaetechrysosporium, Schizophyllum commune, Chondrostereumpurpureum, and Hypsizygusmarmoreus, etc. Furthermore, there is a high demand in the industry for specific proteolytic enzymatic activity. Numerous species of mushrooms have not been explored to date for the optimization and production of enzymes. Therefore, further detailed studies are required to expose the production mechanisms and application of proficient proteolytic enzymes from mushrooms. The present chapter will deliberately deal with proteolytic enzymes downstream processing and their various industrial applications.

Keywords: proteolytic enzymes, Basidiomycetes, macro fungi, mushroom, industrial application

1. Introduction

Enzymes are natural catalysts that evolve or require various biological processes and are utilized in various industrial applications. Scientists have recently focused on detecting new enzymes with various properties and best-suited commercial purposes [1, 2]. There are many advantages associated with industrial enzymes, such as reaction specificity, low energy needs, biodegradable sources such as plants, animals, and microbes used for enzymes production and isolation. Proteases are the best studied and utilized in a group of enzymes that have the best substrate specificity. Total enzymes are produced at the industrial level, of which one-third are hydrolyses, and 65% are proteases. Proteases are hydrolytic enzymes that catalyze the interruption of the polymerization of protein. It evolves in the metabolic processes of biological activities in almost all organisms [3–6].

Different microbial sources have produced different proteases than plants and animals; microbial enzymes are more labor-intensive and best suitable for industrial applications [3, 7, 8]. Approximately two-thirds of commercial protease is produced from microbial origin in the world [6]. Microbial proteases production has advantages: short generation time, high growth rate, high yield, genetic modification is possible, cost-effective, and easy availability. These properties made microbial protease the best choice for biotechnological and industrial applications [9, 10].

Much research has been conducted to isolate and purify proteases from microbial sources and wieldy applied in industrial sectors [1, 11]. Bacteria are the most prominent microbes used for industrial-level protease production. Some groups of basidiomycetes also reported having proteases, and they provide the way for further study, fungal protease is neutral, acidic, or alkaline protease according to the species of fungi [12]. Fungi proteases have easy cell separation techniques, and a study revealed that micromycetes proteases have specific characteristics. Several fungal species include Aspergillus species, *Fusarium graminarum*, *Chrysosporium keratinophilum*, *Penicillium chrysogenum*, *P.griseofulvin, Scedosporium apiosermum,* and *Trametes cingulata*. Of all these species, *Penicillium* and *Aspergillus* are the most widely studied [12–14].

Basidiomycetes are important wood-degrading fungi in biological communities, and some genera of this group have been used as a food source. They are well studied for their extracellular enzymes production properties, such as xylanases, cellulases, and ligninolytic enzymes [15]. Proteases play essential roles in the biochemical process in fungi and the essential completion of the life cycle [16]. Mushrooms are the known Basidiomycetes in the fungi group. They have been used as food products for centuries as well as reported for their biological activity, among which, species of *Pleurotus* are globally known and valued as good food source and ease of cultivation. Mushroom bioactive compounds such as protein, vitamins. and enzymes, etc., are also examined for their biological activity such as antitumor, anti-inflammatory antidiabetic, antiviral, antioxidant, hypocholesterolemic, antitumor, immunomodulatory, and hepatoprotective actions [17, 18]. They are a good source of vitamins, protein, minerals, and very low fat content. They contain a variety of bioactive compounds, including protease, and there are more than 20 proteases that have been isolated [17, 19, 20].

Thus, the isolation of new *proteases* from different mushroom species is a novel area of research that needs much exploration. There is still much progress required for the study of proteases from edible mushrooms and has great future opportunities in the area of genome, proteome, and metabolome of mushroom proteases. Also, open new research relates to exploring downstream processing and economic aspects of mushroom proteases.

2. Classification of proteolytic enzymes or proteases

Proteolytic enzymes significantly participate in the metabolism of organisms such as plants, animals, bacteria, fungi, and viruses. Proteases are not explored and are essential in enzymology because of their substantial physiological significance and broad application in research activities [4]. Since proteolytic enzymes are requisite in providing nitrogen to xylotrophs under natural growth conditions (on living and dead wood), the absence of sufficient systematic information on secreted proteases of higher xylotrophic fungi is unnoticeable yet in biology. Research studies have been conducted to isolate and characterize proteolytic enzymes from the cultured mycelium and fruit bodies of basidiomycetes. Highly diverse types of structures and mechanisms of action, so proteases are not set aside with the rules of enzyme nomenclature [14]. So, the classification of these enzymes is often difficult. The enzyme that enters through the plasma membrane inside the cell is usually called an extracellular enzyme [6]. It must be classified into two categories according to their ability to cleave the peptide bonds as exopeptidases and cleave specific sites of peptide bonds as endopeptidases. They are industrially essential enzymes [21]. The diversity and specificity of these native enzymes are based on their broad characterization and isolation (**Table 1**). Based on active site present on proteases, they are classified as follows:

2.1 Exopeptidases

Exopeptidases are an enzyme that cleaves at the end site and requires free terminal groups close to the bond. It catalyzes the breakdown of specific peptide bonds after the carboxyl or amino terminals in the protein. Based on their efficiency in identifying the active site as either C or N terminal, they are further divided as carboxypeptidases or amino peptidases [36].

2.1.1 Aminopeptidases

Amino peptidases are the class of proteases enzymes that precisely cut at the N-terminal of the amino acid polypeptide chain, breaking it into dimer fragments or a single amino acid residue. After the recognition, they further remove the present methionine N-terminal of the polypeptide chain, which may differ in their expression. It is found in various microbial strains, including basidiomycetes fungi, molds, and bacteria, etc. Overall, amino peptidases work as intracellular enzymes; however, as per a report studied, amino peptidases that originated from *Aspergillus oryzaea* fungal species are extracellular enzymes [3, 4].

2.1.2 Carboxypeptidases

This enzyme performs its catalytic reaction on the C-terminal of the amino acid chain, breaking peptide bonds into monomers form. These are not predominantly recognized as endopeptidases because they leave few amino acid molecules at the target site of the protein. Instead, it can be employed to eliminate the additional tags at the carboxyl-terminal of the target protein. Among specific peptidases, metallocarboxy protease, type A carboxypeptidase, is known primarily for removing amino acid of the aromatic side chain while type B acts on essential amino acids [37].

2.2 Endopeptidases

Endopeptidases act at specific site of the peptide bond of the substrate [36]. It cleaves the internal peptide bonds of proteins influenced by the existing functional group present on the active site of the peptide chain. It is further classified as follows:

S N	Mushroom	Enzyme/protease	Cultivation condition	Method of purification/isolation	References
1.	*Pleurotus ostreatus*	Laccase isoenzymes	Liquid culture	Polyacrylamide gel electrophoresis	[22]
2.	*P. eryngii*	Pleureryn	—	Ion exchange chromatography	[18]
3.	*P. eryngii*	Eryngeolysin	Fruiting body	Ion exchange chromatography	[23]
4.	*P. citrinopileatus*		Fruiting bodies	Ion exchange chromatography	[16]
5.	*P. ostreatus*	Fibrinolytic protease	Mycelia culture	SDS-PAGE	[24]
6.	*Grifola frondosa*	Aminopeptidase	—	SDS-PAGE	[25]
7.	*P. ostreatus*	Metal-dependent proteinases	Fruiting body	Ion exchange chromatography	[26]
8.	*P. nebrodensis*	Nebrodeolysin	Fruiting body	Ion exchange and gel filtration chromatography	[27]
9.	*P.eryngii*	hemolysin	Fruiting body	SDS-PAGE	[28]
10.	*P. eryngii*	Fibrinolytic	solid-state conditions	SDS-PAGE	[29]
11.	*Helvella lacunosa*	Serine protease	—	Ion exchange chromatography	[30]
12.	*Termitomycesalbuminosus*	Alkaline protease	—	Ion exchange chromatography	[31]
13.	*P. ostreatus*	Fibrinolytic enzyme	Submerged culture fermentation	Ammonium sulfate precipitation, hydrophobic interaction, and gel filtration chromatographies	[32]
14.	*Lentinus citrinus*	Alkaline protease	Solid state fermentation	—	[33]
15.	*P. sajor-caju*	Signal Peptide Peptidase	Liquid culture	Ammonium sulphateprecipitation, Ion-exchange chromatography, and HPLC	[34]
16.	*P. sajor-caju*	—	Solid-State Fermentation	Ammonium sulfate precipitation	[35]

Table 1.
Enzymes production from the edible mushroom.

2.2.1 Serine proteases

These classes of proteases are broadly found in nature and present in cellular organisms. Along with all the identified proteolytic enzymes, a significant part is of serine proteases. It generally performs the cleavage action on the bond present in the central part of the amino acid chain. However, few Serine proteases act as exopeptidases by detaching the amino acids from the end terminal of the polypeptide chain. Its name derives from the Ser residue present in the peptide chain, which is nucleophilic and placed in the active site of the chain of amino acids. An intermediate substrate is formed by using the serine residues inform of acyl-enzyme at the C end terminal of the newly structured peptide bond [38].

2.2.2 Cysteine/thiol proteases

This enzyme contains cysteine residues at their active site present both in prokaryotes and eukaryotes microbes. It shows proteolytic activity at the 6–8 pH range with 50–70°C optimum temperature. Hydrogen cyanide is the key component that activates this enzyme, resulting in which SH group is formed in a polypeptide chain. Oxidizing agents can inhibit this kind of proteases and show sensitive action to the sulfhydryl agents, for example, p-CMB [39].

2.2.3 Metalloproteases

Metalloproteases are generally zinc-containing enzymes. In fungi or basidiomycetes, several metal ions such as calcium, cobalt, and zincare are involved in their reactivation. Zinc-containing enzymes and calcium are essential for proteineous activity and structural stability of protein at optimum pH 5–9. These are sensitive to an agent that causes chelation of metal, such as ethylen diamine tetracetic acid (EDTA), but are insensitive to cysteine inhibitors [40].

2.2.4 Aspartic proteases

It is a comparatively small class of endopeptidases that includes aspartic proteases. These proteases are composed of a pair of aspartates bilobed structures, including a leading catalytic site. It functions optimally on acidic pH and is present in nature. This enzyme is secreted by various microorganisms such as bacteria and fungus, as their virulence secretions. Also, it can perform the mutualistic function in the breakdown of proteins yielding nitrogen from urea. These kinds of proteases are primarily biased toward the hydrophobic amino acids nearer to the dipeptides bond. As compared with the other two endoproteases, it utilizes residues present in the active site showing nucleophilic attribute for proteolysis [41].

3. Proteolytic enzymes from mushroom species

As proteolytic enzymes are indispensable in supplying nitrogen to xylotrophs under natural growth conditions (on living and dead wood), the absence of sufficient systematic information on secreted proteases of higher xylotrophic fungi is not much explored [42]. The protein structure contains nitrogen, which is probably the reason for the secretion of extracellular proteolytic enzymes basidiomycetes or mushroom.

The species belong to orders of basidial fungi, *Polyporales, Boletales*, and *Agaricales*, which are reported to secret proteolytic enzymes. Proteases secreted by mushrooms typically have a low molecular mass ranging from 26 to 50 kDa having isoelectric point's up to 3.5–8.8. Acidic pH is usually optimal for these enzymes, ranging from 2.0 to 5.0. Proteinases isolated from *Hypsizygis (H.) marmoreus, P. ostreatus,* and *F. velutipes* are exceptional, as their optimal pH falls under the neutral range. Amino peptidases in the mushroom are usually intracellular enzymes; few reports have studied detecting extracellular aminopeptidases, such as *Tramatellatrogii*. Separately from endopeptidase activity, aminopeptidase, carboxypeptidase, and dipeptidyl aminopeptidase activities were revealed [43]. It is known that there is a high demand for industrial application of proteolytic enzymes with suitable specific properties and must be stable at various temperatures and pH. However, this chapter suggests that the studies of proteases from basidiomycete's fungus or mushroom recommend that further detailed studies are required to explore proteases' mechanisms and physiological effects.

4. Role of proteolytic enzymes in mushroom

Proteases perform complex physiological functions, including protein catabolism; blood clotting, cell growth and migration, morphogenesis, and development [4]. Mushrooms or basidiomycetes fungi are heterotrophic organisms. They can utilize both organic and inorganic nitrogen sources as nutrition. An under natural conditions, they usually secrete various extracellular enzymes to decompose natural organic materials such as ligninolytic enzymes. Protease from mushrooms involves endopeptidases, and exopeptidases act one after another as the former produces many free C and N terminal ends and latter act on the peptide fragments, thus forming the decomposed protein. This broad specificity is a significant property of the fungal secreted proteases and other proteolytic enzymes employed to break down proteins. An investigation reported on fungus *T. rogii* utilizes these enzymes to efficiently break down various peptides in the substrate [43]. Proteases secreted in mushrooms participate in the active regulation of other synthesized enzymes, resulting in regulating some physiological processes in mushrooms species such as *P. ostreatus*, *P. chrysosporium*, etc. The activity of ligninolytic enzymes is regulated via their specific activation or inactivation by the extracellular proteases secreted them [22] and perhaps has the ability to degrade the proteins controlling heat shock response, DNA repair pathway programmed cell death [4]. The metalloprotease plays a significant role in the fruiting body formation in *P. ostreatus*. Previous studies demonstrated that mRNA content is noticeably higher in the primordial stages of fruit body formation than in the vegetative mycelium stage [44]. So the extracellular occurrence of the fungal proteolytic enzymes may help the fungus grow on the host by utilizing its nutritional contents, may act as a pathogenic agent for the host. Under favorable conditions, *P. pulmonarius* grows only on dead decaying wood showing as ubtilisin-like proteolytic activity using proteases [45].

5. Methods used for proteolytic enzymes recovery and production

Enzymes recovery and production from mushrooms were directly influenced by the substrate type, composition, and recovery methods. Various research studies showed that solid-state and submerged fermentation mushrooms had been significantly used for enzyme production [46]. In solid-state cultivation, various agro wastes are utilized and produce fruiting bodies containing various metabolites primarily used for food sources. After harvesting, its by-products, spent mushroom substrate (SMS) that contains plenty of extracellular enzymes can be utilized as animal feeding and for enzymes recovery and production. It reported that laccase (EC 1.10.3.2) was the most prominent and common in *Pleurotus sajor-caju* [47], *P. ostreatus, L. edodes, Flammulina velutipes and Hericium erinaceum* [48], *A. bisporus* [49]. Another enzyme, lignin peroxidase productivity, was found to be the SMS of *P. sajor-caju* [47]. Researchers can significantly explore the production of lipases, pectinases, and phytases. **Table 1** lists all the enzymes found in mushrooms and recovery methods. Convinced enzymes extraction and purification methods were widely applied, for example, dialysis, ultrafiltration, anion-exchange chromatography, and gel [50–52]. It is noteworthy that most of the investigations were carried out only for the fruiting body or mycelium of mushroom, not the SMS; therefore, it is an open possibility for the new finding for enzymes from SMS. Recently works were reported for enzyme recovery from SMS. Mayolo-Deloisa et al. [49] evaluated the use of aqueous two-phase systems to recover laccase from the residual compost of *A. bisporus* mushroom. They observed that valorizations of residual material give a 95% yield and have the potential for value-added products with commercial application [49]. Ko et al. [48] determined the production of amylase, cellulase, glucosidase, laccase, and xylanase from the SMS obtained from four edible mushrooms: *P. ostreatus, L. edodes, F. velutipes* and *H. erinaceum*, and evaluated its potential application using enzymes from SMS as industrial enzymes. It has been reported that a solvent such as water is used for enzymes recover from SMS with good activity; this fact is essential for an industrial application and related environmental concerns. However, the extraction of enzymes from submerged culture supernatant is more straightforward than from SMS because centrifugation is needed, and the obtained supernatant can be used as the crude enzyme. Some physical conditions are essential for optimizing scale-up, such as pH, temperature, extraction medium, incubation time, inoculums density, carbon and nitrogen source, and the impotent parameters for enzyme production.

6. Applications and future prospects

Novel investigation techniques revealed highly specific and selective protein modifications performed by proteases, including activating the zymogenic enzyme forms by limited proteolysis, forming hormones and other physiologically active peptides from precursor proteins, thrombus lysis, or the processing and transport of secreted proteins through the membrane (**Figure 1**). The vital role of proteolytic enzymes in metabolic and regulatory processes explains their occurrence in all living organisms [53].

6.1 In the detergent industry

Proteases were used as a detergent centuries ago as the "Burnus" brand, along with sodium carbonate and pancreatic extract mixed in it [54]. Several industries, such as chemical, pharmaceutical, food processing, detergents, and leather processing, utilize the catalytic properties of proteases. Its application in the bioremediation of pollutants has also been reported. Several factors such as optimum substrate specificity,

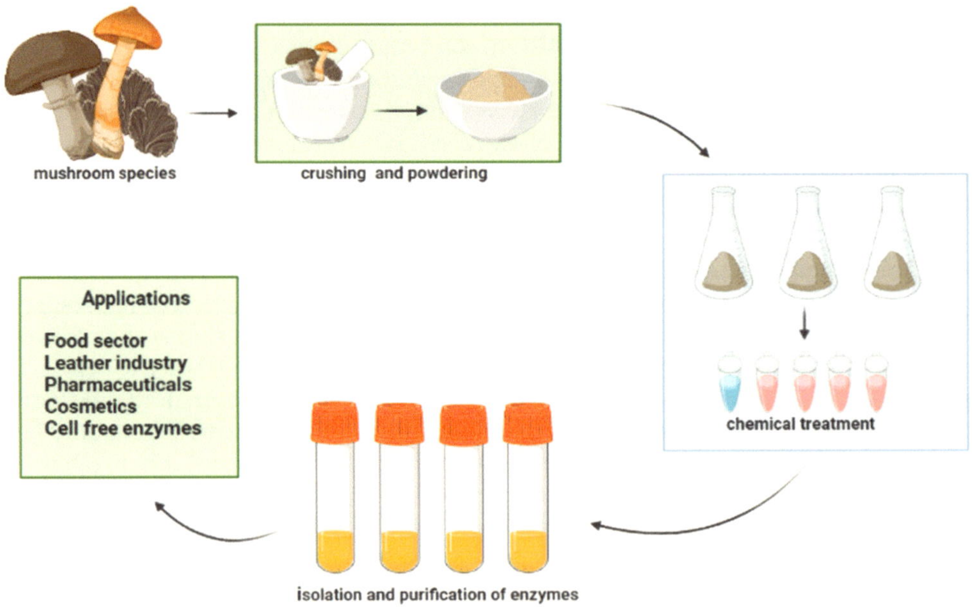

Figure 1.
Application of edible mushroom as a potential source for enzymes production.

temperature, optimum pH, chemical stability, and catalytic activity may vary because of a diverse group and also can affect the production of proteases [21].

6.2 Cell-free enzyme preparation

Immense interest has been grown in proteases due to their thermal ability in a wide range of temperatures. It is also used as detergents in the cell separation process for the production of cell-free enzyme preparations. In these perspective, fungal enzymes have applications as these are extracellularly secreted [55, 56].

6.3 In the pharmaceutical and food industries

Some proteases are also found to produce due to the infection process caused by foreign invaders such as bacteria, fungus, and viruses. A variety of steps regulate the mechanism of proteolytic enzyme reactions, including substrate specificity, ATP-directed protein degradation, restricted access to the active site, highly specific protein modifications. It can activate zymogenic forms of enzymes by restricted proteolysis activity [57]. Including these protease enzymes that cause diseases to host cells has become a good option for developing therapeutic agents for the diseases such as cancer, hepatitis, malaria, and candidiasis. It has also been reported to demonstrate potent immunomodulatory activity [4].

6.4 Leather industry

The leather industry involves various steps to obtain processed leather, for example, soaking, liming, hair removal, bating, deliming, and degreasing. These steps are applied using poisonous chemicals such as salt, lime, solvents, and sodium sulfide, resulting in pollution. The exclusion of non-collagenous particles is required in leather processing, which decides the softness and durability of leather products [58, 59]. It can be controlled by applying enzymes such as proteases in the place of chemicals [60].

7. Conclusion

Most of the industrial proteases used are of microbial origin, especially of bacteria. These enzymes are preferentially selected because of their desired characteristics and lower cost. The bioengineering manufacture of microbial proteases is favored as they have short generation periods, high yield, ease of genetic desired modification, and diverse species available. Future opportunities are high in cutting-edge research from the pharmaceutical perspective of the protease gene. By the help via recombinant DNA technology, respective genes must have been cloned and sequenced to determine the function of enzymes that cause changes in the attributes of protease enzymes and enhance enzyme production for their commercial usage. In industries, proteases contribute to the high value-added products development, and the same way biological catalysts offer advantages over the use of chemical catalysts for numerous reasons, such as high catalytic activity, high specificity, and their availability in economically viable quantities.

Conversely, cost associated with the production of proteases from mushrooms or basidiomycetes is the major obstacle to their application in industries and pharmaceuticals. For that reason, further research studies should have been implemented to discover novel low-cost proteases from mushrooms and their application in commercial and industrial sectors. So, a great extent of the study of proteases from the mushroom requires further investigations.

Acknowledgements

The authors are thankful to the Junior Research Fellowship (DBT/JRF/BET-18/I/2018/AL/123), Department of Biotechnology, Biotech Consortium of India Limited, and Pt. Ravishankar Shukla University Research scholarship award (797/Fin/Sch./2021) for providing funding support. The authors are also are thankful to the Head, School of Studies in Biotechnology, Pt. Ravishankar Shukla University, Raipur.

Conflict of interest

The authors declare no conflict of interest.

Author details

Nagendra Kumar Chandrawanshi[1*], Deepali Koreti[1], Anjali Kosre[1] and Ashish Kumar[2]

1 School of Studies in Biotechnology, Pt. Ravishankar Shukla University, Raipur, Chhattisgarh, India

2 Department of Biotechnology, Sant Gahira Guru Vishwavidyalaya, Sarguja, Ambikapur, Chhattisgarh, India

*Address all correspondence to: chandrawanshi11@gmail.com

References

[1] Inacio DF, Ferreira RO, CA VA, Brugnari T, Castoldi R, Peralta RM, et al. Proteases of wood rot fungi with emphasis on the genus *Pleurotus*. BioMed Research International. 2015;**2015**:1-10

[2] Daroit DJ, Correa APF, Brandelli A. Production of keratinolytic proteases through bioconversion of feather meal by the Amazonian bacterium *Bacillus* sp. P45. International Biodeterioration and Biodegradation. 2011;**65**:45-51

[3] Rao CS, Sathish T, Ravichandra P, Prakasham R. Characterization of thermo- and detergent stable serine protease from isolated *Bacillus circulans* and evaluation of eco-friendly applications. Process Biochemistry. 2009;**44**:262-268

[4] Rao MB, Tanksale AM, Ghatge MS, Deshpande VV. Molecular and biotechnological aspects of microbial proteases. Microbiological Reviews. 1998;**62**:597-635

[5] Singhal P, Nigam V, Vidyarthi A. Studies on production, characterization and applications of microbial alkaline proteases. International Journal of Advanced Biotechnology and Research. 2012;**3**:653-669

[6] Beg QK, Gupta R. Purification and characterization of an oxidation-stable, thiol-dependent serine alkaline protease from *Bacillus mojavensis*. Enzyme and Microbial Technology. 2003;**32**:294-304

[7] Gupta R, Beg Q, Lorenz P. Bacterial alkaline proteases: Molecular approaches and industrial applications. Applied Microbiology and Biotechnology. 2002;**59**:15-32

[8] Kalaiarasi K, Sunitha P. Optimization of alkaline protease production from *Pseudomonas fluorescens* isolated from meat waste contaminated soil. African Journal of Biotechnology. 2009;**8**:24

[9] Nisha N, Divakaran J. Optimization of alkaline protease production from *Bacillus subtilis* NS isolated from sea water. African Journal of Biotechnology. 2014;**13**(65):24

[10] Ali N, Ullah N, Qasim M, Rahman H, Khan SN. Molecular characterization and growth optimization of halo-tolerant protease producing *Bacillus subtilis* strain BLK-1.5 isolated from salt mines of Karak, Pakistan. Extremophiles. 2016;**20**:395-402

[11] Adrio J, Demain A. Microbial enzymes: Tools for biotechnological processes. Biomolecules. 2014;**4**:117-139

[12] Said D, Pietro RCLR. Enzimascomoagents Bio Tecnologicos. Ribeirao Preto, Brazil: Legis Summa; 2004

[13] Kang SG, Kim IS, Rho YT, Lee KJ. Production dynamics of extracellular proteases accompanying morphological differentiation of *Streptomyces albidoflavus* SMF301. Microbiology. 1995;**141**:3095-3103

[14] Kudryavtseva OA, Dunaevsky YE, Kamzolkina OV, Belozersky MA. Fungal proteolytic enzymes: Features of the extracellular proteases of xylotrophic basidiomycetes. Microbiology. 2008;77: 643-653

[15] Nakamura M, Iketani A, Shioi Y. A survey of proteases in edible mushrooms with synthetic peptides as substrates. Mycoscience. 2011;**52**:234-241

[16] Cui L, Liu QH, Wang HX, Ng TB. An alkaline protease from fresh fruiting

bodies of the edible mushroom *Pleurotus citrinopileatus*. Applied Microbiology and Biotechnology. 2007;**75**:81-85

[17] Patel Y, Naraian R, Singh VK. Medicinal properties of *Pleurotus* species (oyster mushroom): A review. World Journal of Fungal and Plant Biology. 2012;**3**:1-12

[18] Wang H, Ng TB. Pleureryn, a novel protease from fresh fruiting bodies of the edible mushroom *Pleurotus eryngii*. Biochemical and Biophysical Research Communications. 2001;**289**:750-755

[19] Iandolo D, Piscitelli A, Sannia G, Faraco V. Enzyme production by solid substrate fermentation of *Pleurotus ostreatus* and *Trametes versicolor* on tomato pomace. Applied Biochemistry and Biotechnology. 2011;**163**:40-51

[20] AbdWahab NA, Abdullah N, Aminudin N. Characterisation of potential antidiabetic-related proteins from *Pleurotus pulmonarius* (Fr.) Quel. (grey oyster mushroom) by MALDITOF/TOF mass spectrometry. BioMed Research International. 2014;**2014**:1-9

[21] Sabotic J, Trcek T, Popovi T, Brzin T. Basidiomycetes harbour a hidden treasure of proteolytic diversity. Journal of Biotechnology. 2007;**128**:297-307

[22] Palmieri G, Giardina P, Bianco C, Fontanella B, Sannia G. Copper induction of laccase isoenzymes in the ligninolytic fungus *Pleurotus ostreatus*. Applied and Environmental Microbiology. 2000;**66**:920-924

[23] Ngai PHK, Ng TB. A hemolysin from the mushroom *Pleurotus eryngii*. Applied Microbiology and Biotechnology. 2006;**72**:1185-1191

[24] Shen MH, Kim JS, Sapkotaetal K. Purification, characterization, and cloning of fibrinolytic metalloprotease from *Pleurotus ostreatus* mycelia. Journal of Microbiology and Biotechnology. 2007;**17**:1271-1283

[25] Islam MM. Molecular cloning, expression and characterization of a serine proteinase from Japanese edible mushroom, *Grifolafrondosa*: Solving the structure–function anomaly of a reported aminopeptidase. Electronic Journal of Biotechnology. 2008;**11**:1-12

[26] Lebedeva GV, Proskuryakov MT. Purification andcharacterization of milk-clotting enzymes from oyster mushroom (*Pleurotus ostreatus* (Fr.) Kumm). Applied Biochemistry and Microbiology. 2009;**45**:623-625

[27] Lv H, Kong Y, Yaoetal Q. Nebrodeolysin, a novel hemolytic protein from mushroom *Pleurotus nebrodensis* with apoptosis inducing and anti-HIV-1effects. Phytomedicine. 2009;**16**:98-205

[28] Shibata T, Kudou M, Hoshi Y, Kudo A, Nanashima N, Miyairi K. Isolation and characterization of a novel two-component hemolysin, erylysin A and B, from an edible mushroom, *Pleurotus eryngii*. Toxicon. 2010;**56**:1436-1442

[29] Cha WS, Park SS, Kim SJ, Choi D. Biochemicaland enzymatic properties of a fibrinolytic enzyme from *Pleurotus eryngii* cultivated under solid-state conditions using corncob. Bioresource Technology. 2010;**101**:6475-6481

[30] Zhang G, Wang H, Zhang X, Ng T. Helvellisin, a novel alkaline protease from the wild ascomycete mushroom *Helvella lacunosa*. Journal of Bioscience and Bioengineering. 2010;**109**:20-24

[31] Zheng S, Wang H, Zhang G. A novel alkaline protease from wild edible mushroom *Termitomyces*

albuminosus. Acta Biochimica Polonica. 2011;**58**:269-273

[32] Liu XL, Zheng XQ, Qian PZ. Purification and characterization of a novel fibrinolytic enzyme from culture supernatantof *Pleurotus ostreatus*. Journal of Microbiology and Biotechnology. 2014;**24**:245-253

[33] Machado ARG, Teixeira MFS, de Souza KL, Campelo MCL, de Aguiar Oliveira IM. Nutritional value and proteases of *Lentinus citrinus* produced by solid state fermentation of lignocellulosic waste from tropical region. Saudi Journal of Biological Sciences. 2016;**23**:621-627

[34] Benmrad MO, Mechri S, Jaouadi NZ, Elhoul MB, Rekik H, Sayadi S, et al. Purification and biochemical characterization of a novel thermostable protease from the oyster mushroom *Pleurotus sajor-caju* strain CTM10057 with industrial interest. BMC Biotechnology. 2019;**19**:43

[35] Ravikumar G, Gomathi D, Kalaiselvi M, Uma C. A protease from the medicinal mushroom *Pleurotus sajor-caju*; production, purification and partial characterization. Asian Pacific Journal of Tropical Biomedicine. 2012;**2**:411-417

[36] Sawant R, Nagendran S. Protease: An enzyme with multiple industrial applications. World Journal of Pharmacy and Pharmaceutical Sciences. 2014;**3**:568-579

[37] Motyan JA, Toth F, Tozser J. Research applications of proteolytic enzymes in molecular biology. Biomolecules. 2013;**3**:923-942

[38] Page M, Di Cera E. Serine peptidases: Classification, structure and function. Cell and Molecular Life Sciences. 2008;**65**:1220-1236

[39] Thakur N, Goyal M, Sharma S, Kumar D. Proteases: Industrial applications and approaches used in strain improvement. Biological Forum–An International Journal. 2018;**10**:158-167

[40] Ellaiah P, Srinivasulu B, Adinarayana K. A review on microbial alkaline proteases. Journal of Scientific and Industrial Research. 2002;**61**:690-704

[41] Turk B, Turk D, Turk V. Protease signalling: The cutting edge. The EMBO Journal. 2012;**31**:1630-1643

[42] Abraham L, Breuil C, Bradshaw ED, Morris PI, Byrne T. Proteinases As potential targets for new generation anti-Sapstain chemicals. Forest Products Journal. 1997;**47**:57-63

[43] Caporale C, Garzillo AMV, Caruso C, Buonocore V. Characterisation of extracellular proteases from *Trametes trogii*. Phytochemistry. 1996;**41**:385-393

[44] Joh JH, Kim BG, Kong WS, Yoo YB, Kim NK, Park HR, et al. Cloning and developmental expression of a metzincin family metalloprotease CDNA from oyster mushroom *Pleurotus ostreatus*. FEMS Microbiology Letters. 2004;**239**:57-62

[45] Barsukova TN, Garibova LV, Ivanov AI. Ecologo biological characterization of *Pleurotus pulmonarius* (Fr.) Quel. Mikologiya i Fitopatologiya. 1989;**23**:14-19

[46] Pandey A, Soccol CR, Mitchell D. New developments in solid state fermentation. I bioprocesses and products process. The Biochemist. 2000;**35**:1153-1169

[47] Singh AD, Abdullah N, Vikineswary S. Optimization of

extraction of bulk enzymes from spent mushroom compost. Journal of Chemical Technology & Biotechnology. 2003;**78**:743-752

[48] Ko HG, Park SH, Kim SH, Park HG, Park WM. Detection and recovery of hydrolytic enzymes from spent compost of four mushroom species. Folia Microbiologica. 2005;**50**:103-106

[49] Mayolo-Deloisa K, Trejo-Hernández MDR, Rito-Palomares M. Recovery of laccase fromthe residual compost of Agaricusbisporusin aqueous two-phase systems. Process Biochemistry. 2009;**44**:435-439

[50] Chen S, Ge W, Buswell JA. Biochemical and molecular characterization of a laccase from the edible straw mushroom, Volvariellavolvacea. European Journal of Biochemistry. 2004;**271**:318-328

[51] Ullrich R, Huong LM, Dung NL, Hofrichter M. Laccase from the medicinal mushroom Agaricusblazei: Production, purification and characterization. Applied Microbiology and Biotechnology. 2005;**67**:357-363

[52] Quaratino D, Federici F, Petruccioli M, Fenice M, D'Annibale A. Production, purification and partial characterisation of a novel laccase from the white-rot fungus Panustigrinus CBS 577.79. Antonie Van Leeuwenhoek. 2007;**91**:57-69

[53] Sani JT, Gharibi SOS, Shariati MA. The importance of alkaline protease commercial applications: A short review. Indian Journal of Research in Pharmacy and Biotechnology. 2017;**5**:534-538

[54] Landolo D, Piscitelli A, Sannia G, Faraco V. Enzyme production by solid substrate fermentation of *Pleurotus ostreatus* and *Trametes versicolor* on tomato pomace. Applied Biochemistry and Biotechnology. 2011;**163**:40-51

[55] Sumantha A, Larroche C, Pandey A. Microbiology and industrial biotechnology of food-grade proteases: A perspective. Food Technology and Biotechnology. 2006;**44**(2):211-220

[56] Shaba AM, Baba J. Screening of *Pleurotus ostreatus* and *Gleophylumsepiarium* strains for extracellular protease enzyme production. Bayero Journal of Pure and Applied Sciences. 2014;**5**(1):187-190

[57] Vandeputte-Rutten L, Gros P. Novel proteases: Common themes and surprising features. Current Opinion in Structural Biology. 2002;**12**(6):704-708

[58] Khan F. New microbial proteases in leather and detergent industries. Recent Innovations in Chemical Engineering. 2013;**1**:1-6

[59] Wahab WAA, Ahmed SA. Response surface methodology for production, characterization and application of solvent, salt and alkali-tolerant alkaline protease from isolated fungal strain *aspergillus Niger WA*. International Journal of Biological Macromolecules. 2017;**115**:447-458

[60] Sharma M, Gat Y, Arya S. A review on microbial alkaline protease: An essential tool for various industrial approaches. Industrial Biotechnology. 2019;**15**:69-78

Chapter 3

Cholinesterases and Their Inhibitors

Mesut Işık

Abstract

The main focus of this section is to review the available information on ChEs (ChEs) and their inhibitors. The ChE enzymes cause damage to the cholinergic system by hydrolyzing the neurotransmitter acetylcholine (ACh). ChE inhibitors, playing an important role in the cholinergic system, are used in the treatment of Alzheimer's disease (AD) because of their effects on maintaining ACh levels in brain regions and preventing Aβ accumulation by inhibiting ChE. In this context, it is important to develop many synthetic and natural origin ChE inhibitors for the treatment of abnormalities in the cholinergic system and disorders with neuropsychiatric symptoms. In this section, firstly, general information about ACh and its synthesis in the cholinergic system is given, then ChEs and their catalytic properties, their roles in AD, and their molecular forms are explained. In the following section, the active site of Cantis was defined. The anti-ChE activity of the developed inhibitors was discussed, and then the mechanism of their binding to the ChE active site was explained by molecular docking. In the final section, many types of ChE inhibitors are described and discussed in detail in this section, and the properties and binding mechanism of these inhibitors are summarized.

Keywords: Alzheimer's disease, acetylcholine, acetylcholinesterase, butyrylcholinesterase, cholinesterase inhibitors, molecular docking

1. Introduction

Acetylcholine (ACh), one of the most important neurotransmitters, is the primary substrate for cholinesterases (ChEs). First identified in autonomic ganglia, neuromuscular junctions, and many synapses in the central nervous system (CNS), the ACh is also the primary neurotransmitter in preganglionic sympathetic and parasympathetic neurons and in the adrenal medulla. In the CNS, ACh is mainly found in interneurons and long axon cholinergic pathways [1, 2]. The ACh is formed as a result of esterification of acetic acid with quaternary ammonium alcohol choline. This ACh found in cholinergic neurons and other cell types is biosynthesized by the transfer of acetyl group from acetyl-coenzyme A to choline catalyzed by choline acetyl transferase (ChAT, EC 2.3.1.6). The synthesis of ACh by ChAT is presented in the following reaction.

$$\text{Acetyl} - \text{CoA} + \text{Choline} \overset{\text{ChAT}}{\rightarrow} \text{Acetylcholine} + \text{CoA} - \text{SH} \quad (1)$$

The synthesis of acetyl-CoA, which is the precursor for ACh synthesis from various metabolic pathways with carbohydrate, protein, and fat metabolism, has been reported by many researchers [3].

The cholinergic system is based on ACh, which was first recognized by Loewi in the 1920s and is widely found in both the peripheral and central nervous systems. There are two types of receptors in the nervous system and neuromuscular junctions: ACh receptors (muscarinic and nicotinic). The cholinergic receptors are also known to be expressed in many cells, including immune system and endothelial cells [4]. The receptors are identified based on their response to specific antagonists and agonists [5]. Although three types (M1–M3) of muscarinic receptors have been identified pharmacologically, five types have been reported based on molecular cloning experiments [6]. Muscarinic receptors are metabotropic and use G proteins for signal transduction, while nicotinic receptors are ionotropic and use ligand-gated ion channels for signal transduction [7]. The M1 receptor, one of the muscarinic receptor types and most common in the cerebral cortex, has its highest concentrations in the anterior olfactory nucleus, cerebral cortex, olfactory tubercle, dentate gyrus, hippocampus, and nucleus accumbens. The M2 receptor, a presynaptic autoreceptor that directs cholinergic release, is located in brain regions with abundant cholinergic neurons [8]. M3 and M4 receptors are located in the mainly in the olfactory tubercle, diencephalic, striatum, and brainstem regions [9]. Two types of the nicotinic receptor have also been identified in the CNS. The receptors are commonly found in the thalamus, substantia nigra, and periaqueductal gray [6, 10].

2. Cholinesterases

The existence of ChEs, one of the serine hydrolase class enzymes, was first suggested by Henry Dale in 1914, and research on the subject has been continuing since the 1930s. Research on the determination of the sequence of amino acids in the active site of ChEs has been started since the early 1959 [11]. The ChEs are divided into two groups due to their substrate selectivity to ACh and butyrylcholine. AcetylChE (AChE, acetylcholine acetyl hydrolase, EC 3.1.1.7) is known as true ChE, while butyrylChE (BChE, acylcholine acyl hydrolase, EC 3.1.1.8) is known as pseudoChE, serum ChE, or nonspecific ChE. While AChE inhibits at high substrate concentrations, BChE is activated. These enzymes also differ from each other in terms of selective inhibition of enzymes. The enzymes, which show 65% similarity in their amino acid sequence contents, are encoded in the chromosomes (7 and 3) [12–14].

The AChE is found in erythrocyte membranes and many tissues such as central and peripheral tissues, nerve and muscles, motor and sensory fibers, and noncholinergic and cholinergic fibers [1]. Another enzyme, BChE, is mainly synthesized in the liver and released into the plasma and is known to show high activity in many tissues [12, 13]. AChE, one of the fastest known enzymes with catalytic activity, is responsible for the hydrolysis of ACh in cholinergic synapses [11, 14]. Although the basic physiological function of BChE is not yet fully known, it is important both pharmacologically and toxicologically due to its ability to break down ester drugs such as carbamates, aspirin, succinylcholine, cocaine, antidepressant drugs such as sertraline, amitriptyline, pesticides, organophosphate, and chemical warfare agents. It has been reported that the expression of AChE and BChE is increased in many

Enzyme	Substrate	Product
AChE	Acetyl-β-methyl-thiocholine	β-metyl-thiocholine + acetate
AChE	Acetyl-β-methyl-choline	β-methyl-choline + acetate
AChE > BChE	Acetylcholine	acetate + choline
AChE > BChE	Acetylthiocholine	acetate + thiocholine
BChE > AChE	Butyrylcholine	butyrate + choline
BChE > AChE	Butyrylthiocholine	butyrate + thiocholine
BChE	Succinylcholine	succinate + choline
BChE	Benzoylcholine	benzoiate + choline

Table 1.
Selected substrates and products of AChE and BChE hydrolysis [16].

pathological conditions such as glioma, lung cancers, meningioma, leukemia, and ovarian tumors [12, 15].

The ChEs differ in their catalytic activities according to the artificial substrate preference. Since ACh is the more selective substrate for AChE, it is more hydrolyzed by AChE than by BChE. As expressed in **Table 1**, the catalytic activity of the ChEs may vary depending on the type of substrates [16]. AChE and BChE are responsible for the hydrolysis of different types of choline esters.

AChE, playing an important role in impulse transmission by hydrolyzing ACh in the central and peripheral nervous system, catalyzes the hydrolysis of the neurotransmitter ACh to choline and acetate. BChE is responsible for the hydrolysis of its more specific substrate (butyrylcholine) to butyrate and choline. Hydrolysis of ACh/butyrylcholine by the ChEs is presented in the following reaction.

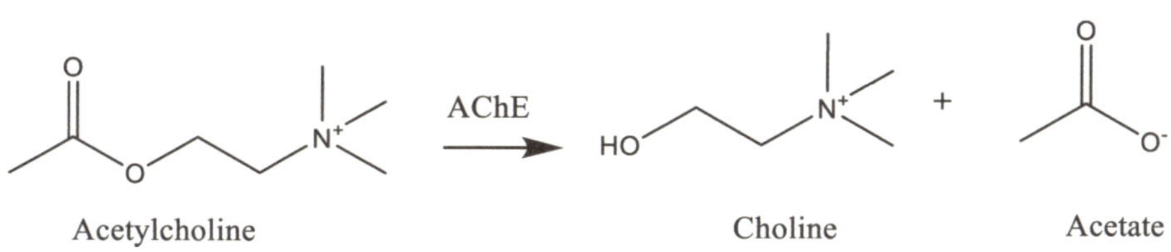

(2)

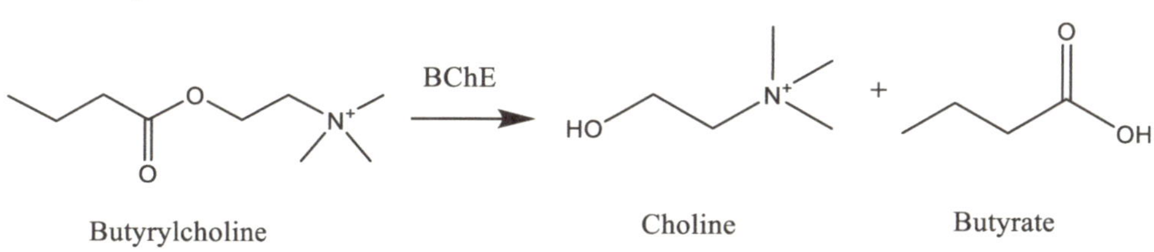

(3)

The ChE activity method is based on the formation of 5-thio-2-nitrobenzoic acid, a yellow compound, with 5,5-ditiyobis(2-nitrobenzoik) asit (DTNB) of thiocholine, which is formed by the hydrolysis of ACh and BCh as substrate. The ChE activities were determined using the Ellman method. Reaction mixture contains acetylthiocholine iodide (10 mM) for AChE and butyrylthiocholine iodide (0,5 mM) for BChE as substrate, 5,5-ditiyobis(2-nitrobenzoik) asit (DTNB, 10 mM), and Tris–HCl buffer (pH 8, 1 M) [17–19].

2.1 The role of cholinesterases in Alzheimer's disease

Alzheimer's disease (AD) is the most common type of dementia seen in the elderly population. Its etiology is not known exactly. This disease has been expressed as a neurodegenerative disease characterized by irreversible loss of nerve cells, difficulties in cholinergic nerve conduction, memory and mental functions, thinking and interpretation, and personality and behavioral disorders. It has been reported that amyloid β-peptide (Aβ) aggregation, neurofibrillary network formation originating from hyperphosphorylated tau proteins, oxidative stress, and low ACh levels are effective in the pathology of AD [20–22]. Some hypotheses have been proposed to explain the pathogenesis of AD.

Cholinergic hypothesis: It is one of the first hypotheses. In this hypothesis, it is thought that AD occurs due to the reduction of acetylcholine, an important neurotransmitter. Degeneration of cholinergic neurons in the basal forebrain of AD patients, as well as a significant decrease in cholinergic receptors and choline acetyltransferase (ChAT) levels in the cerebral cortex, was detected. Although many of the previous treatment approaches are based on this hypothesis, clinical studies have revealed that treatment strategies to increase ACh levels provide only symptomatic relief but are not effective in the development and treatment of the disease [23, 24]. However, there is evidence that the use of ChE inhibitors in new approaches based on this hypothesis may affect Aβ aggregation [25].

Amyloid cascade hypothesis: The hypothesis proposed in 1991 reports that Aβ deposits are the primary factor in AD. In AD, there is a series of neurodegenerative events triggered by the production, aggregation, storage, and toxicity of derivatives that occur after the production of amyloid precursor protein (APP) in the brain [26], and this study has been supported by molecular genetic, neuropathological, and biochemical studies. In this hypothesis, it is suggested that Aβ proteins, which accumulate in certain parts of the brain and form insoluble fibrils, later damage nerve cells by forming senile plaques; these cells break their connections with each other and reduce the amount of neurotransmitters [27].

Dysfunction of cholinergic neurotransmission in the brain causes an increase in AD symptoms. In the brain, choline-ester-based neurotransmitters are catalyzed by ChEs such as AChE and BChE. Therefore, ACh level decreases in the perisynaptic region of AD. Inhibition of AChE to prevent this reduction in ACh level is considered an important therapeutic target for the disease. The role of BChE in the progression of AD has been stated, and inhibition of BChE gains importance for treatment purposes. In addition, the presence of two types of the ChEs was detected between neuritic plaques and neurofibrillary tangles in AD brain. Thus, it has been suggested that both AChE and BChE may be involved in the formation of aggregates of the Aβ peptide in the AD brain [28]. Toxicity in AD brains varies depending on the amount of AChE and BChE, which have the potential to form complexes with Aβ aggregates [29, 30]. Depending on the disease, changes are seen in the amounts of AChE forms. The tetrameric form of AChE (G4), which predominates in the healthy brain, decreased at the onset of AD, while its monomeric and dimeric forms (G1 and G2) remained unchanged. In addition, studies have reported that the level of G1 and G2 forms is elevated in the plasma of AD patients [28, 31, 32].

2.2 Molecular forms of cholinesterases

The ChEs exist as amphiphilic or soluble molecular forms in tissues and body fluids [1, 12]. Monomeric, dimeric, and tetrameric molecular forms of ChE arise from the posttranslational modification of the expressed protein. The soluble monomeric and tetrameric membrane-bound forms are the predominant enzyme species in humans [10, 33] The forms are summarized below:

Amphiphilic dimers (Type 1): The dimeric form of AChE covalently bound to the membrane by a glycophosphatidylinositol extension, soluble only by detergents and aggregates in the absence of detergent. This form is found in erythrocytes, muscle, and lymphocytes of mammals.

Amphiphilic monomers and dimers (Type 2): Unlike Type 1, the form does not contain glycolipids anchor, does not aggregate in the absence of detergent, and can only be dissolved in salt solutions. The forms of the ChEs are frequently encountered in the muscle, brain, and intestine.

Hydrophobic-tailed tetramers (Type 3): This form of AChE, hydrophobically attached to the plasma membrane by a 20 kDa polypeptide anchor, is widely found in the CNS of mammals.

Collagen-like tailed or asymmetrical forms (Type 4): These forms can be identified by the presence of a collagen-like tail that allows them to attach to the basal lamina. This tail consists of collagen 3 helical subunits, each associated with a ChE tetramer. It finds it more common for AChE rather than BChE in neuromuscular junctions.

Soluble tetrameric form (G4): This form consists of four identical monomers and is stabilized by the interactions of hydrophobic amino acids at the C-terminus of the monomers. These forms are common in mammalian body fluids and soluble fractions of tissue homogenates. The form of ChE in the mammalian brain, mammalian body fluids, and soluble fractions of tissue homogenates is the tetrameric form (G4) [1].

2.3 Active site of cholinesterases

Homo sapiens AChE belonging to the serine hydrolases is known to have a very high catalytic activity, with each molecule of AChE degrading about 25,000 molecules of ACh per second [34, 35]. The enzyme's active site contains the esteratic subsite and anionic subsite in catalytic center. There is a glutamate residue in the anionic region and a serine residue with a functional -OH group in the esteratic region (ES) and an imidazole ring. In addition to the ES, there is an acyl region and a choline-binding subsite at the catalytic active site (CAS) [34, 36]. The enzyme also contains the peripheral anionic site (PAS). The sites serve as sites for binding AChE and other quaternary ligands and involved in substrate-based inhibition [37]. The anionic site contains many amino acid residues such as Phe 330, Trp 84, and Tyr 121 for the electric eel AChE or Phe 338, Trp 86, and Tyr 337 for murine AChE. The PAS, localized on the AChE surface around the cavity entrance, was recognized as a target for multiple AChE activity modulators. Asp70 and Tyr332 residues in the PAS are involved in the binding of positively charged substrate/ligands and in the activity control of the enzyme. The site contains the most significant amino acids residues such as Asp 72, Tyr 121, Tyr 70, Trp 279, and Tyr 334 [16].

How the substrate binds to the active site of AChE (PDB code: 4EY5) via different types of amino acid residues is predicted by molecular docking. The quaternary nitrogen in the ACh interacts with the anionic site formed by the amino acid

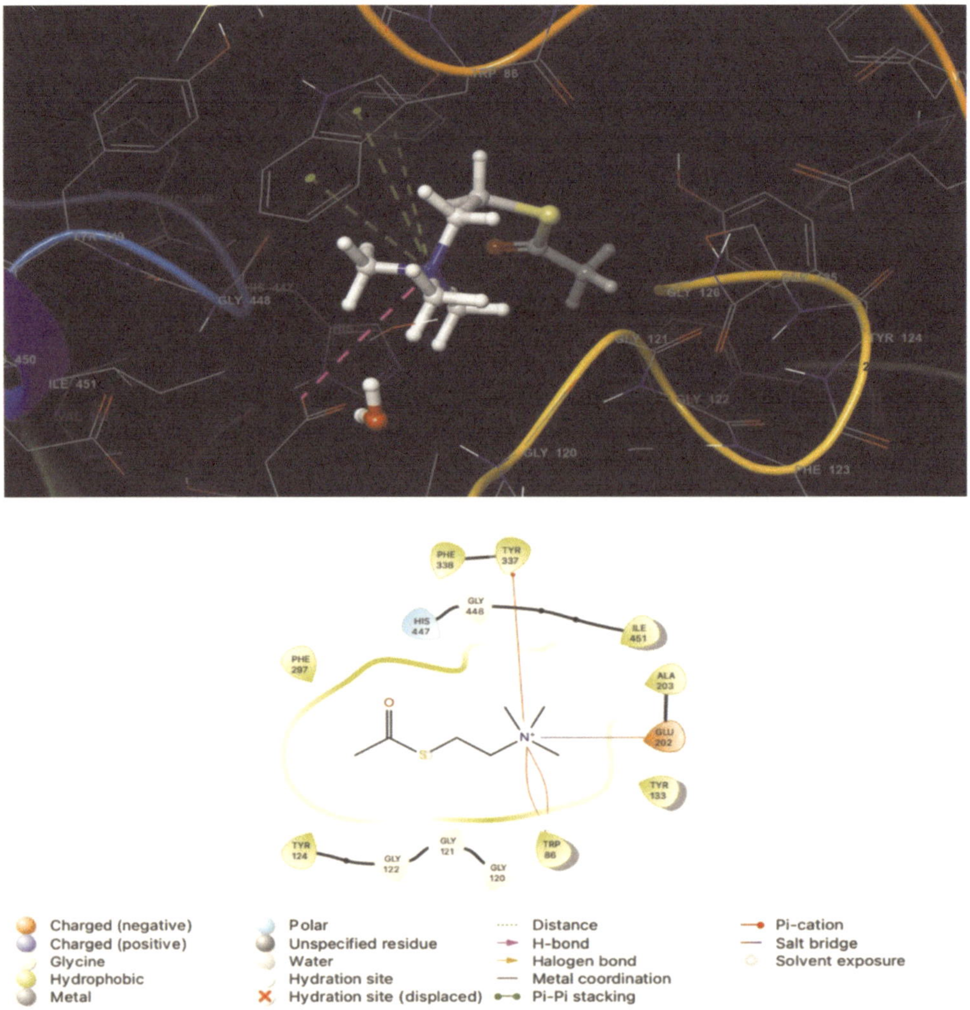

Figure 1.
2D and 3D interaction diagrams of ACh in the active site of human AChE (PDB ID: 4EY5).

tryptophan (Trp), and through the aspartic acid and tyrosine located at the entrance of the active center, the substrates are directed to interact with the active center [38]. The molecular docking results show that ACh has Pi-cation interaction with Tyr337 and Trp86 in the active site of human AChE. In addition, nitrogen in the ACh formed a salt bridge with Glu202 (**Figure 1**). The AChE is composed of a hydrophobic cloud with Ala203, His447, Gly120-Gly122, Gly448,Tyr133, Tyr124, and Phe297.

3. Molecular docking

Molecular docking has gained importance with the increase in studies on drug development and design. Such studies have gained momentum with the creation of protein data banks. Molecular docking is a computer-generated tool for finding the optimal configuration and energy between the two if the protein and ligand structures are known. The basis of the method is to identify possible conformational states between the protein-ligand complex and to choose the complex with the lowest free energy. There are many applications such as Schrodinger, AutoDock, DOCK, and MolDock developed to simulate interactions (hydrogen bonding, hydrophobic interactions, Van der Waals) between two or more molecules [39]. Computational docking of test ligands to predict binding sites to the ChE active site becomes one approach to predict potential inhibitors against the target [40]. The docking studies were carried out using the GLIDE software [41].

Previous studies reported interactions with the AChE amino acid residues with various antipsychotic drugs. Recent studies reported interactions with critical residues such as Leu288, Ser292, Thr237, Val238, Gln368, Pi-cation interaction with Arg295, Pi-alkyl and alkyl interactions with Pro289, Val299, Pro367, 234 of AChE [42, 43] Donepezil and Pimozide ligands showed the lowest energy and the best favorable interactions. The Galanthamine (−7.3 kcal/mol) and Risvagtimine (−7.95 kcal/mol) showed higher binding energies than Donepezil (−8.5 kcal/mol) ligands [44].

In addition, in another study, binding free energy analysis was performed to understand that some drugs are potential candidates. It is a more suitable drug candidate with lower binding energies. Brexpiprazole and pimavanserin have binding free energies of −212.690 kcal/mol and − 108.626 kcal/mol, respectively, whereas Donepezil has binding energy of −180.517. The values, providing the highest contribution in terms of energy, are obtained from Van Der Waals forces. With significantly lower binding energy, these results indicate that Brexipyrazole with lower binding energy binds to the active site and other binding pockets in the protein with higher affinity than Donepezil and Pimavanserin [13]. In the study, it was stated that reference tacrine has higher binding affinity to AChE than to the natural substrate ACh [40]. In the present study, the tacrine was predicted to interact with AChE-binding site residues Trp84, Phe330, Tyr334, Ile 439, Trp432, Tyr442, and Gly441.The molecular docking results show that tacrine has Pi-stacking interaction with Phe 330 and Trp84 in the active site of human AChE. In addition, the AChE is composed of a hydrophobic cloud with Tyr334, Ile 439, Trp432, Tyr442, and Gly441 (**Figure 2**).

4. Cholinesterase inhibitors

ChE inhibitors constitute a wide group of compounds with different physicochemical properties, including drugs used in the treatment of many diseases, natural toxins, pesticides, and chemical warfare agents. The ChEs, which play an important role in the hydrolysis of pesticides, are also of great importance in food and agriculture research. Specific AChE inhibitors play an important role in the regulation of ACh metabolism due to the physiological importance of AChE, especially in the brain and blood. Since ACh is a more selective substrate for AChE, it takes a more active role in this metabolism than BChE. Specific BChE inhibitors such as tetraisopropyl pyrophosphoramide (iso-OMPA) are mainly of diagnostic importance [45, 46]. AChE inhibitors, one of the ChE inhibitors used in the treatment of many diseases, are more widely known than BChE. Drugs that suppress the development of AD through action on the cholinergic system are predominantly selective inhibitors of AChE. It is stated that these inhibitors also suppress Aβ aggregate formation and oxidative stress. Selective inhibitors of BChE for this disease have also been investigated as potential drugs for AD, but to a much lesser extent than AChE [47–50]. Inhibition of AChE also plays an important role in nerve agent toxicology. However, BChE can sometimes replace AChE temporarily inhibited by these agents and slowly hydrolyze accumulated ACh [51].

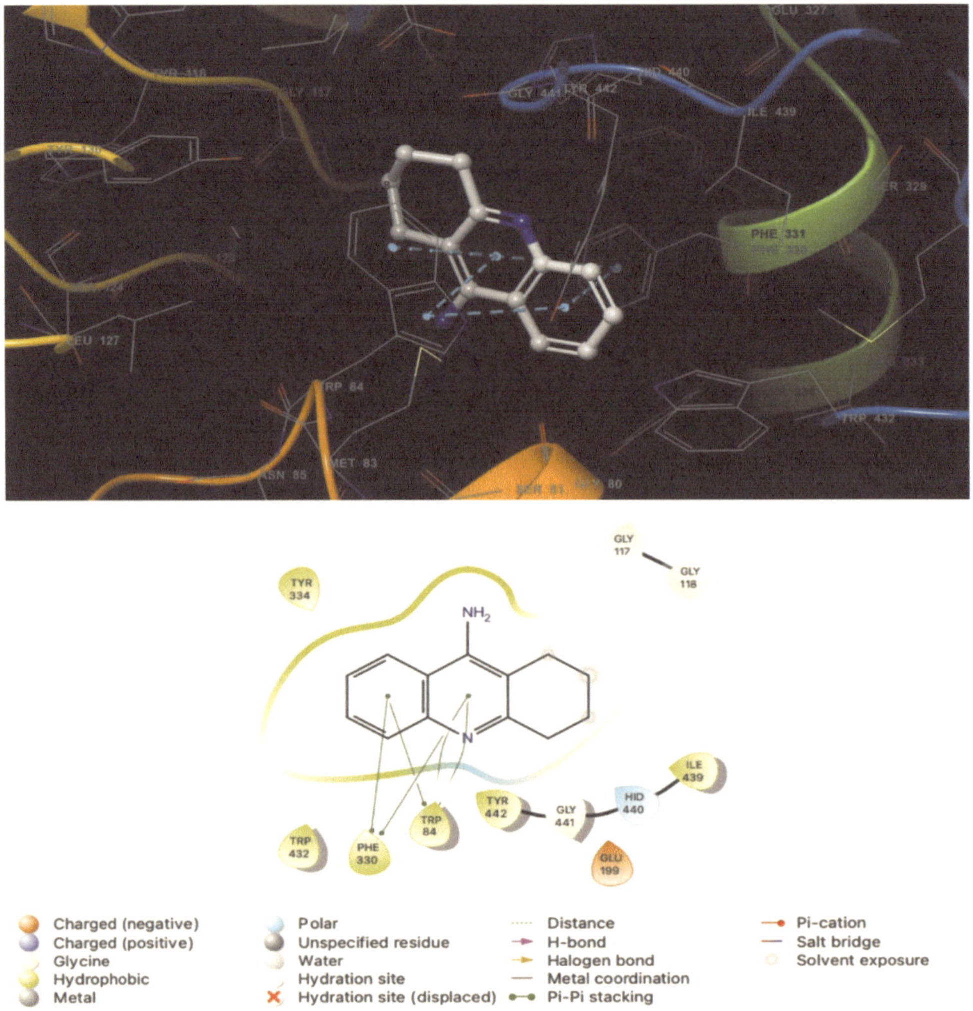

Figure 2.
2D and 3D interaction diagrams of tacrine in the active site of human AChE (PDB ID: 4EY5, Schrodinger 2017).

Compounds with AChE inhibitory potential can be divided into three main groups as follows [52]:

1. Compounds binding at the active site interact with either anionic site (e.g., tacrine) or esteratic (e.g., nerve agents).

2. Compounds interacting with the aromatic gorge (e.g., decamethonium).

3. Compounds bound at the PAS (e.g., propidium, huperzine).

In vitro inhibition studies: The inhibition effects of compounds against AChE are determined with at least five different inhibitor concentrations. The IC_{50} values of the compounds are calculated from activity (%)-[compound] graphs for each compound. The inhibition types and K_I constants are found by Lineweaver and Burk's curves according to previous works [53–55].

4.1 ChE inhibitors with therapeutic potential

Many compounds such as donepezil, galantamine, metrifonate, physostigmine, rivastigmine and tacrine, and memantine have been developed as ChE inhibitors. ChE inhibitors, having a regulatory effect on ACh metabolism, have been developed for the treatment of many diseases. For example, donepezil, galantamine, rivastigmine, tacrine, and memantine are drugs used in the treatment of AD [56]. However, the efficacy of these drugs is limited, and they are known to have various dose-related side effects, especially at higher doses. Galantamine and donepezil have AChE inhibitory potential [57, 58], whereas rivastigmine has a reversible inhibitory effect for both AChE and BChE. In particular, donepezil is a highly selective inhibitor for AChE rather than BChE. In a study, the inhibitory potentials (IC_{50} values) of donepezil, physostigmine, tacrine, and rivastigmine for AChE were found to be 6.7, 0.67, 77, and 4.3 nM, respectively [58].

Donepezil: Donepezil is a drug approved for the mild to moderate treatment of AD (**Figure 3A**). However, donepezil has many side effects such as muscle weakness loss of appetite, diarrhea, muscle cramps, nausea, and insomnia. In addition, when high

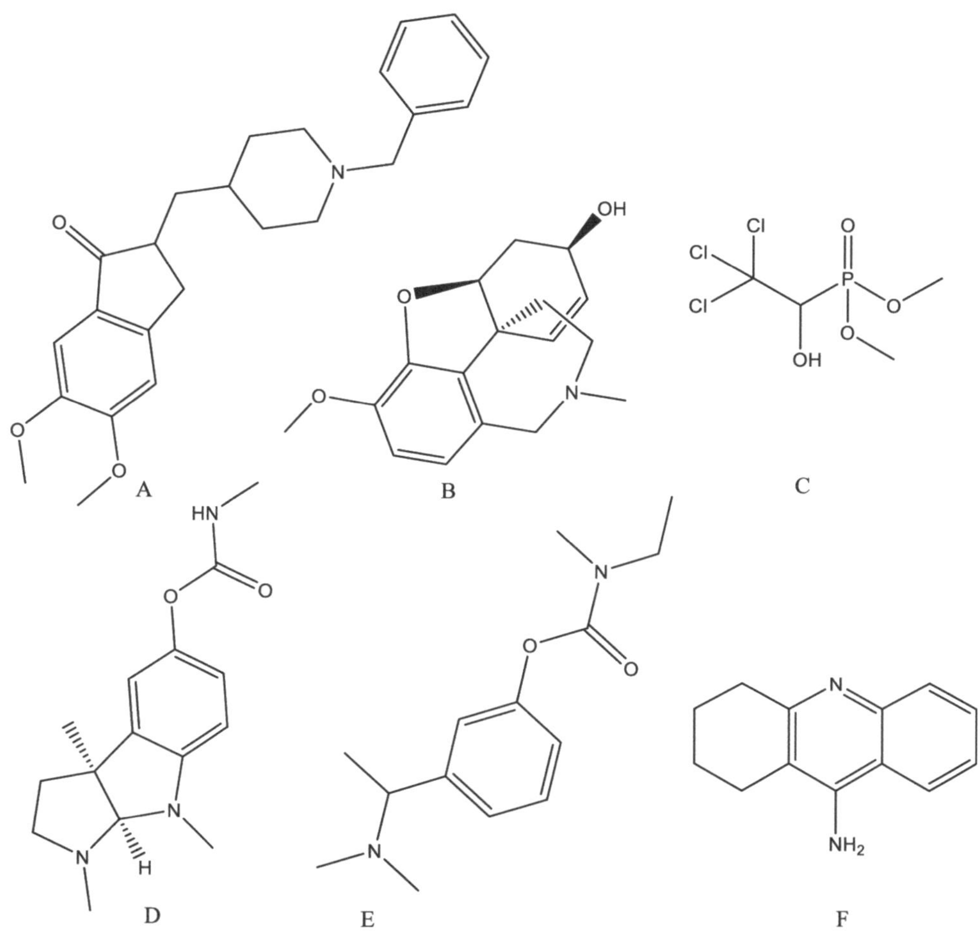

Figure 3.
The molecular structures of ChE inhibitors such as donepezil (a), galantamine (B), metrifonate (C), physostigmine (D), rivastigmine (E), and tacrine (F).

doses of the drug are given to patients, many symptoms such as muscle weakness, low blood pressure, severe nausea, severe vomiting, and respiratory problems occur [59]. In addition to being a selective inhibitor of ChE, donepezil may have multiple mechanisms of action. They act at molecular and cellular levels in many pathways, including inhibition of various aspects of glutamate-induced excitotoxicity, stimulation of the neuroprotective isoform of AChE and reduction of oxidative-stress-dependent effects, and in the pathogenesis of AD [60]. The donepezil has a molecular structure that causes simultaneous inhibition by binding to the active and PAS of AChE. However, it has been stated that it does not interact directly with the catalytic triad or the oxyanion hole [61].

Galantamine: Galantamine, one of the alkaloid groups, is found in many plants and has been used as a medicine for decades in Russia and Eastern European countries for many purposes such as the treatment of myasthenia, myopathy, and CNS-related sensory and motor defects. The galantamine has the appropriate molecular structure to bind to nicotinic receptors in the cholinergic system (**Figure 3B**). As its effectiveness against ChE was revealed in the 1950s, it started to be used in the treatment of various neurological diseases [56]. It was approved for use after it was found to be effective in the treatment of cognitive and many symptoms related to AD. However, it also has many side effects such as severe nausea, convulsions, vomiting, stomach cramps, irregular breathing, and muscle weakness.

Metrifonate: Metrophonate, a prodrug of dichlorvos (2,2-dichlorovinyl dimethyl phosphate), is an irreversible AChE inhibitor with biphasic action (**Figure 3C**). Initially, it interacts competitively with the enzyme and then turns into a noncompetitive type of inhibition by phosphorylation of the enzyme esteratic site. In other words, metrifonate from the organophosphate group interacts with the esteratic site of the enzyme (**Table 2**). Having close inhibition effects on both AChE and BChE, metrifonate can be defined as a pseudo-irreversible ChE inhibitor [62].

Physostigmine: Physostigmine (also known as Eserine) with AChE inhibitory effect is one of the compounds first isolated from Calabar bean in 1864. Although this compound is able to cross the blood–brain barrier (BBB), its use is limited due to its short half-life and many side effects such as stomach cramps, diarrhea, increased saliva production, and excessive sweating. Therefore, physostigmine is not approved for the treatment of AD. Physostigmine from the organophosphate group has a molecular structure to interact with the ES of the enzyme (**Figure 3D**). Based on this structure, many of its derivatives have been designed and synthesized. Among the derivatives, eseroline, tolserine, and phenserine have ChE inhibitory potential [56, 63].

Inhibitors	Class	Enzyme	Enzymatic active site
Donepezil	Piperidine	BChE > AChE	Anionic
Galantamine	Phenanthrene alkaloid	AChE > BChE	—
Metrifonate	Organophosphate	BChE > AChE	Esteratic
Physostigmine	Carbamate	BChE > AChE	Esteratic
Rivastigmine	Carbamate	AChE > BChE	Esteratic
Tacrine	Acridine	BChE > AChE	Anionic

Table 2.
Pharmacological properties of some selective ChE inhibitors [10].

Rivastigmine: Rivastigmine has been used for the treatment of moderate AD and dementia associated with Parkinson's disease [64]. Although it is stated that rivastigmine may exert its pharmacological effect through the cholinergic system, its mechanism of action has still not been clarified. Rivastigmine binds reversibly to both AChE and BChE and can cause inhibition. Also, rivastigmine is a carbamate class compound that is converted to various phenolic derivatives that are rapidly excreted from the body. The compound has the potential to bind with high affinity to the ES of AChE during ACh hydrolysis [56, 65], and has the appropriate molecular structure to bind to ChE in the cholinergic system (**Figure 3E**). The rivastigmine can cause many side effects such as weight loss, diarrhea, stomach pain, and loss of appetite, and in overdose, it can cause numerous symptoms such as irregular heartbeat and chest pain [66].

Tacrine: Tacrine, synthesized in the 1930s, was first used as a muscle relaxant antagonist and respiratory stimulant, moreover, due to its therapeutic effect on the cholinergic system in AD patients, and was approved by the FDA in 1993. The amine group of tacrine, whose molecular structure is presented in **Figure 3F**, interacts with amino acid residues Phe330 and Trp84 located in the "anionic region" of AChE (**Table 2**). As a result of the interaction, it has been stated that it is an effective inhibitor developed for the ChEs [67]. However, the use of tacrine has been limited due to liver toxicity, short half-life, and many side effects such as loss of appetite, nausea, vomiting, and diarrhea [56, 68].

4.2 Naturally derived inhibitors

ChE inhibitors used in the treatment of Alzheimer's, Parkinson's, and many diseases have short half-lives and many side effects. For this reason, there has been an increased interest in studies on the determination of natural origin inhibitors. Although these types of compounds have less ChE inhibitory activity than synthetics, they have much less side effects due to their natural origin.

Phenolic compounds: A number of flavonoid compounds, which have free radical scavenging properties and important roles in the prevention of oxidative stress, are natural ChE inhibitors in vitro. Galangin, a flavonol group from Alpiniae officinarum, showed a strong inhibitory effect against AChE [69]. Many studies have reported that phenolic compounds, reducing oxidative stress due to antioxidant properties, have inhibitory effects on AChE. The inhibitory effect on the AChE of phenolic compounds strongly depends on the structure of a particular compound, especially the position and/or number of the C=O and OH groups [70]. Chlorogenic acid has an inhibition effect on AChE in the hippocampus and frontal cortex (IC_{50}: 98.17 μg/ml). *In vitro*, caffeic acid has an activation effect in the cerebral cortex, cerebellum, hypothalamus, whole blood, and lymphocytes, while it has an inhibition effect at the concentrations of 0.5, 1.0, 1.5, and 2 mM in the muscles [71]. Hydroquinone, chlorogenic acid, and 4-hydroxybenzoic acid have inhibitory potential against the AChE with IC_{50} and K_I values in the range of 0.26 ± 0.01–36.34 ± 2.72 mM and 0.72 ± 0.00–29.23 ± 2.62 mM, respectively. The effectiveness of the compounds has been associated with its structure [49]. Consequently, as phenolic compounds have both AChE inhibitory effect and antioxidant properties, they can be considered as alternative drugs in the treatment of AD.

Cardanol: In 2009, various non-isoprenoid phenolic lipids obtained from Anacardium occidentale were investigated for their inhibitory activity against AChE. In particular, cardanol, a phenolic lipid, has shown promising results [56]. In one

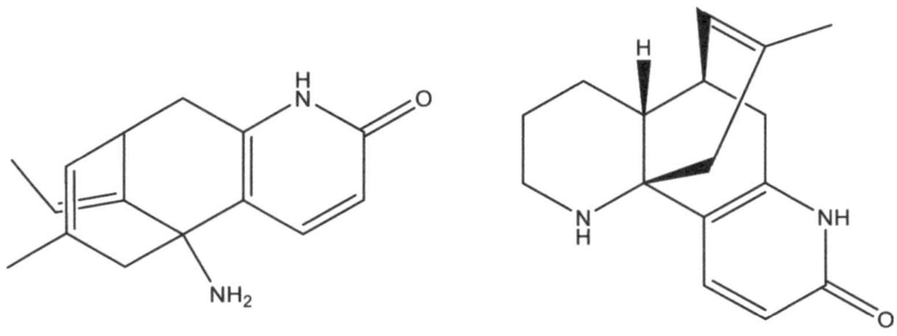

Figure 4.
The molecular structures of Hup-A and Hup-B.

study, a novel series of cardanol derivatives was designed as AChE inhibitors and tested for their inhibitory effects against ChEs. The derivatives showed the highest inhibitory activity against AChE, with IC_{50} of 6.6 μM [72].

Huperzine (Hup): Hup, obtained from the *Huperzia serrata*, is available in two types, Hup-A and Hup-B (**Figure 4A** and **B**, respectively). Hup-A is used in the treatment of AD and age-related memory loss and in improving cognitive functions by regulating ACh level. It is more effective than tacrine, galantamine, and rivastigmine and is a highly selective and potent inhibitor of AChE. However, it is less active against BChE compared with AChE. Many hybrids of Hup-A also have an AChE inhibitory effect [56, 73].

4.3 Synthetic analogs

In order to prevent or reduce the side effects and toxicity of ChE inhibitors, which are known to be used as drugs, their synthetic analogs have been developed as ChE inhibitors [74]. However, the main problem of synthetic analogs is their inability to penetrate the blood–brain barrier (BBB), and their inhibitory effect may be lower compared with reference inhibitors [75]. The derivatives of tacrine have shown improved AChE-inhibitory activities compared with the tacrine used as drug [76]. Various analog compounds were synthesized and tested by Ali et al. in 2009. Most of them showed moderate AChE-inhibitory effects. Ali et al. suggested that the presence of methoxy groups on the phenyl ring significantly improved the inhibition of AChE [75]. Sulfonamide compounds, which have a wide range of biological applications such as antimicrobial, antiviral, diuretic, and anticancer agents, are found as active ingredients in many drugs. In recent years, many studies have been carried out on the design and synthesis of sulfonamide-derived compounds with ChE inhibitory potential. Many of the compounds have been reported to be selective inhibitors for AChE and BChE [19, 77–80].

5. Conclusions

In conclusion, many synthetic and natural ChE inhibitors have been discovered recently. Due to the side effects of ChE inhibitors used in the treatment of many neurodegenerative diseases such as AD, studies conducted within the scope of the discovery and design of alternative inhibitors have gained importance. Although the side effects of natural origin inhibitors are much less than those of synthetic compounds, their effect is low. In some studies, derivatives of known inhibitors have been synthesized to increase the effectiveness of inhibitors and reduce their toxicity. The binding affinities of these derivatives to the enzyme active site vary depending on the structure. Therefore, the design of targeted inhibitors suitable for the enzyme active site by molecular docking method sheds light on the development of alternative drugs for treatment. The design of compounds with the ChE inhibitor potential by molecular docking method provides a significant advantage in terms of financial and workload since ineffective syntheses are not made.

Acknowledgements

I would like to thank Dr. Cüneyt Türkeş for his kindly helpful suggestions during the preparation of the book chapter.

Conflict of interest

The authors declare no conflict of interest.

Appendices and nomenclature

AD	Alzheimer disease
ACh	acetylcholine
AChE	acetylcholinesterase
Aβ	beta-amyloid
BBB	blood-brain-barrier
BChE	butyrylcholinesterase
CAS	catalytic active site
CNS	central nervous system
ChE	cholinesterases
ChAT	choline acetyl transferase
DTNB	5,5-ditiyobis(2-nitrobenzoik) asit
ES	esteratic region
iso-OMPA	tetraisopropyl pyrophosphoramide
PAS	peripheral anionic site

Author details

Mesut Işık
Bilecik Seyh Edebali University, Bilecik, Turkey

*Address all correspondence to: mesut.isik@bilecik.edu.tr

References

[1] Wang R, Xi CT. Neuroprotective effects of huperzine a: A natural cholinesterase inhibitor for the treatment of Alzheimer's disease. Neuro-Signals. 2005;**14**:71-82

[2] Perry E, Walker M, Grace J, Perry R. Acetylcholine in mind: A neurotransmitter correlate of consciousness? Trends in Neurosciences. 1999;**22**:273-280

[3] Jankowska-Kulawy A, Bielarczyk H, Pawelczyk T, Wróblewska M, Szutowicz A. Acetyl-CoA and acetylcholine metabolism in nerve terminal compartment of thiamine deficient rat brain. Journal of Neurochemistry. 2010;**115**:333-342

[4] Wessler I, Kirkpatrick CJ. Acetylcholine beyond neurons: The non-neuronal cholinergic system in humans. British Journal of Pharmacology. 2008; **154**:1558-1571

[5] BL S, SI D, NR S, SM R, CH B, RB R. The effects of prolonged administration of an alpha7 nicotinic cholinergic agonist intervention on neurocognitive function in schizophrenia. Neuropsychopharmacology. 2011;**36**:2774-2781

[6] Hurst R, Rollema H, Bertrand D. Nicotinic acetylcholine receptors: From basic science to therapeutics. Pharmacology and Therapeutics. 2013; **137**:22-54

[7] Turner AJ. Proteins, transmitters and synapses. Biochemical Education. 1995: 223-223

[8] Flynn DD, Ferrari-DiLeo G, Mash DC, Levey AI. Differential regulation of molecular subtypes of muscarinic receptors in Alzheimer's disease. Journal of Neurochemistry. 1995;**64**:1888-1891

[9] Schliebs R, Roßner S. Distribution of muscarinic acetylcholine receptors in the CNS. CNS Neurotransmitters and Neuromodulators. 1995. pp. 67-83

[10] Cummings JL. Cholinesterase inhibitors: A new class of psychotropic compounds. The American Journal of Psychiatry. 2000;**157**:4-15

[11] Massoulié J, Pezzementi L, Bon S, Krejci E, Vallette FM. Molecular and cellular biology of cholinesterases. Progress in Neurobiology. 1993;**41**:31-91

[12] Cokugras AN. Butyrylcholinesterase: Structure and physiological importance. Turkish Journal of Biochemistry. 2003; **28**:54-61

[13] Kundu D, Dubey VK. Potential alternatives to current cholinesterase inhibitors: An in silico drug repurposing approach. Drug Development and Industrial Pharmacy. 2021;**47**: 919-930

[14] Işık M, Beydemir Ş, Yılmaz A, Naldan ME, Aslan HE, Gülçin İ. Oxidative stress and mRNA expression of acetylcholinesterase in the leukocytes of ischemic patients. Biomedicine & Pharmacotherapy. 2017;**87**:561-567

[15] Özturan Özer E, Unsal Tan O, Ozadali K, Küçükkilinç T, Balkan A, Uçar G. Synthesis, molecular modeling and evaluation of novel N′-2-(4-benzylpiperidin‒/piperazin-1-yl) acylhydrazone derivatives as dual inhibitors for cholinesterases and Aβ aggregation. Bioorganic Med Chem Lett. 2013;**23**:440-443

[16] Pohanka M. Cholinesterases, a target of pharmacology and toxicology. Biomed Pap. 2011:155

[17] Ellman GL, Courtney KD, Andres V, Featherstone RM. A new and rapid colorimetric determination of acetylcholinesterase activity. Biochemical Pharmacology. 1961;**7**:88-95

[18] Türkeş C, Arslan M, Demir Y, Çoçaj L, Rifati Nixha A, Beydemir Ş. Synthesis, biological evaluation and in silico studies of novel N-substituted phthalazine sulfonamide compounds as potent carbonic anhydrase and acetylcholinesterase inhibitors. Bioorganic Chemistry. 2019;**89**:103004

[19] Türkeş C, Akocak S, Işık M, Lolak N, Taslimi P, Durgun M, et al. Novel inhibitors with sulfamethazine backbone: Synthesis and biological study of multi-target cholinesterases and α-glucosidase inhibitors. Journal of Biomolecular Structure & Dynamics. 2021:1-13

[20] Banks WA. Drug delivery to the brain in Alzheimer's disease: Consideration of the blood-brain barrier. Advanced Drug Delivery Reviews. 2012; **64**:629-639

[21] Maltsev AV, Bystryak S, Galzitskaya OV. The role of β-amyloid peptide in neurodegenerative diseases. Ageing Research Reviews. 2011;**10**:440-452

[22] Işık M, Beydemir Ş. AChE mRNA expression as a possible novel biomarker for the diagnosis of coronary artery disease and Alzheimer's disease, and its association with oxidative stress. Archives of Physiology and Biochemistry. 2019:1-8

[23] Auld DS, Kornecook TJ, Bastianetto S, Quirion R. Alzheimer's disease and the basal forebrain cholinergic system: Relations to β-amyloid peptides, cognition, and treatment strategies. Progress in Neurobiology. 2002;**68**:209-245

[24] Bartus RT, Emerich DF, Davis KL, Marin D, Purohit DP, Perl D, et al. Cholinergic markers in Alzheimer disease. Journal of the American Medical Association. 1999;**282**:2208-2209

[25] Shen ZX. Brain cholinesterases: III. Future perspectives of AD research and clinical practice. Medical Hypotheses. 2004;**63**:298-307

[26] Hardy JA, Higgins GA. Alzheimer's disease: The amyloid cascade hypothesis. Science. 1992;**256**:184-185

[27] Robinson SR, Bishop GM. Aβ as a bioflocculant: Implications for the amyloid hypothesis of Alzheimer's disease. Neurobiology of Aging. 2002;**23**: 1051-1072

[28] Kandasamy S, Loganathan C, Sakayanathan P, Karthikeyan S, Stephen AD, Marimuthu DK, et al. In silico, theoretical biointerface analysis and in vitro kinetic analysis of amine compounds interaction with acetylcholinesterase and butyrylcholinesterase. International Journal of Biological Macromolecules. 2021;**185**:750-760

[29] Alvarez A, Alarcón R, Opazo C, Campos EO, Muñoz FJ, Calderón FH, et al. Stable complexes involving acetylcholinesterase and amyloid-β peptide change the biochemical properties of the enzyme and increase the neurotoxicity of Alzheimer's fibrils. The Journal of Neuroscience. 1998;**18**: 3213-3223

[30] Mesulam -Marsel, Geula C. Butyrylcholinesterase reactivity differentiates the amyloid plaques of aging from those of dementia. Annals of Neurology. 1994;**36**:722-727

[31] Atack JR, Perry EK, Bonham JR, Perry RH, Tomlinson BE, Blessed G, et al. Molecular forms of acetylcholinesterase in senile dementia of Alzheimer type: Selective loss of the intermediate (10S) form. Neuroscience Letters. 1983;**40**:199-204

[32] García-Ayllón MS, Riba-Llena I, Serra-Basante C, Alom J, Boopathy R, Sáez-Valero J. Altered levels of acetylcholinesterase in Alzheimer plasma. PLoS One. 2010;**5**:e8701

[33] Schegg KM, Harrington LS, Neilsen S, Zweig RM, Peacock JH. Soluble and membrane-bound forms of brain acetylcholinesterase in Alzheimer's disease. Neurobiology of Aging. 1992;**13**: 697-704

[34] Colovic MB, Krstic DZ, Lazarevic-Pasti TD, Bondzic AM, Vasic VM. Acetylcholinesterase inhibitors: Pharmacology and toxicology. Current Neuropharmacology. 2013;**11**:315-335

[35] Taylor P, Radić Z. The cholinesterases: From genes to proteins. Annual Review of Pharmacology and Toxicology. 1994;**34**:281-320

[36] Tougu V. Acetylcholinesterase: mechanism of catalysis and inhibition. Current Medicinal Chemistry-Central Nervous System Agents. 2001;**1**:155-170

[37] Lopes JPB, Silva L, da Costa FG, Antonio Ceschi M, Seibert Lüdtke D, Ferreira Dantas R, et al. Design, synthesis, cholinesterase inhibition and molecular modelling study of novel tacrine hybrids with carbohydrate derivatives. Bioorganic. Medicinal Chemistry. 2018;**26**:5566-5577

[38] Masson P, Legrand P, Bartels CF, Froment MT, Schopfer LM, Lockridge O. Role of aspartate 70 and tryptophan 82 in binding of succinyldithiocholine to human butyrylcholinesterase. Biochemistry. 1997;**36**:2266-2277

[39] Novikov FN, Chilov GG. Molecular docking: Theoretical background, practical applications and perspectives. Mendeleev Communications. 2009;**19**: 237-242

[40] Mohan V, Gibbs A, Cummings M, Jaeger E, DesJarlais R. Docking: Successes and challenges. Current Pharmaceutical Design. 2005;**11**:323-333

[41] Release S. 2: Maestro. New York, NY: Schrödinger, LLC; 2017

[42] Larik FA, Shah MS, Saeed A, Shah HS, Channar PA, Bolte M, et al. New cholinesterase inhibitors for Alzheimer's disease: Structure activity relationship, kinetics and molecular docking studies of 1–butanoyl–3–arylthiourea derivatives. International Journal of Biological Macromolecules. 2018;**116**:144-150

[43] Ujan R, Saeed A, Channar PA, Larik FA, Abbas Q, Alajmi MF, et al. Drug-1,3,4-thiadiazole conjugates as novel mixed-type inhibitors of acetylcholinesterase: Synthesis, molecular docking, pharmacokinetics, and ADMET evaluation. Molecules. 2019;**24**:860

[44] Kumar S, Chowdhury S, Kumar S. In silico repurposing of antipsychotic drugs for Alzheimer's disease. BMC Neuroscience. 2017;**18**:1-16

[45] Adler M, Filbert MG. Role of butyrylcholinesterase in canine tracheal smooth muscle function. FEBS Letters. 1990;**267**:107-110

[46] Akdag A, Işık M, Göktaş H. Conducting polymer-based electrochemical biosensor for the detection of acetylthiocholine and

pesticide via acetylcholinesterase. Biotechnology and Applied Biochemistry. 2021;**68**:1113-1119

[47] Villarroya M, García A, Marco J. New classes of AChE inhibitors with additional pharmacological effects of interest for the treatment of Alzheimers disease. Current Pharmaceutical Design. 2004;**10**:3177-3184

[48] Giacobini E. Selective inhibitors of butyrylcholinesterase: A valid alternative for therapy of Alzheimer's disease? Drugs and Aging. 2001;**18**: 891-898

[49] Işık M, Beydemir Ş. The impact of some phenolic compounds on serum acetylcholinesterase: Kinetic analysis of an enzyme/inhibitor interaction and molecular docking study. Journal of Biomolecular Structure & Dynamics. 2021;**39**:6515-6523

[50] Işık M. The binding mechanisms and inhibitory effect of intravenous Anesthetics on AChE In vitro and In vivo: Kinetic analysis and molecular docking. Neurochemical Research. 2019; **44**:2147-2155

[51] Giacobini E. Cholinesterase inhibitors: New roles and therapeutic alternatives. Pharmacological Research. 2004;**50**:433-440

[52] Weiner L, Shnyrov VL, Konstantinovskii L, Roth E, Ashani Y, Silman I. Stabilization of torpedo californica acetylcholinesterase by reversible inhibitors. Biochemistry. 2009;**48**:563-574

[53] Türkeş C, Beydemir Ş. Inhibition of human serum Paraoxonase-I with Antimycotic drugs: In vitro and In silico studies. Applied Biochemistry and Biotechnology. 2020;**190**:252-269

[54] Demir Y, Işık M, Gülçin İ, Beydemir Ş. Phenolic compounds inhibit the aldose reductase enzyme from the sheep kidney. Journal of Biochemical and Molecular Toxicology. 2017;**31**:e21936

[55] Lineweaver H, Burk D. The determination of enzyme dissociation constants. Journal of the American Chemical Society. 1934;**56**:658-666

[56] Sharma K. Cholinesterase inhibitors as Alzheimer's therapeutics (review). Molecular Medicine Reports. 2019;**20**: 1479-1487

[57] Weinstock M. Selectivity of cholinesterase inhibition. CNS Drugs. 1999;**12**:307-323

[58] Ogura H, Kosasa T, Kuriya Y, Yamanishi Y. Comparison of inhibitory activities of donepezil and other cholinesterase inhibitors on acetylcholinesterase and butyrylcholinesterase in vitro. Methods and Findings in Experimental and Clinical Pharmacology. 2000;**22**:609-613

[59] Rogers SL, Farlow MR, Doody RS, Mohs R, Friedhoff LT. A 24-week, double-blind, placebo-controlled trial of donepezil in patients with Alzheimer's disease. Neurology. 1998;**50**:136-145

[60] Jacobson SA, Sabbagh MN. Donepezil: Potential neuroprotective and disease-modifying effects. Expert Opinion on Drug Metabolism and Toxicology. 2008;**4**:1363-1369

[61] Kryger G, Silman I, Sussman JL. Structure of acetylcholinesterase complexed with E2020 (Ariceptρ): Implications for the design of new anti-Alzheimer drugs. Structure. 1999;**7**: 297-307

[62] Nordberg A, Svensson AL. Cholinesterase inhibitors in the

treatment of Alzheimer's disease. A comparison of tolerability and pharmacology. Drug Safety. 1998;**19**: 465-480

[63] Coelho F, Birks JFC. Physostigmine for Alzheimer's disease. Cochrane Database of Systematic Reviews. 2001: Cd001499

[64] Inglis F. The tolerability and safety of cholinesterase inhibitors in the treatment of dementia. International Journal of Clinical Practice, Supplement. 2002;**127**: 45-46

[65] Onor ML, Trevisiol M, Aguglia E. Rivastigmine in the treatment of Alzheimer's disease: An update. Clinical interventions in aging. 2007;**2**:17-32

[66] Corey-Bloom J, Anand R, Veach J. A randomized trial evaluating the efficacy and safety of ENA 713 (rivastigmine tartrate), a new acetylcholinesterase inhibitor, in patients with mild to moderately severe Alzheimer's disease. International Journal of Geriatric Psychopharmacology. 1998;**1**:55-65

[67] Harel M, Schalk I, Ehret-Sabatier L, Bouet F, Goeldner M, Hirth C, et al. Quaternary ligand binding to aromatic residues in the active-site gorge of acetylcholinesterase. Proceedings of the National Academy of Sciences of the United States of America. 1993;**90**:9031-9035

[68] Farlow M, Gracon SI, Hershey LA, Lewis KW, Sadowsky CH, Dolan UJ. A controlled trial of Tacrine in Alzheimer's disease. JAMA J Am Med Assoc. 1992; **268**:2523-2529

[69] Guo AJY, Xie HQ, Choi RCY, Zheng KYZ, Bi CWC, Xu SL, et al. Galangin, a flavonol derived from Rhizoma Alpiniae Officinarum, inhibits acetylcholinesterase activity in vitro. Chemico-Biological Interactions. 2010; **187**:246-248

[70] Szwajgier D. Anticholinesterase activity of phenolic acids and their derivatives. Zeitschrift fur Naturforsch - sect C Journal of Biosciences. 2013;**68**: 125-132

[71] Anwar J, Spanevello RM, Thomé G, Stefanello N, Schmatz R, Gutierres J, et al. Effects of caffeic acid on behavioral parameters and on the activity of acetylcholinesterase in different tissues from adult rats. Pharmacology, Biochemistry, and Behavior. 2012;**103**: 386-394

[72] Lemes LFN, De Andrade RG, De Oliveira AS, Da Silva FMR, De Castro CG, Da Silva BM, et al. Cardanol-derived AChE inhibitors: Towards the development of dual binding derivatives for Alzheimer's disease. European Journal of Medicinal Chemistry. 2016; **108**:687-700

[73] Camps P, El Achab R, Morral J, Munoz-Torrero D, Badia A, Eladi Banos J, et al. New tacrine-huperzine a hybrids (huprines): Highly potent tight-binding acetylcholinesterase inhibitors of interest for the treatment of Alzheimer's disease. Journal of Medicinal Chemistry. 2000;**43**:4657-4666

[74] Korabecny J, Musilek K, Holas O, Binder J, Zemek F, Marek J, et al. Synthesis and in vitro evaluation of N-alkyl-7-methoxytacrine hydrochlorides as potential cholinesterase inhibitors in Alzheimer disease. Bioorganic Med Chem Lett. 2010;**20**:6093-6105

[75] Ali MA, Yar MS, Hasan MZ, Ahsan MJ, Pandian S. Design, synthesis and evaluation of novel 5,6-dimethoxy-1-oxo-2,3-dihydro-1H-2-indenyl-3,4-substituted phenyl methanone

analogues. Bioorganic Med Chem Lett. 2009;**19**:5075-5077

[76] De La Torre P, Saavedra LA, Caballero J, Quiroga J, Alzate-Morales JH, Cabrera MG, et al. A novel class of selective acetylcholinesterase inhibitors: Synthesis and evaluation of (E)-2-(benzo[d]thiazol-2-yl)-3-heteroarylacrylonitriles. Molecules. 2012;**17**:12072-12085

[77] Lolak N, Akocak S, Türkeş C, Taslimi P, Işık M, Beydemir Ş, et al. Synthesis, characterization, inhibition effects, and molecular docking studies as acetylcholinesterase, α-glycosidase, and carbonic anhydrase inhibitors of novel benzenesulfonamides incorporating 1,3,5-triazine structural motifs. Bioorganic Chemistry. 2020;**100**:103897

[78] Taslimi P, Işık M, Türkan F, Durgun M, Türkeş C, Gülçin İ, et al. Benzenesulfonamide derivatives as potent acetylcholinesterase, α-glycosidase, and glutathione S-transferase inhibitors: Biological evaluation and molecular docking studies. Journal of Biomolecular Structure & Dynamics. 2021;**39**: 5449-5460

[79] Işık M, Demir Y, Durgun M, Türkeş C, Necip A, Beydemir Ş. Molecular docking and investigation of 4-(benzylideneamino)- and 4-(benzylamino)-benzenesulfonamide derivatives as potent AChE inhibitors. Chemical Papers. 2020;**74**:1395-1405

[80] Akocak S, Taslimi P, Lolak N, Işık M, Durgun M, Budak Y, et al. Synthesis, characterization, and inhibition study of novel substituted Phenylureido Sulfaguanidine derivatives as α-glycosidase and cholinesterase inhibitors. Chemistry & Biodiversity. 2021;**18**:e2000958

Chapter 4

Trypsins: Structural Characterization and Inhibition Focus in Insects

Yaremis Beatriz Meriño-Cabrera and Maria Goreti de Almeida Oliveira

Abstract

Serine proteases are considered the main class of protein digestive enzymes present in the midgut of many lepidopteran species and are the focus of the review in this chapter. Among them, trypsin and chymotrypsin are the most studied and participate in a great diversity of physiological processes that include, in addition to digestion, activation of specific proteins, such as in the coagulation cascades, in the immune system of insects and plants, in the development and production of biologically active peptides, in signal transduction, hormone activation, and development. In this chapter, a review was made of the structural characteristics of trypsins, specifically of Lepidoptera insects, main experimental and theoretical techniques for the study of their function and structure, and interaction with other proteins and ligands as protease inhibitors. Finally, it was described how this type of hydrolases can be a focus of inhibition in pests to the detriment of the development and death of the target insect. Until now, the main strategies of agricultural crop management, especially of large crops, consist of the use of inorganic pesticides and transgenic cultivars containing *Bacillus thuringiensis* toxins. Therefore, new and ecologically friendly strategies are necessary, such as the use of protease inhibitors.

Keywords: 3D structure of trypsin, catalytic triad, inhibitor protein, larvae, pest

1. Introduction

Enzymes have extraordinary catalytic power, often greater than synthetic or inorganic catalysts. They have a high degree of specificity for their respective substrates, accelerate chemical reactions, and act in aqueous solutions under mild temperature and pH conditions. Few non-biological catalysts have this set of properties. Except for a small group of catalytic RNA molecules, all enzymes are proteins. Enzymes are at the heart of every biochemical process. Acting in organized sequences, they catalyze each of the reactions of the hundreds of steps that degrade nutrient molecules, that conserve and transform chemical energy, and that build biological macromolecules from elementary precursors [1, 2].

One type of enzyme is the proteases or peptidases, molecules that promote cleavage through the hydrolysis of peptide bonds present in proteins and polypeptides, transforming them into a smaller amino acid or polypeptide residues [3]. The term protease appeared in the German literature of physiological chemistry in the latter part of the nineteenth century about proteolytic enzymes and was used as a general term embracing all the hydrolases that act on proteins or further degrade the fragments of them [4].

Proteolytic cleavage of peptide bonds is one of the most frequent and important enzymatic modifications of proteins. Historically, enzymatic proteolysis has generally been associated with protein digestion and early drew the attention of physiologists and biochemists who were interested in the process of protein digestion in animals and humans [5]. Hence, the digestive proteases of the pancreatic and gastric secretions are among the best-characterized enzymes, and the current knowledge of protein structure and enzyme function has been derived from a study of these proteases. Investigations of the kinetics, specificity, and inhibition, together with detailed analyses of their amino acid sequence and X-ray structure, have led to the identification of the components and geometry of their active sites, and from these, the mechanism of action of these digestive proteases was deduced [6]. As a result, it became evident that proteases can be classified into families, members of each family having similar structures and mechanisms of action.

Proteases are classified according to the Enzyme Commission of the International Union of Biochemistry and Molecular Biology (IUBMB) within group 3 (hydrolases), subgroup 4. They are also classified based on three criteria: (1) type of reaction catalyzed, (2) chemical nature of the catalytic site, and (3) evolutionary relationship according to structure [7].

Proteases are subdivided into two main groups, the exopeptidases, and the endopeptidases, depending on their site of action. The exopeptidases cleave peptide bonds close to the amino or carboxy-terminal group in the substrate, while endopeptidases cleave peptide bonds far from the terminal group of the protein substrate [5, 8].

Insect digestive proteases are classified based on the functional group present in the active site into serine, cysteine, aspartyl, and metalloproteases [5, 8, 9]. Serine proteases have a serine residue at their active center, while aspartyl proteases have two aspartic acid units at their catalytic center. Cysteine-proteases have an amino acid cysteine, and metalloproteases have a metal ion in their catalytic mechanism [8]. Some insects often have multiple digestive proteases in their intestinal tract, belonging to different or the same mechanic group, although they usually use one principal type in their digestive function [10, 11].

Serine proteases are considered the main class of digestive protein enzymes present in the midgut of many lepidopteran species [9] and are the review focus in this chapter. Among them, trypsin and chymotrypsin are the most studied and participate in a great diversity of physiological processes that include, in addition to digestion, activation of specific proteins, such as in the coagulation cascades, in the immune system of insects and plants, in the development and production of biologically active peptides, in signal transduction, hormone activation, and development [12–14].

Serine proteases belong to the largest gene families in the animal kingdom, are widely distributed in nature, and are found in all assemblies of cellular life, as well as in several viral genomes, indicating vital participation in the metabolism of these organisms [14]. In insects, a study with *Helicoverpa armigera* demonstrated the existence of at least 28 different genes belonging to the serine-proteases family whose products are expressed in the gut [15]; Furthermore, a proteomic analysis of the gut of the soybean pest *Anticarsia gemmatalis* where proteases were characterized by LC/MS, 54 expressed antigens were found for gut protease, suggesting multiple important isoforms involved in the digestion process and other functions in the larval gut [16].

In Ref. [8] it is described that serine proteases are generally active at neutral and alkaline pH, with a pH optimum between 7.0 and 11.0. They have broad specificity, including amidase and esterase activities. The molecular mass of serine proteases is generally in the range of 18–35 kDa [17, 18]. However, several organisms have serine proteases with higher molecular masses, such as Melolontha melolontha (Coleoptera: ɛelolonthidae), whose molecular mass for two trypsin-like enzymes is 56 and 63 kDa [19]. The isoelectric point of serine proteases is generally in the pH range 4.0 and 6.0 [16, 20].

The catalytic function of serine proteases is realized through the action of the catalytic triad (Serine 195 reactive, Histidine 57, and Aspartic acid 102 to trypsin bovine). The degree and type of substrate specificity are determined by the nature of the active center region [21]. When residues in the catalytic triad are altered, separately or together, large changes in the enzyme turnover rate (Kcat) occur, changing the enzyme mechanism, with little effect on KM. The residues of the triad act in perfect synergism and contribute to an optimized catalytic activity [22]. Serine proteases generally act in a two-step hydrolysis reaction, where a covalently linked intermediate, acyl-enzyme, is formed.

This acylation is followed by the deacylation, a process in which a water-mediated nucleophilic attack occurs, resulting in hydrolysis of the peptide. The nucleophilic attack of the hydroxyl group of the serine residue 195 on the carboxylic carbon atom of the peptide bond, catalyzed by the histidine residue, which acts as a base, leads to the formation of a tetrahedral intermediate and an imidazole ion. The intermediate decomposes via acid-base catalysis by the action of the polarized groups of aspartate and histidine into an acyl-enzyme intermediate, an imidazole base, and an amine. This mechanism involves close contact between the tetrahedral intermediate and the imidazole ion, which inhibits proton release to the solvent medium before acid-base catalysis, regenerating the active enzyme and releasing the degradation product [24]. Each step occurs through the formation of a tetrahedral intermediate, whose structure resembles a high-energy transition state in both reactions. This mechanism is capable of accelerating the speed of peptide bond hydrolysis more than 109-fold compared with the uncatalyzed reaction [14, 23–25].

The ratio between the speed of acylation and deacylation depends on the type of substrate used. For an amide substrate, the velocity of acylation is slower than for deacylation, and for an ester substrate, can be one to three times faster. Therefore, in the amidase activity, the acylation step is slow and deacylation fast, while in esterase enzymes the acylation step is fast and the deacylation step is slow, and the slow step is the limiting step in hydrolysis [26].

2. Trypsin-*like* enzymes in insects

The occurrence of different digestive enzymes in the insect alimentary canal is usually associated with the chemical composition of the ingested diet. However, the theory of dietary adaptation should not disregard phylogenetic aspects in determining the type (not quantity) of enzymes present in the insect gut. The possibility that insects possess a wide variety of digestive enzymes, whose relative amounts present may change in response to diet, is considered. This change could occur during an individual's feeding or result from the adaptation of a taxonomic group of insects to a particular diet, resulting in the presence of enzymes whose activities are permanently greater than others [27].

In Ref. [28] it was demonstrated that the intestinal proteolytic profile changes during larval development of *Anticarsia gemmatalis* caterpillars, the activity of cysteine proteases is more intense in the first instar. On the contrary, the serine proteases showed major activities in the late stages of the larval phase. Furthermore, Zymogram analysis and protein identification by liquid chromatography–mass spectrometry indicated serine protease as the main protease class expressed in the fifth instar.

Protein digestion in lepidopterans is performed mainly by serine proteases, the increase in the structural and functional diversity of genes that code for this sub-subclass can be attributed to the insect's response machinery to circumvent of protease inhibitors (PIs). Therefore, these constitutively expressed proteins represent an adaptive advantage [16].

Trypsins (EC 3.4.21.4) are serine proteases and are considered the most important digestive proteases of most insects. Trypsins are involved in the initial phase of protein digestion and are characterized by containing a catalytic triad consisting of the amino acid residues Hys, Asp, and Ser; in *A. gemmatalis* trypsin are Ser 229, Hys 85, and Asp 132 [16, 29, 30].

Every trypsin-like serine protease has a preference for substrates with a basic residue at P1, Lys, or Arg. This is mainly caused by the presence of a negatively charged Asp 189 at the bottom of the S1 pocket (numbering used in chymotrypsinogen). The architecture of the S1 site among these proteases is highly conserved. A striking difference is found at position 190, which can be an Ser or Ala, and serves as an identification point for subfamilies [31]. The nomenclature for substrate amino acid residues is Pn, ..., P2, P1, P1′, P2′, ..., Pn′, where P1-P1' denotes the hydrolyzed bond. Sn, ..., S2, S1, S1′, S2′, ..., Sn′ denotes the corresponding binding subsites of the enzyme [22].

Insect trypsins have similar primary specificity; except Lepidoptera, these proteases hydrolyze more efficiently substrates containing Arg than Lys at the P1 position [32, 33]. Lepidopteran trypsins have higher specificity for Lys-containing substrates than for Arg-containing substrates. Protease inhibitors produced by plants present a region known as the reactive site, which interacts with its target enzyme. The sequence alignment of several plant protease inhibitors indicated that the reactive sites of most of these inhibitors have a Lys residue at the P1 position [32]. The presence of a Lys at the P1 site in the reactive site is a survival strategy, as these inhibitors would act by inhibiting the trypsins of most insects that preferentially hydrolyze Arg at this position [30, 34].

Trypsins isolated from the midgut of various insects typically exhibit molecular mass between 20 and 35 kDa and optimal activity at alkaline pH [17, 18]. Trypsins from lepidopterans typically exhibit higher pH optimum corresponding to the high pH value of the midgut contents. Most lepidopteran serine proteinases sequences do not contain lysine residues. The lysine residue at position 188 is conserved in mammalian and dipteran trypsins. However, in Lepidoptera, this residue is replaced by arginine. It has been suggested that this substitution is a necessary adaptation for the stability of the enzyme in the gut of these insects, where the pH value is very high, associated with the need for the digestive enzymes to remain protonated [15, 35]; However, the presence of lysine in the sequence of the digestive trypsins of *Sesamia nonagroides* and *Helicoverpa armigera* has already been described [36].

Although the primary specificity of trypsins from most insects is similar to that of vertebrates, and they show sequence homology in the catalytic site region, their properties contrast in several respects. Insect trypsins, in most cases, are unstable at acidic pH, exhibit different sensitivities to inhibitors, and typically contain fewer cysteine pairs in their sequences than do vertebrate trypsins [37].

In vertebrates, calcium prevents the aggregation of enzyme molecules, protecting it from autolysis and denaturation by heat, inducing a conformational change in its structure to a more compact form, which is necessary for catalytic activity [38]. According to the literature, insect trypsins are not influenced in their tryptic activity by calcium ions, which has been demonstrated in several studies on serine proteases from various insects such as *Spodoptera litura* (Lepidoptera) [39], in *Helicoverpa armigera* (Lepidoptera) [40]; soluble and membrane trypsin-like proteins of *Musca domestica* (Diptera) [41]; trypsin of *Spodoptera littoralis* (Lepidoptera) [42], and in Ref. [11] Pilon and coworkers analyzed the trypsins of *A. gemmatalis* (Lepidoptera), verifying that trypsins found in these insects are not stabilized or activated by calcium ions.

The autolysis of insect trypsins typically differs from that of mammalian trypsins. The stabilization of mammalian trypsins depends on calcium binding to the binding motif in their structure; this motif is not present in the sequences of insect trypsins. The autocatalytic sites in mammalian trypsins (Lys 61-Ser 62, Arg 117-Val 118, and Lys 145-Ser 146) are not conserved in insects, having their cleavage sites [33]. However, several authors found that the trypsins of the studied insects suffered a drop in tryptic activity on the substrate BApNA in the presence of EDTA, a result that led the authors to suggest that calcium maybe, in some way, acting on the enzymes of these insects [19, 43, 44].

The processing and secretion mechanisms of insect trypsins also appear to include aspects that differ from other animals. There is evidence that the soluble form of insect trypsin is derived from a membrane-bound form [45]. The presence of trypsin in the soluble form whose kinetic and physical properties were identical to the microvillar membrane-associated form in *M. domestica* [46] led the authors to propose a mechanism of trypsin secretion, where trypsin is synthesized in the midgut cells of insects in the active form but associated with the vesicle membrane, which will be processed in the microvilli becoming soluble before being secreted [45].

Through immunolocalization studies, trypsins also were detected in more than one form in the gut of *Spodoptera frugiperda* (Lepidoptera) [47]. Trypsins were observed associated with the microvillar membrane of midgut cells, the membrane of secretory vesicles, and within the microvilli. A secretion model was proposed, where trypsin is synthesized bound to the membrane via a hydrophobic anchor peptide. Subsequently, the enzyme would be processed in the Golgi complex and transported in secretory vesicles. These vesicles migrate through the microvilli and either before or after fusing with the microvillar membrane are released into the intestinal lumen in the form of double- or single-membrane vesicles, respectively. The double-membrane vesicles fuse to form a single membrane. The trypsin present on the luminal surface of these vesicles is then solubilized by limited proteolysis or by the dissolution of the vesicles due to the highly alkaline pH of the midgut contents. Remnant membranes, with some trypsin bound, are finally incorporated into the peritrophic membrane [48].

In *A. gemmatalis*, a study provided evidence of the presence of membrane-bound trypsin-like proteases in midgut preparations of the velvet bean caterpillar, a key soybean pest in warm climates, and the likely occurrence of members of other protease families [43]. The detection of proteolytic activity in the insoluble fraction from the midgut of *A. gemmatalis*, after treatment with Triton X-100 and centrifugation at

100,000 g, indicates the occurrence of membrane-bound proteases that may be least partially transferred to the peritrophic membrane [43].

3. Methodologies and techniques for the study of digestives trypsin-*like* enzyme of lepidoptera larvae

3.1 In vitro

3.1.1 Gel-filtration chromatography

Gel-filtration chromatography is performed on a Superose 12 HR10/30 column (10 mm × 30 cm) in an FPLC system equilibrated with 0.1 M Tris-HCl pH 7.5 and 0.1 M NaCl. A flow rate of 0.5 mL/min is used and fractions of 1.5 mL, collected in each tube, and before collection, 1.5 M Phomic acid is added to keep the collected samples at pH 3.0. The sample applied to the gel-filtration column is the enzyme extract obtained from the intestines of the insect larvae. The enzymatic activity of each fraction collected in the chromatographs is monitored using L-BApNA as substrate. For calibration of the Superose 12 HR10/30 column, the following molecular mass standards are used: Blue Dextran (2 × 106 Da), amylase (205,000 Da), alcohol dehydrogenase (150,000 Da), BSA (66,000 Da), ovalbumin (45,000 Da), chymotrypsinogen (25,000 Da), cytochrome C (12,327 Da), aprotinin (6500 Da), which are applied to the column; and the retention time and Kav obtained for each standard applied, through these data, it is possible to calculate the molecular mass ranges of the samples of interest.

3.1.2 Ion exchange chromatography

The samples after gel-filtration chromatography are dialyzed, in membranes with a cutoff of 12,000–14,000 Da, 150 times against 5 mM Tris–HCl pH 7.5 buffer, and subsequently applied (10 mL per run) to the ion exchange column. Ion exchange chromatography is performed on a Mono-Q HR 5/5 column (5 mm × 5 cm). Equilibrated with 0.1 M Tris–HCl pH 7.5 and eluted with a NaCl gradient. A flow rate of 1 mL/min is used and 1.5 ml fractions are collected. The enzyme activity of each fraction is monitored using L-BApNA as substrate.

3.1.3 Affinity chromatography

Affinity chromatography by column HiTrap Benzamidine is an efficient method for partial purification of trypsin-like proteases and has been used in other studies getting similar results [11, 18]. The efficiency of the method may be due to the fact that the benzamidine is a potent competitive inhibitor of trypsin-like proteases that occupies the S1 subsite (site of specificity) of the enzyme; benzamidine is stabilized by hydrophobic interactions in its hydrophobic pocket and electrostatic interactions between the amidine group and carboxyl group belonging to the residue aspartic acid present in the bottom of the S1 site [11, 49].

The samples, after gel-filtration or ion-exchange chromatography, have their pH set to 7.5 and were applied (5 mL) to affinity chromatography. This chromatography is done on a HiTrap Benzamidine column equilibrated with 0.1 M Tris-HCl, pH 7.5, and eluted with 10–3 M HCl. A flow rate of 0.5 mL/min is used, and 1.5 mL fractions are collected. The enzymatic activity of each fraction is monitored, using L-BApNA as substrate.

3.1.4 Enzyme activity

The amidase activity is carried out according to the method already described [50], using sample (extract of the intestate or enriched trypsin fraction after chromatography), chromogenic substrate L-BApNA at a final concentration of 0.5 mM, and 0.1 M Tris-HCl buffer, pH 8.2. The initial velocities are determined by the formation of the product p-nitroanilide, by measuring its absorbance at 410 nm as a function of time (120 s), using for the calculations of the molar extinction coefficient of 8800 $M^{-1} \times cm^{-1}$ for the product. The experiment is performed in triplicates.

3.1.5 Determination of protein concentration

The determination of the total protein concentration of the samples is estimated by the method described [51], using a 0.2 mg/mL BSA solution to obtain a standard curve for quantification.

3.1.6 SDS-PAGE

The efficiency of the purification method was also confirmed by the results of SDS-PAGE with a reduction of protein species in the purified sample compared with the crude extract. Electrophoresis is performed by the method already described [51]. Using a 12.5% polyacrylamide gel in the presence of SDS (0.1%), the experiment is performed at a constant voltage of 100 V for 1 h20 at room temperature. The gel is stained by silver or Coomassie blue staining. The molecular mass standard used is purchased from Invitrogen ("BenchMark ™ Protein Ladder").

3.1.7 Zymogram

Zymogram is performed using 12.5% SDS-PAGE containing 0.1% copolymerized gelatin. Electrophoresis occurred at 50 V at 4°C, and the gel was subsequently incubated in 2.5% (v/v) Triton X-100 for 1 h at room temperature and under stirring to remove SDS. After incubation, the gels are washed and again incubated with 0.1 M Tris-HCl buffer, pH 8.0, for 2 h at 37°C. The activity is revealed by staining with "Coomassie Blue" R-250 (0.25%).

3.2 In silico methods

The interaction between PIs and insect trypsin-like enzymes is an example of ligand-macromolecule recognition, required in the plant defense process. The understanding of these recognition mechanisms is one of the central aspects for the success in the discovery of new promissory compounds. The characterization of binding mode between inhibitor and protease can be obtained from several methods, among them the in silico studies that allow the reduction of time and costs, specifically docking-molecular analyzes allow to identify the binding conformation and affinity quantification [52]. Below we will show the main methodologies.

3.2.1 Molecular modeling

Molecular modeling encompasses all the computational techniques used to simulate the behavior of molecules. These techniques are widely used in the fields of computational chemistry and drug development to study biological systems and can therefore be applied to the discovery of enzyme inhibitors.

3.2.1.1 Comparative protein modeling

Functional characterization of protein sequences is a frequent problem in the biological field. It is now well established that knowledge of molecular structure is a powerful tool to understand, control, and alter functions of biomolecules. Although three-dimensional structures of proteins can be determined by X-ray crystallography and nuclear magnetic resonance (NMR), these experiments demand time and large quantities of proteins in self-purify and have some limitations. The NMR technique is difficult to apply to large proteins (greater than 250 amino acid residues), or very flexible proteins, while X-ray crystallography depends on obtaining crystals with good diffraction ability, a process performed by trial and error, and solving the phase problem [53]. However, protein sequences can be determined much more easily by using molecular biology and protein sequencing techniques. Therefore, in cases where the structure cannot be determined experimentally, homology modeling can often produce a useful three-dimensional model of a target sequence based on its similarity to a protein with a known structure used as a template protein [54].

The principle of molecular homology modeling is based on the fact that throughout evolution the structures of proteins are more conserved than their sequence [53, 55]. The biological evolution of proteins has some rules such as homology between amino acid residue sequence implies structural and functional similarity; homologous proteins have conserved internal regions (mainly consisting of secondary structure elements: α-helices and β-sheets); structural changes between homologous proteins occur in the loop regions [55]. Furthermore, proteins are grouped into a limited number of three-dimensional families making it possible to model the proteins of interest if there is a member of the family that already has its structure determined. A model built by comparative modeling needs that at least one 3D structure of the family in question has been elucidated by experimental techniques. Another important point is the identity between the sequences (target and template), this value should be above 25% so that the generated model can be reliable [53, 56]. Homology molecular modeling features four main steps: the search for homologous protein sequences, the alignment of the sequences, the construction and optimization of the models, and finally, the evaluation and validation of the generated structures [56].

3.2.1.2 Phyre2 and protein modeling

The Phyre2 system is a combination of software created and written in several languages by a researcher's group in London, England. The system runs on a Linux program with an approximately 300-core CPU. The Phyre2 server can be used in several ways, depending on the user's research focus. The most commonly used facility is the prediction of the 3D structure of a single submitted protein sequence [57].

The Phyre and Phyre2 servers predict the three-dimensional structure of a protein sequence using the principles and techniques of homology modeling. A protein sequence of interest (the target) can be modeled with reasonable accuracy using a sequence far removed from the known structure (the template) since the relationship between the target and template can be discerned through sequence alignment. Currently, the most powerful and accurate methods for detecting and aligning remotely related sequences rely on profiles or hidden Markov models (HMMs). These profiles/HMMs capture

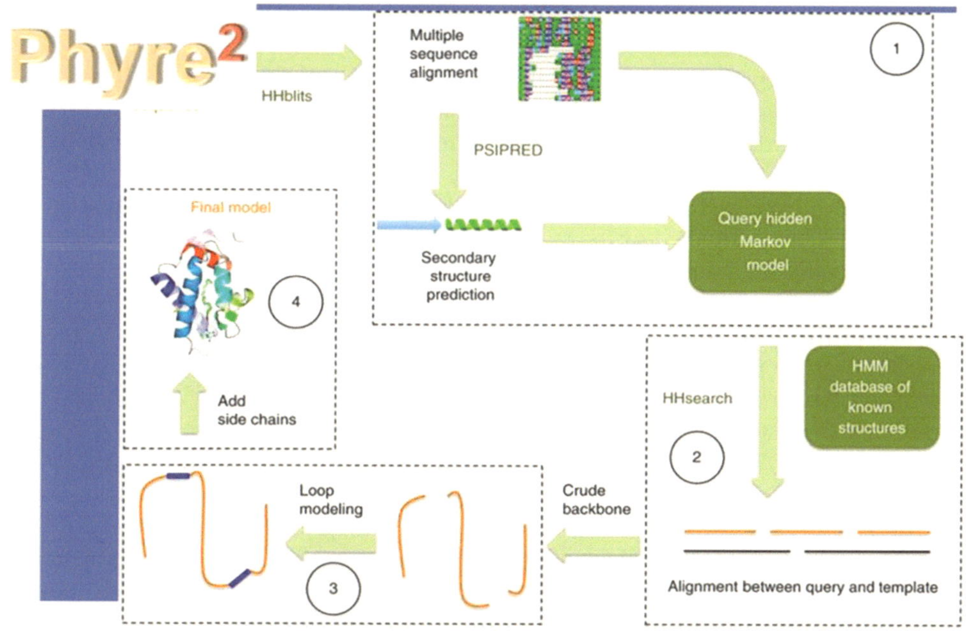

Figure 1.
Normal-mode Phyre2 pipeline showing algorithmic stages taken to the Phyre2 web portal for protein modeling, prediction, and analysis. Stage numbers are shown in circles, and elements within a stage are surrounded by a dashed box. Stage 1 (gathering homologous sequences): A query sequence is scanned against the specially curated nr20 (no sequences with >20% mutual sequence identity) protein sequence database with HHblits. The resulting multiple-sequence alignment is used to predict secondary structure with PSIPRED and both the alignment and secondary structure prediction combined into a query hidden Markov model. Stage 2 (fold library scanning): This is scanned against a database of HMMs of proteins of known structure. The top-scoring alignments from this search are used to construct crude backbone-only models. Stage 3 (loop modeling): Indels in these models are corrected by loop modeling. Stage 4 (side-chain placement): amino acid side chains are added to generate the final Phyre2 model [58].

the mutational propensity of each position in an amino acid residue sequence based on mutations observed in related sequences and can be thought of as an "evolutionary fingerprint" of a specific protein (**Figure 1**) [57, 58].

Typically, the amino acid residue sequences of a representative set of all known three-dimensional protein structures are compiled, and these sequences are processed by scanning into a large protein sequence database. The result is a database of profiles or HMMs, one for each known 3D structure. A user sequence of interest is processed in a similar way to form a profile/HMM. This user profile is verified in the profile database using profile-profile or HMM-HMM alignment techniques. These alignments can also take into account predicted or known secondary structure element patterns and can be scored using various statistical models [57, 58].

3.2.2 Quality and validation of three-dimensional protein models

The quality of the generated models depends mainly on the existence of appropriate templates, i.e., with good resolution, high coverage, and high identity. For close homologs, the most commonly used programs in most cases generate resolutive models with RMSD (root mean square deviation) of approximately 2 Å from the experimental structure. Generally, a sequence identity above 35% is sufficient to produce good models for proteins above approximately 100 amino acid residues, and as the similarity between target and template decreases, the model error increases [59].

Among the programs used for validation of stereochemical features of protein structures is the PROCHECK program, which uses selected stereochemical information from high-quality structures to provide an overall assessment of the structure and to highlight regions that need refinement.

The program can be used independently of experimental data and applied to structures already published or to structures generated by comparative modeling. The program also analyzes the torsion angles of the main (phi and psi) and side chains of the molecule informing the bad contacts and planarity of peptide bonds [60]. One of the most well-known output files generated by the program is the Ramachandran plot that presents a correlation between the torsional angles of the main chain. Analysis of the rotation of these angles led to the identification of allowed and disallowed regions where collisions between atoms occur [60].

Another widely used program for validation is ProSa, a tool that relies on a statistical analysis of all protein structures deposited in the PDB. Soluble protein structures whose z-scores, a score used by the program to evaluate the quality of three-dimensional protein models, depart from the averages obtained for experimentally determined structures are uncommon and usually stem from various structural errors. This tool uses a knowledge-based function of the Potential of Mean Force type, which describes the preferred geometries of a given sequence of amino acid residues by statistically analyzing the interaction geometries between atoms of structures deposited in the PDB [61].

3.2.3 Molecular docking

Understanding the mechanisms of protein-ligand molecular recognition is one of the central aspects for the discovery and planning of new compounds. Obtaining an accurate and automated description of the molecular recognition process, using computational methodologies, can allow the reduction of the time and high costs involved in the development of new drugs [62].

Among these methodologies, receptor-ligand docking has contributed significantly to advances in drug development and is employed in the refinement and optimization of previously identified prototype compounds, virtual database screening, and estimation of protein-ligand binding affinities. The molecular docking methodology aims to predict the binding orientation of two molecules forming a stable complex and to estimate the binding affinity between them. Therefore, the success of the technique is measured by comparing the predicted results with binding modes determined by X-ray crystallography of the complexes and affinity measurements determined in vitro assays [62].

To perform molecular docking, basically, three steps are required: definition of the structure of the target molecule, location of the binding site, and prediction of the binding mode and affinity of a ligand using specific algorithms. The structure can be obtained by X-ray crystallography, NMR techniques, or computationally predicted by comparative modeling as described earlier. The application of these models for docking is already well established and represents an important alternative when experimental structures are not yet available [63].

The prediction of binding mode and affinity is performed using search algorithms and evaluation functions, two main aspects that differentiate docking programs. Search algorithms are used to sample the possible orientations of the ligands bound to the protein target, considering translational, rotational, and conformational degrees of freedom (which evaluate the dihedral angles associated with simple covalent bonds).

Evaluation functions can be divided into three main classes: force-field-based functions, empirical functions, and knowledge-based functions. Force-field-based functions use a force field to calculate the binding energy between the ligand and the protein target. Empirical functions use empirical and semiempirical methods whose coefficients have been pre-optimized based on experimental results of receptor-ligand structures and their respective inhibition constants. Knowledge-based functions also use experimental data but use crystallographic structures to describe the receptor-ligand interaction geometries, instead of using data from the inhibition constants. Through statistical analysis of these geometries, "Mean Force Potentials" are derived, which evaluate the change in free energy as a function of interatomic coordinates. Because of the errors associated with these three-class evaluation functions, newer programs are using combinations of these functions to produce consensus functions, which appear to increase the quality of the results [64].

3.2.4 Protein-protein docking

Protein–protein docking has immense applications (**Figure 2**). It is particularly important in predicting metabolic pathways, macromolecular interactions, and macromolecular assemblies. Due to the difficulty in determining macromolecular assemblies experimentally, computational prediction of possible binding modes is one of the main goals of this type of docking [64]. Protein-protein docking simulates molecular recognition and is the most complex docking task. This is because the number of degrees of freedom is enormous, and it is not a possibility to do an

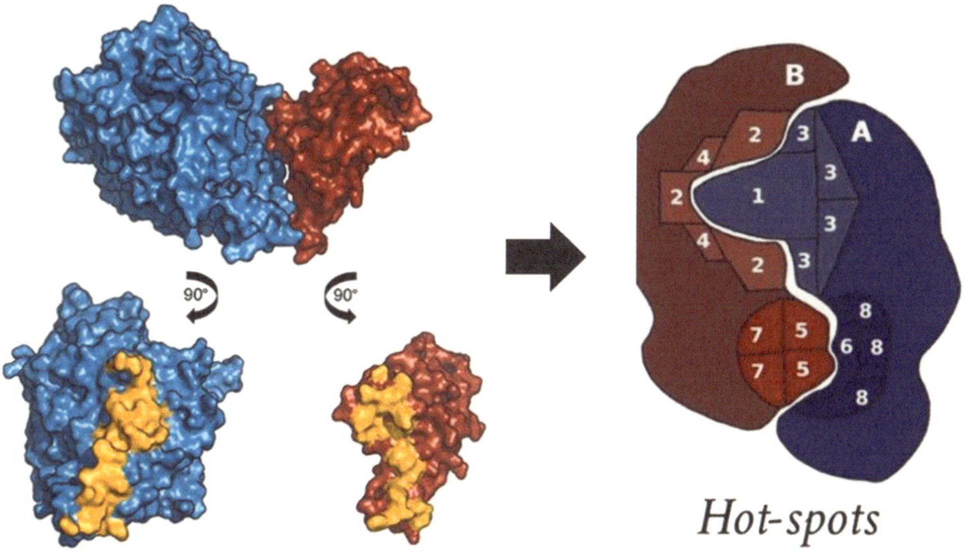

Figure 2.
Protein-protein docking, the receptor protein in blue surface and ligand protein in red surface. The docking study is determined the interface region (yellow), i.e., the residues of amino acids of the receptor and ligand that participate in the interaction or formation of the protein-protein complex, these residues (hot-spots) can serve to design new molecules with an inhibitory character, for example, in the case of the study of the trypsin-inhibitor complex.

exhaustive search of the conformational space. This is why many docking algorithms treat proteins as rigid bodies [65].

The Cluspro 2.0 server could be employed for protein-protein docking. This server performs rigid body docking and generates 109 complexes by performing rotation and translation movements of one protein ("ligand") relative to another ("receptor", held fixed). Docking conformations are classified according to the properties of their clusters. To rank the docking conformations, the program considers the pairwise interaction potential, the solvation energy, as well as the van der Waals (attraction and repulsion) and electrostatic contributions.

4. Trypsin-inhibition focus for the pest control

Insects represent one of the most important biotic stresses in agriculture worldwide. They are responsible for reducing crop yields, despite the use of chemical pesticides. These not only cause yield losses directly due to herbivore attack but also indirectly by acting as vectors for various plant pathogens. The estimated losses in crops around the world, without the use of pesticides or other non-chemical control strategies, reach about 70% of production, representing a loss of US$ 400 billion to the agricultural sector [66].

In the attempt to control the attack of insects on cultivars, new methods have been sought that are not based on agrochemical strategies. Although today the methods of pest control still concentrate primarily on the use of these substances, the high cost of developing new products whose formulations must suit pests that are increasingly resistant to their use, the unacceptable environmental consequences, and consumer pressure against this practice have caused a revolution in pest control in modern agriculture [67].

Thus, the study of molecules that can help in the control of herbivorous insects is very important. Until now, the main strategies of agricultural crop management, especially of large crops, consisted of the use of inorganic pesticides and transgenic cultivars containing *Bacillus thuringiensis* toxins. However, in addition to the damage caused to the environment and human health by inorganic insecticides, several insects of agronomic importance have developed resistance to these molecules.

For example, the frequent applications of diamide-type insecticides have already selected resistant individuals of important lepidopteran pests, in several locations around the world, since the beginning of their use 10 years ago [68]. Moreover, the other, more sustainable control method, based on the expression of Bt toxins, also suffers similar problems as insecticides. There have already been reported cases of Bt resistance since 1990, from *Plutella xylostella* (Lepidoptera: Plutellidae) larvae that required dosages of the dispel toxin (Abbott Laboratories North Chicago, OL), e.g., one of the first Bt toxin formulations for field spraying; 2x higher than susceptible populations to be controlled at acceptable levels [69]. In the following years, reports of resistance involving Bt insecticides and transgenic cultivars containing Bt toxin variants began to emerge in several other parts of the globe as well [70]. For these reasons, searches for new, preferably more sustainable, molecules have been ongoing in science.

Among possible new molecules are plant-derived protease inhibitors (PIs). Until the early 1980s, it was known that molecules that reduced the activity of proteolytic enzymes, the PIs, were present in plant tissues. It was then, in 1992, that Dr. Terry Green, a postdoctoral fellow at the University of Washington, showed

DOI: http://dx.doi.org/10.5772/intechopen.102632

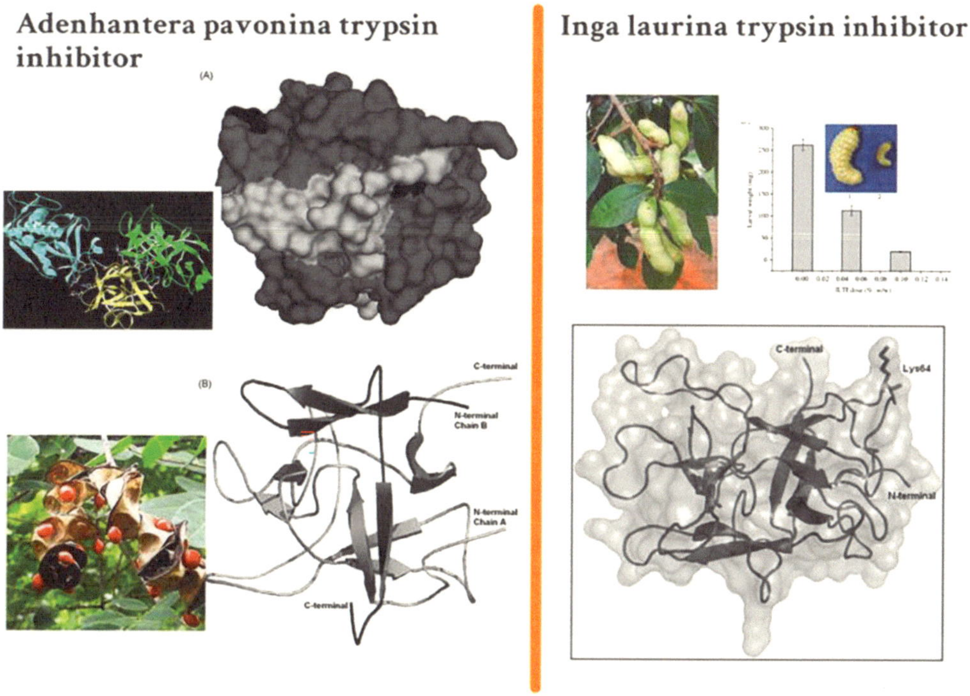

Figure 3.
APTI (Adenanthera pavonina trypsin inhibitor) and ILTI (Inga Laurina trypsin inhibitor) are an example of protein inhibitors of lepidoptera trypsins.

that tomato and potato plants accumulated large amounts of protease inhibitors when exposed to herbivory by the potato beetle, Leptinotarsa Decemliata (Coleoptera: Chrysomelidae). From this moment on, the IPs and digestive proteases of herbivorous insects became the target of the study of many researchers. Several works show that chronic exposure of insects, mainly those of the order Lepidoptera, to plant-derived protease inhibitors such as SKTI (Soybean Kunitz Trypsin Inhibitor), SBBI (Soybean Bowman-Birk Inhibitor), APTI (Adenanthera pavonina Trypsin Inhibitor), ILTI (Inga Laurina Trypsin Inhibitor) showed negative effects on the larval cycle of these herbivores (**Figure 3**). These adverse effects include reduced weight, survival, and delayed larval cycle [17, 29, 71–75]. However, the use of these molecules so far in agriculture has not proven effective, mainly due to the long exposure period required to cause effective control rates. In this regard, even efforts to express exogenous IPs constitutively (e.g., as is done with Bt toxins) in plants have not proven effective yet, since the initial attempt in 1987, when researchers expressed an inhibitor found in peas (Cowpea Trypsin Inhibitor) in tobacco (*Nicotiana tabacum*) leaves and observed that several orders of insects adapted quickly, returning to normal weight in a short period of time [76].

The commonly accepted mode of action of IPs in herbivores is that these molecules inhibit digestive proteases in the gut, resulting in a deficiency of free amino acids and consequently slowing the larval cycle and reducing survival and fecundity [77]. However, the effect of IPs may be more complex than just reducing proteolytic activity in the gut. It has been shown that feedback mechanisms in response to IPs lead to hyperproduction of proteases to compensate for the activity of the inhibited enzymes.

The nutritional stress imposed by this mechanism, which needs to divert amino acids important for insect development, slows development and reduces survival [78].

The main mechanisms of adaptation to IPs by herbivores so far recorded involve the following strategies: overexpression of the target protease, expression of proteases insensitive to the IP, and degradation of the molecule by endogenous proteolytic cleavage [79]. Thus, although naturally produced protease inhibitors are supplanted by the high gene plasticity of herbivores, these molecules are undoubtedly a defense mechanism against insects and thus a valuable target for the development of new insecticides. Plant tissues have a suboptimal protein content. Thus, nitrogen often becomes the limiting factor in the nutrition of many if not most phytophagous insects. Efficient hydrolysis of plant proteins to obtain essential amino acids is crucial for the survival of herbivores.

Indeed, for these molecules to be efficient against herbivores, more complex studies using more current bioinformatics tools, omics, and protease kinetics can be used to further explore and understand the mechanisms of adaptations to IPs. It is known that the set of proteolytic enzymes present in the midgut of herbivores can be composed of serine, threonine, cysteine, aspartic, and metalloproteases [80]. However, in the case of insects of the order Lepidoptera, the vast majority use protein digestion systems based on serine-proteases, trypsin-like, and chymotrypsin-like [81]. In addition, different isoforms of serine-proteases are known to be present in the gut of the same species. These different isoforms assist in the complex mechanism of insect response to IPs [76]. Insect trypsins share similar, but not identical, specificities with vertebrate trypsins. For example, some insect trypsins, unlike those of mammals, are calcium-stabilized and others not [11].

The mechanisms of response to IPs in herbivorous insects are not yet fully understood, and therefore protease inhibitors with characteristics that can prevent the insect from supplanting the effect of these molecules may generate more promising results. To date, much of the work testing the anti-insect effects of IPs has used molecules extracted and purified from plants [76]. However, due to the long periods of close association/interaction between insects and plants, possibly the mechanisms of herbivores are more prepared to counteract these kinds of molecules. Therefore, exposing herbivores to protease inhibitors that were not closely evolved, such as mammalian IPs and designed peptides, may elevate the anti-insect effects of these molecules. In *Spodoptera gregaria* larvae (Lepidoptera: Noctuidae) exposed to inhibitors of the pacifastin multidomain fam¬ily (115 kDa), e.g., (inhibitors of non-vegetable origin) was shown further growth suppression than plant inhibitors early in the insect cycle. However, the effect was gradually supplanted by the herbivores, which normalized their growth by the end of the cycle. Possibly, due to the high number of residues of the pacifastin IPs, a high amount of cleavage sites may be present, causing them to undergo endogenous proteolysis in the midgut.

Several works unravel the mechanisms behind the interactions between protease inhibitors (natural or otherwise) and proteases in the digestive tract of herbivorous insects, mainly the soybean caterpillar Anticarsia gemmatalis. Until then, much information has been generated regarding the specificities of A. gemmatalis proteases, mainly cysteine-proteases and serine-proteases. In 2005, trypsin-like enzymes from the gut of *A. gemmatalis* were purified and characterized. They showed the potential effect of several protease inhibitors on the activity of these enzymes. Inhibitors of serine proteases, including Benzamidine, TLCK, PMSF, and BPTI, reduced the activity of purified trypsins by more than 50% at relatively low concentrations in the micromolar and millimolar range [44]. And in the reference [11] the in vivo effects of synthetic trypsin inhibitors such as Benzamidine were evaluated.

Benzamidine was able to reduce protein digestibility, which affects the survival and formation of *A. gemmatalis* adults. The reference [11] evaluated the effect of this inhibitor throughout the insect cycle and concluded that although it caused an increase in the larval cycle and a higher percentage of mortality, most larvae were able to adapt to the inhibitor by remodeling the amount and type of enzyme present during digestion of the artificial diet. In addition, negative effects of other protease inhibitors on A. gemmatalis, mainly the organic ones [29, 30], were shown. Peptide protease inhibitors have also been tested, which were developed from molecular minimization using docking-molecular techniques in developing insects of the order Lepidoptera [81]. Even after molecule minimization, the conserved domains of trypsin IPs were able to reduce survival and important biological parameters of A. gemmatalis and S. cosmioides. An important work opened the vision for another approach in the attempt to use protease inhibitors in herbivore control.

Smaller, rationally designed molecules, based on enzyme kinetics and bioinformatics results, can help in the development of molecules that present better levels of control [82]. Besides generating increased molar concentration of protease inhibitors in the gut lumen and presenting higher stability than the conventionally tested PIs. In addition, protease inhibitor molecules without close association with the herbivore are less likely to find adaptation mechanisms. Transgenic plants can also be generated by generating proteins based on tandem sequences of previously tested peptides.

5. Conclusion

We concluded that the trypsin enzymes are serine proteases and are considered the most important digestive proteases of most insects. Trypsins are involved in the initial phase of protein digestion and are characterized by containing a catalytic triad consisting of the amino acid residues Hys, Asp, and Ser; in A. gemmatalis trypsin is Ser 229, Hys 85, and Asp 132. In the gut, insects exhibit the potential expression of various trypsin isoforms, but the proteolytic metabolism can be targeted by protease inhibitors, such as SKTI, ILTI, ApTI, BPTI, which offer possibilities for the development of novel biorational-based insect control approaches in silico methodologies such as molecular modeling and docking. The action of the protease inhibitors on the development of Lepidoptera larvae shows that these inhibitors influence larval survival, indicating that these proteins may have great toxic potential.

Acknowledgements

The authors would like to thank the National Council for Scientific and Technological Development (Conselho Nacional de Desenvolvimento Científico e Tecnológico—CNPq), the Brazilian Federal Agency for the Support and Evaluation of Graduate Education (Coordenação de Aperfeiçoamento de Pessoal de Ensino Superior—CAPES), and the Protection Foundation for Research in Minas Gerais (Fundação de Amparo à pesquisa de Minas Gerais, FAPEMIG).

Conflict of interest

The authors declare no conflict of interest.

Author details

Yaremis Beatriz Meriño-Cabrera* and Maria Goreti de Almeida Oliveira
Laboratory of Enzymology and Biochemistry of Proteins and Peptides, Department of Biochemistry and Molecular Biology, Institute of Applied Biotechnology to Agriculture—BIOAGRO, Federal University of Viçosa (UFV), Brazil, Vicosa, Minas Gerias, Brazil

*Address all correspondence to: yaremis.cabrera@ufv.br

References

[1] Nelson DL, Cox MM. Lehninger Principles of Biochemistry. 7th ed. New York, New York, United States: W.H. Freeman; 2017

[2] Robinson PK. Enzymes: principles and biotechnological applications. Essays in Biochemistry. 2015;**59**:1

[3] Santos JAND, Marques NAM, França AFJD, Santos YQD, Carelli GSC, Cruz JIOD, et al. Proteases and their inhibitors in coagulation and inflammation. Brain Research. 2020;**1407**:107-122

[4] Phillips J. Fundamentals of Enzymology. 1 st ed. Delhi: Scientific e-Resources; 2019

[5] Mótyán JA, Tóth F, Tőzsér J. Research applications of proteolytic enzymes in molecular biology. Biomolecules. 2013;**3**:923-942. DOI: 10.3390/biom3040923

[6] Verhamme IM, Leonard SE, Perkins RC. Proteases: Pivot points in functional proteomics. Methods in Molecular Biology. 1871;**2019**:313-392. DOI: 10.1007/978-1-4939-8814-3_20

[7] Barret AJ. Proteolytic enzymes: Nomenclature and classification. In: Beynon R, Bond JS, editors. Proteolytic Enzymes: A Practical Approach. 2nd ed. Oxford University Press: Oxford; 2001. pp. 1-21. DOI: 10.1016/0307-4412(90)90038-P

[8] Rao MB, Tanksale AM, Ghatge MS, Deshpande VV. Molecular and biotechnological aspects of microbial proteases. Microbiology and Molecular Biology Reviews. 1998;**62**:597-635

[9] Terra WR, Ferreira C. Insect digestive enzymes: compartimentalization and function. Comparative Biochemistry & Physiology. 1994;**109B**:1-62. DOI: 10.1016/0305-0491(94)90141-4

[10] Henriques BS, Gomes B, Costa SGD, Moraes CDS, Mesquita RD, Dillon VM, et al. Genome wide mapping of peptidases in Rhodnius prolixus: Identification of protease gene duplications, horizontally transferred proteases and analysis of peptidase A1 structures, with considerations on their role in the evolution of hematophagy in Triatominae. Frontiers in Physiology. 2017;**8**:1051. DOI: 10.3389/fphys.2017.01051

[11] Pilon FM, Silva CDR, Visôtto LE, Barros RDA, da Silva Júnior NR, Campos WG, et al. Purification and characterization of trypsin produced by gut bacteria from *Anticarsia gemmatalis*. Archives of Insect Biochemistry and Physiology. 2017;**96**:e21407. DOI: 10.1002/arch.21407

[12] Hillyer JF. Insect immunology and hematopoiesis. Developmental and Comparative Immunology. 2016;**58**: 102-118. DOI: 10.1016/j.dci.2015.12.006

[13] Herrero S, Combes E, Van Oers MM, Vlak JM, Maagd RA, Beekwilder J. Identificacion and rcombinat expression of a novel chymotrypsin from Spodoptera exígua. Insect Biochemistry and Molecular Biology. 2005;**35**:1073-1082. DOI: 10.1016/j.ibmb.2005.05.006

[14] Page MJ, Di Cera E. Serine peptidases: Classification, structure and function. Cellular and Molecular Life Sciences. 2008;**65**:1220-1236. DOI: 10.1007/s00018-008-7565-9

[15] Bown DP, Wilkinson HS, Gatehouse JA. Differentially regulated inhibitor- sensitive and insensitive

protease genes from the phytophagous insect pest, Helicoverpa armigera, are members of complex multigene families. Insect Biochemistry and Molecular Biology. 1997;**27**:625-638. DOI: 10.1016/s0965-1748(97)00043-x

[16] Silva-Júnior NR, Cabrera YM, Barbosa SL, Barros RDA, Barros E, Vital CE, et al. Intestinal proteases profiling from Anticarsia gemmatalis and their binding to inhibitors. Archives of Insect Biochemistry and Physiology. 2021;**107**:e21792. DOI: 10.1002/arch.21792

[17] Meriño-Cabrera Y, de Oliveira Mendes TA, Macedo MLR, de Almeida Oliveira MG. Inhibition of digestive trypsins by plant Kunitz proteins reduces the viability of Spodoptera cosmioides larvae. Annals of Applied Biology. 2019;**175**:336-349. DOI: 10.1111/aab.12539

[18] Patarroyo-Vargas AM, Merino-Cabrera YB, Zanuncio JC, Rocha F, Campos WG, de Almeida Oliveira MG. Kinetic characterization of anticarsia gemmatalis digestive serine-proteases and the inhibitory effect of synthetic peptides. Protein and Peptide Letters. 2017;**24**:1040-1047. DOI: 10.2174/0929866524666170918103146

[19] Wagner W, Möhrlen F, Schnetter W. Characterization of the proteolytic enzymes in the midgut of the European cockchafer, Melolontha melolontha (coleoptera: Scarabaeidae). Insect Biochemistry and Molecular Biology. 2002;**32**:803-814. DOI: 10.1016/s0965-1748(01)00167-9

[20] Hofer F, Kraml J, Kahler U, Kamenik AS, Liedl KR. Catalytic site pK a values of aspartic, cysteine, and serine proteases: Constant pH MD simulations. Journal of Chemical Information and Modeling. 2020;**60**:3030-3042. DOI: 10.1021/acs.jcim.0c00190

[21] Blow DM. Structure and mechanism of chymotrypsin. Accounts of Chemical Research. 1976;**9**:145-152. DOI: 10.1021/ar50100a004

[22] Perona JJ, Craik CS. Structural basis of substrate specificity in the serine proteases. Protein Science. 1995;**4**:337-360. DOI: 10.1002/pro.5560040301

[23] Hedstrom L. Serine protease mechanism and specifcity. Chemical Reviews. 2002;**102**:4501-4523. DOI: 10.1021/cr000033x

[24] Blay V, Pei D. Serine proteases: How did chemists tease out their catalytic mechanism? ChemTexts. 2019;**5**:1-7. DOI: 10.1007/s40828-019-0093-4

[25] Radisky ES, Lee JM, Lu CJK, Koshland DE. Insights into the serine protease mechanism from atomic resolution structures of trypsin reaction intermediates. Proceedings of the National Academy of Sciences. 2006;**103**:6835-6840. DOI: 10.1073/pnas.0601910103

[26] Fastrez J, Fersht AR. Demonstration of the acyl-enzyme mechanism for the hydrolysis of peptides and anilides by chymotrypsin. Biochemistry. 1973;**12**:2025-2034. DOI: 10.1021/bi00735a001

[27] Levy SM, Falleiros ÂM, Moscardi F, Gregório EA, Toledo LA. Morphological study of the hindgut in larvae of Anticarsia gemmatalis Hübner (Lepidoptera: Noctuidae). Neotropical Entomology. 2004;**33**:427-431. DOI: 10.1590/S1519-566X2004000400005

[28] da Silva Júnior NR, Vital CE, de Almeida BR, Faustino VA, Monteiro LP, Barros E, et al. Intestinal proteolytic profile changes during larval development of Anticarsia gemmatalis

caterpillars. Archives of Insect Biochemistry and Physiology. 2020;**103**:e21631. DOI: 10.1002/arch.21631

[29] Meriño-Cabrera Y, de Oliveira Mendes TA, Castro JG, Barbosa SL, Macedo ML, de Almeida Oliveira MG. Noncompetitive tight-binding inhibition of Anticarsia gemmatalis trypsins by Adenanthera pavonina protease inhibitor affects larvae survival. Archives of Insect Biochemistry and Physiology. 2020;**104**:e21687. DOI: 10.1002/arch.21687

[30] de Almeida BR, Meriño-Cabrera Y, Vital CE, da Silva Júnior NR, de Oliveira CN, Lessa Barbosa S, et al. Small peptides inhibit gut trypsin-like proteases and impair Anticarsia gemmatalis (L epidoptera: N octuidae) survival and development. Pest Management Science. 2021;**77**:1714-1723. DOI: 10.1002/ps.6191

[31] Sichler K, Hopfner KP, Kopetzki E, Huber R, Bode W, Brandstetter H. The influence of residue 190 in the S1 site of trypsin-like serine proteases on substrate selectivity is universally conserved. FEBS Letters. 2002;**530**:220-224. DOI: 10.1016/s0014-5793(02)03495-6

[32] Lopes AR, Juliano MA, Juliano L, Terra WR. Coevolution of insect trypsins and inhibitors. Archives of Insect Biochemistry and Physiology. 2004;**55**:140-152. DOI: 10.1002/arch.10134

[33] Lopes AR, Juliano MA, Marana SR, Terra WR. Substrate specificity of insect trypsins and the role of their subsites in the catalysis. Insect Biochemistry and Molecular Biology. 2006;**36**:130-140. DOI: 10.1016/j.ibmb.2005.11.006

[34] de Oliveira G, de Almeida BR, da Silva Júnior NR, Vital CE, Cordeiro G, da Silva CR, et al. Inhibitory effects of tripeptides to enzymatic activity and life cycle parameters of Anticarsia gemmatalis. Phytoparasitica. 2020;**48**:823-831. DOI: 10.1007/s12600-020-00837-0

[35] Pereira ME, Dörr FA, Peixoto NC, Lima-Garcia JF, Dörr F, Brito GG. Perspectives of digestive pest control with proteinase inhibitors that mainly affect the trypsin-like activity of Anticarsia gemmatalis Hübner (lepidoptera: Noctuidae). Brazilian Journal of Medical and Biological Research. 2005;**38**:1633-1641. DOI: 10.1590/S0100-879X2005001100010

[36] Mazumdar-Leighton S, Broadway RM. Identification of six chymotrypsin cDNAs from larval midguts of Helicoverpa zea and Agrotis ipsilon feeding on the soybean (Kunitz) trypsin inhibitor. Insect Biochemistry and Molecular Biology. 2001;**31**:633-644. DOI: 10.1016/s0965-1748(00)00168-5

[37] Magalhães CP, Fragoso RR, Souza DS, Barbosa AE, Silva CP, Finardi-Filho F, et al. Molecular and structural characterization of a trypsin highly expressed in larval stage of Zabrotes subfasciatus. Archives of Insect Biochemistry and Physiology: Published in Collaboration with the Entomological Society of America. 2007;**66**:169-182. DOI: 10.1002/arch.20208

[38] Eich T, Ståhle M, Gustafsson B, Horneland R, Lempinen M, Lundgren T, et al. Calcium: A crucial potentiator for efficient enzyme digestion of the human pancreas. Cell Transplantation. 2018;**27**:1031-1038. DOI: 10.1177/0963689718779350

[39] Ahmad Z, Saleemuddin M, Siddi M. Purification and characterization of three alkaline proteases from the gut of the larva of armyworm Spodoptera litura.

Insect Biochemistry. 1980;**10**:667-673. DOI: 10.1016/0020-1790(80)90056-6

[40] Johnston KA, Gatehouse JA, Anstee JH. The partial purification and characterization of serine protease activity in midgut of larval Helicoverpa armigera. Insect Biochemistry. 1991;**21**:389-397. DOI: 10.1016/0020-1790(91)90005-Y

[41] Lemos JA, Terra WR. Soluble and membrane-bound forms of trypsin-like enzymes in Musca domestica larval midgut. Insect Biochemistry and Molecular Biology. 1992;**22**:613-619

[42] Lee MJ, Anstee JH. Endoproteases from the midgut of larval Spodoptera littoralis includes a chymotrypsin-like enzyme with an extended binding site. Insect Biochemistry and Molecular Biology. 1995;**25**:49-61. DOI: 10.1016/0965-1748(94)00042-G

[43] Xavier LP, Oliveira MGA, Guedes RNC, Santos AV, De Simone SG. Trypsin-like activity of membrane-bound midgut proteases from Anticarsia gemmatalis (Lepidoptera: Noctuidae). European Journal of Entomology. 2005;**102**:147-153. DOI: 10.14411/eje.2005.023

[44] Oliveira MGA, De Simone SG, Xavier LP, Guedes RNC. Partial purification and characterization of digestive trypsin-like proteases from the velvet bean caterpillar, Anticarsia gemmatalis. Comparative Biochemistry and Physiology Part B: Biochemistry and Molecular Biology. 2005;**140**:369-380. DOI: 10.1016/j.cbpc.2004.10.018

[45] Lazarević J, Janković-Tomanić M. Dietary and phylogenetic correlates of digestive trypsin activity in insect pests. Entomologia Experimentalis et Applicata. 2015;**157**:123-151. DOI: 10.1111/eea.12349

[46] Reis DTC, dos Mares-Guia TR, de Oliveira JS, Santos AMC, Santoro MM, Oliveira MGA. Purification of a membrane-bound trypsin-like enzyme from the gut of the velvetbean caterpillar (Anticarsia gemmatalis Hübner). Acta Scientiarum:Biological Sciences. 2012;**34**:263-270. DOI: 10.4025/actascibiolsci.v34i3.9239

[47] Jordão BP, Lehane MJ, Terra WR, Ribeiro AF, Ferreira C. An immunocytochemical investigation of trypsin secretion in the midgut of the stablefly, Stomoxys calcitrans. Insect Biochemistry and Molecular Biology. 1996;**26**:445-453. DOI: 10.1016/0965-1748(95)00112-3

[48] Fuzita FJ, Pimenta DC, Palmisano G, Terra WR, Ferreira C. Detergent-resistant domains in Spodoptera frugiperda midgut microvillar membranes and their relation to microapocrine secretion. Comparative Biochemistry and Physiology Part B: Biochemistry and Molecular Biology. 2019;**235**:8-18. DOI: 10.1016/j.cbpb.2019.05.008

[49] Oliveira MG, Rogana E, Rosa JC, Reinhold BB, Andrade MH, Greene LJ, et al. Tyrosine 151 is part of the substrate activation binding site of bovine trypsin. Identification by covalent labeling with p-diazoniumbenzamidine and kinetic characterization of Tyr-151-(p-benzamidino)-azo-beta-trypsin. Journal of Biological Chemistry. 1993;**268**:26893-26903. DOI: 10.1016/S0021-9258(19)74195-7

[50] Erlanger BF, Kokowsky N, Cohen W. The preparation and properties of two new chromogenic substrates of trypsin. Archives of Biochemistry and Biophysics. 1961;**95**:271-278. DOI: 10.1016/0003-9861(61)90145-X

[51] Bradford MM. A rapid and sensitive method for the quantification of

microgram quantities of proteins utilizing the principle of protein dye binding. Analytical Biochemistry. 1976;**72**: 248-254. DOI: 10.1006/abio.1976.9999

[52] Laemmli UK. Cleavage of structural protein during the assembly of the head of bacteriophage T4. Nature. 1976;**227**:680-685. DOI: 10.1038/227680a0

[53] Pagadala NS, Syed K, Tuszynski J. Software for molecular docking: A review. Biophysical Reviews. 2017;**9**:91-102. DOI: 10.1007/s12551-016-0247-1

[54] Schneidman-Duhovny D, Hammel M. Modeling structure and dynamics of protein complexes with SAXS profiles. Methods in Molecular Biology (Clifton, NJ). 2018;**1764**:449. DOI: 10.1007/978-1-4939-7759-8_29

[55] Ovchinnikov S, Kamisetty H, Baker D. Robust and accurate prediction of residue–residue interactions across protein interfaces using evolutionary information. eLife. 2014;**3**:e02030. DOI: 10.7554/eLife.02030

[56] Branden C, Tooze J. Prediction, Engineering, and Design of Protein Structures. 2nd ed. New York: Garland Publishing Inc; 1991. p. 26

[57] Cavasotto CN, Phatak SS. Homology modeling in drug discovery: Current trends and applications. Drug Discovery Today. 2009;**14**:676-683. DOI: 10.1016/j.drudis.2009.04.006

[58] Basyuni M, Wati R, Sulistiyono N, Hayati R, Oku H, Baba S, et al. Protein modelling of triterpene synthase genes from mangrove plants using Phyre2 and Swiss-model. In. Journal of Physics: Conference Series. 2018;**978**:012095. DOI: 10.1088/1742-6596/978/1/012095, 10.1088/1742-6596/978/1/012095

[59] Kelley LA, Mezulis S, Yates CM, Wass MN, Sternberg MJ. (). The Phyre2 web portal for protein modeling, prediction and analysis. Nature Protocols. 2015;**10**:845-858. DOI: 10.1038/nprot.2015.053

[60] Rayan A. New tips for structure prediction by comparative modeling. Bioinformation. 2009;**3**:263. DOI: 10.6026/97320630003263

[61] Brereton AE, Karplus PA. Native proteins trap high-energy transit conformations. Science Advances. 2015;**1**:e1501188. DOI: 10.1126/sciadv.1501188

[62] Rasheed M, Bettadapura R, Bajaj C. Computational refinement and validation protocol for proteins with large variable regions applied to model HIV env spike in CD4 and 17b bound state. Structure. 2015;**23**:1138-1149. DOI: 10.1016/j.str.2015.03.026

[63] Meng XY, Zhang HX, Mezei M, Cui M. Molecular docking: A powerful approach for structure-based drug discovery. Current Computer-Aided Drug Design. 2011;**7**:146-157. DOI: 10.2174/157340911795677602

[64] Lionta E, Spyrou GK, Vassilatis D, Cournia Z. Structure-based virtual screening for drug discovery: Principles, applications and recent advances. Current Topics in Medicinal Chemistry. 2014;**14**:1923-1938. DOI: 10.2174/1568026614666140929124445

[65] Han X, Liu Z, Sun M. Neural knowledge acquisition via mutual attention between knowledge graph and text. In: Thirty-second AAAI Conference on Artificial Intelligence. New Orleans, Louisiana, USA: AAAI-18 Conference Committee; 2018

[66] Macalino S, Basith S, Clavio S, Chang N, Kang H, Choi S. Evolution of in silico strategies for protein-protein

interaction drug discovery. Molecules. 2018;**23**:1963. DOI: 10.3390/molecules23081963

[67] War AR, Paulraj MG, Ahmad T, Buhroo AA, Hussain B, Ignacimuthu S, et al. Mechanisms of plant defense against insect herbivores. Plant Signaling & Behavior. 2012;7:1306-1320. DOI: 0.4161/psb.21663

[68] Parra JRP. Biological control in Brazil: an overview. Scientia Agricola. 2014;**71**:420-429. DOI: 10.1590/0103-9016-2014-0167

[69] Richardson EB, Troczka BJ, Gutbrod O, Davies TE, Nauen R. Diamide resistance: 10 years of lessons from lepidopteran pests. Journal of Pest Science. 2020;**93**:911-928. DOI: 10.1007/s10340-020-01220-y

[70] Melo ALDA, Soccol VT, Soccol CR. Bacillus thuringiensis: Mechanism of action, resistance, and new applications: A review. Critical Reviews in Biotechnology. 2016;**36**:317-326. DOI: 10.3109/07388551.2014.960793

[71] Reisig DD, Kurtz R. Bt resistance implications for Helicoverpa zea (Lepidoptera: Noctuidae) insecticide resistance management in the United States. Environmental Entomology. 2018;**47**:1357-1364. DOI: 10.1093/ee/nvy142

[72] Macedo M, Garcia V, Freire M, Richardson M. Characterization of a Kunitz trypsin inhibitor with a single disulfide bridge from seeds of Inga laurina (SW.) Willd. Phytochemistry. 2007;**68**:1104-1111. DOI: 10.1016/j.phytochem.2007.01.024

[73] Macedo MLR, Durigan RA, Silva DS, Marangoni S, Machado MF, Postali JRP. Adenanthera pavonina trypsin inhibitor retard growth of Anagasta kuehniella (Lepidoptera: Pyralidae). Archives of Insect Biochemistry and Physiology. 2010;**73**:213-231. DOI: 10.1002/arch.20352

[74] Macedo MLR, Machado MF, Franco OL, Migliolo L, Ramalho CFO. Practical and theoretical characterization of Inga laurina Kunitz inhibitor on the control Homalinotus coriaceus. Comparative Biochemistry and Physiology. B. 2011;**158**:164-172. DOI: 10.1016/j.cbpb.2010.11.005

[75] Macedo RML, das Graças Machado Freire M. Insect digestive enzymes as a target for pest control. Invertebrate Survival Journal. 2011;8:190-198. DOI:

[76] Machado S, de Oliveira C, Zério N, Parra J, Macedo M. Inga laurina trypsin inhibitor (ILTI) obstructs Spodoptera frugiperda trypsins expressed during adaptive mechanisms against plant protease inhibitors. Archives of Insect Biochemistry and Physiology. 2017;**95**:21393. DOI: 10.1002/arch.21393

[77] Zhu-Salzman K, Zeng R. Insect response to plant defensive protease inhibitors. Annual Review of Entomology. 2015;**60**:233-252. DOI: 10.1146/annurev-ento-010814-020816

[78] Narita K, Nagao K, Bannai M, Ichimaru T, Nakano S, Murata T, et al. Dietary deficiency of essential amino acids rapidly induces cessation of the rat estrous cycle. PLoS One. 2011;**6**:e28136. DOI: 10.1371/journal.pone.0028136

[79] Gatehouse AJ. Prospects for using proteinase inhibitors to protect transgenic plants against attack by herbivorous insects. Current Protein and Peptide Science. 2011;**12**:409-416

[80] Singh S, Singh A, Kumar S, Mittal P, Singh IK. Protease inhibitors:

Recent advancement in its usage as a potential biocontrol agent for insect pest management. Journal of Insect Science. 2018;**27**:186-201. DOI: 10.1111/1744-7917.12641

[81] Rawlings ND, Bateman A. How to use the MEROPS database and website to help understand peptidase specificity. Protein Science. 2021;**30**:83-92. DOI: 10.1002/pro.3948

[82] Meriño-Cabrera Y, Castro JGS, Diez JDR, Macedo MLR, de Oliveira Mendes TA, de Almeida Oliveira MG. Rational design of mimetic peptides based on the interaction between Inga laurina inhibitor and trypsins for Spodoptera cosmioides pest control. Insect Biochemistry and Molecular Biology. 2020;**122**:103390. DOI: 10.1016/j.ibmb.2020.103390

Chapter 5

The Hydrolysates from Fish By-Product, An Opportunity Increasing

Jose Edgar Zapata Montoya and Angie Franco Sanchez

Abstract

The fishery industries have continuously increased over the last decade. This growth comes accompanied by a high volume of by-products released to environment, because these industries discard between 60 and 70% of their production as waste. This waste includes fish whole or part from these such as fillet remains (15–20%), skin and fins (1–3%), bones (9–15%), heads (9–12%), viscera (12–18%) and scales (5%). This by-products are rich in proteins and lipids which of several nature, which can be recovered to obtain compounds of high added value. In this chapter, some methods to recover compounds from fish by-products will be discussed. Among others, will be discussed topics about postharvest of fish, by-product releasing, enzymatic hydrolysis of by-product and bioactive peptide obtaining from fish waste.

Keywords: bioactives peptides, enzymatic hydrolysis, protein revaluation

1. Introduction

Production levels of fishery and aquaculture have been increasing for the last 30 years, as fish is an important protein source for human consumption and it is expected to reach a production of 196 mt by 2025 [1]. As a result, more and more people depend on fish or other fisheries production, capture, processing and marketing. By 2018, aquaculture production in the world was estimated to reach over 178 million tons [2], whereas marine capture fisheries have been around half the global production [3].

A huge waste volume has been produced along with that production increase, too. Around 70% of fish is processed before final sale, producing between 20 and 80% of fish waste, depending on the fish type and its transformation technology [4]. Furthermore, important amounts of water are required for those processes [5]. That situation represents a challenge from an environmental perspective because around 50% of that fish waste is discarded without being used [6]. Most of it is buried or deposited in water sources, either in the ocean, rivers, or streams. In the case of landfills, it can lead to saturations that cause odor and leachate problems. As for dumping in water sources, aerobic bacteria use organic matter by the action of oxygen, releasing large amounts of phosphorus, nitrogen and ammonium, affecting pH, causing algae growth, and

Process	Organic byproduct	%	Goal
Stunning	N/A		Decrease agony time to reduce undesirable compound production
Classification	Whole fish		Separate by size or species
Slime removal	Aqueous fluid		Reduce microbial contamination surface
Scaling	Scales	5	Reduce bacterial contamination
Washing	Washing water	100	Remove micro-organisms and contaminants
Head removal	Heads	9–32	Remove non-edible or low-value parts
Evisceration	Viscera	12–18	Remove internal organs to reduce microbial contamination
Fin Cutting	Fins	1–2	Remove non-edible parts
Skinning	Skin	3	Remove non-edible parts
Filleting	Fillet remains	15–20	Separation of dorsal and abdominal meat from fish
Bone/meat separation	Bones and skeletons	9–15	Separate meat from ribs and bones

Table 1.
Processes used for fish preparation after capture.

turbidity. The absence of oxygen in water results in the release of hydrogen sulfide, carbon dioxide, organic acids, methane, and ammonium [7].

These wastes contain important nutrient levels [8] and their composition depends on species, source organs or obtaining processes, as seen in **Table 1**. On the other hand, some of those nutrients represent an opportunity from an economic perspective, as in the case of the protein, which can be recovered to obtain high added-value compounds.

Among the methods used to add value to fish residues, there are protein hydrolysis, silage, and collagen recovery [9]. In the first hydrolysis tests evaluated, chemical processes and extraction with organic solvents were used, showing that they affected the nutritional quality of proteins and amino acids. For this reason, commercial enzymes have been increasingly applied to intend to obtain hydrolyzed protein of this substrate type [10]. These latter processes have moderate operating conditions, show greater reproducibility, and are more controllable and selective than chemical processes. Besides, they deliver products with techno-functional properties, excellent digestibility, rapid absorption, and amino acid balance, in addition to high levels of bioactive peptides [11].

This chapter will address the issue of protein residues used in fish processing aiming to obtain bioactive peptides through enzymatic hydrolysis using commercial enzymes. The basic concepts of fish processing, the characteristics of the waste generated, their use by enzymatic hydrolysis, and bioactive and functional peptide production will be addressed.

2. Fish post-harvest

Once the fish is harvested, it undergoes different processes intending to improve conservation conditions, separate the non-edible or low commercial value parts, and leave the product ready to deliver to the consumer. **Table 1** lists, in general terms, the stages of fish processing, many of which release some type of organic by-product [3, 6].

3. Bromatological characteristics of the Main fish-farmed by-products

Fish by-products are made up of different compound types with food importance [12]. The major components are moisture, fat, and protein. However, the bromatological composition varies depending on the species, age, and gender of the fish, in addition to the part of the fish from which the by-product comes, or the processes to which it has been subjected [13]. Thus, **Table 2** presents the bromatological composition of different fish by-products, for different species, fish parts, and processes.

As **Table 2** shows, these residues contain mainly proteins, fats and water, but they may also contain high added-value compounds such as collagen and gelatin, polyunsaturated fatty acids (EPA and DHA), monounsaturated such as palmitic and oleic, in addition to minerals and enzymes such as pepsin, trypsin, chymotrypsin and collagenase [3]. Because of their nutrient richness, inappropriate dumping of these residues affects not only the area where they are directly discharged, but it can also alter natural ecosystems in a wider area. In this sense, phosphorus and dissolved nitrogen release can be favored and thus increase biochemical demand for oxygen (BDO), because at least 80% of the nutrients in fish residues are potentially eutrophic substances. This leads to the higher growth of macroalgae in aquifers [31].

In some regions of the world, alternatives to use by-products have been sought. That is how the demand for complete fish heads and skeletons as food for humans has increased in Asia and Africa [32]. Bones, which contain the highest protein levels among the residues (41–84%), are a good source of collagen and gelatin. Besides, their mineral content has been used in the manufacture of food products for schoolchildren (85 mg/kg zinc, 350 mg/kg iron, and 84 g/kg calcium) [32]. Whereas skeletons contain significant amounts of meat remaining after filleting, whose protein is highly digestible and can be extracted for different purposes since it has better nutritional properties than plant proteins, and better essential amino acids balance than other animal protein sources [33] but are more sensitive to heat [34]. On the other hand, fish skin, provides gelatin [32], such as, Nile Tilapia skin has been used to produce collagen [35], which can be used for tissue regeneration [36].

A fish by-product that has gained the most attention in recent years is the viscera (12–20% of the fish), which comprise all organs of the main body cavity, including gills, heart, liver, spleen, swim bladder, stomach, gonads, intestines and their contents [6]. This residue has an average composition of 8–21% proteins, 2–12% lipids, 60–81% humidity and 1–5% ash [6]. The high protein content, in addition to being an excellent enzyme source, makes them a potential source of added-value products with exceptional properties for different industrial applications [37].

Between 70 and 80% of fish muscle is a structural protein, between 20 and 30% sarcoplasma proteins, and the remaining 2–3% of proteins are insoluble connective tissue. The main food protein is myofibrillar, with 66–77% of the total in fish meat. This protein comprises between 50 and 60% myosin and 15–30% actin [38]. Myosin fibers are connected by actin molecules and can be cut at one end by trypsin and chymotrypsin, while at the other end by papain, to form they divide into two forms of meromyosin, heavy and light, with different functional properties [39].

Type of waste	Protein	Fat	Moisture	Ash	Reference
Freeze-dried Viscera of Yamú (Brycon siebenthalae)	19.19	79.49	0.48	—	[14]
Argentine hake (Merluccius hubbsi) gonad	21.95	10.92	68.72	11.61	[15]
Raw Viscera Tilapia of (Oreochromis spp.)	4.03	32.93	61.36	0.67	[16]
Tilapia (Oreochromis nilotica) skeleton (D.B)	50.6	30.6	65.3	15.3	[17]
Raw Viscera of Trucha (Oncorhynchus mykiss)	9.14	31.15	56.93	1.51	[10]
Argentine hake (M. hubbsi) liver	16.38	29.71	55.79	1.61	[15]
Viscera of Catla Catla	8.52	12.46	76.25	2.50	[18]
Cape hake (Merluccius capensis) by products	18.0	1.1	78.5	1.9	[19]
Tilapia (Oreochromis spp.) scales (D.B)	67.96	—	15.18	32.08	[20]
Greenland halibut (Reinhardtius hippoglossoides) skin	15.95	10.62	55.44	17.63	[21]
Tilapia (Oreochromis spp.) viscera	7.87	26.08	62	1.19	[22]
Blue shark (Prionace glauca) skin	22.79	0.24	76.03	4.24	[21]
Rainbow trout (O. mykiss) viscera	15	13	71.7	2.7	[23]
Atlantic salmon (Salmo salar) viscera	8	44	60	1	[24]
Tilapia (Oreochromis spp.) defatted viscera	10.04	1.88	83.21	1.71	[22]
Yellowfin tuna (Thunnus albacares) skin	32.38	3.22	0.67	62.57	[21]
Red tilapia (Oreochromis niloticus) head, skeleton, and tail	14.6	5.5	66.6	8.9	[25]
Tilapia del Nilo (O. niloticus) skin	29.68	13.89	54.91	1.61	[26]
Black Sea anchovy (Engraulis encrasicholus) head	13.39	10.02	70.94	5.00	[27]
Atlantic salmon (S. salar) head (D.B.)	13	22	39	44	[24]
Tilapia (Oreochromis spp.) spines (D.B.)	55.54	—	53.46	22.91	[20]
Tilapia (Oreochromis nilótica) skeleton	50.6	30.6	65.3	15.3	[17]
Black Sea anchovy (Engraulis encrasicholus) frame	16.47	15.50	59.72	7.60	[27]
Tilapia (Oreochromis spp.) viscera	4.574	33.602	62.693	0.732	[28]
Atlantic salmon (S. salar) skeleton	15	27	42	4	[24]
Tilapia (Oreochromis spp.) defatted viscera	12.644	2.525	82.607	1.462	[28]
Black Sea anchovy (Engraulis encrasicholus) viscera	12.05	23.90	61.50	2.09	[27]
Cuttlefish (Sepia officinalis) viscera	17.45	4.78	74.99	1.95	[29]
Salmon (S. salar) head	13	22	39	4	[24]
Tilapia (Oreochromis spp.) scales	83.9	0.9	—	15.1	[30]
Salmon (S. salar) skeleton	15	27	42	4	[24]

Table 2.
Bromatological composition of fish by-products D.B.: Dry base.

Fish proteins contain between 16 and 18 amino acids, which have an excellent balance, usually 8 essential and 8 non-essential. This makes this type of protein very widely used for animal feed, although they are also used for fertilizer production, silage and in recent decades, bioactive peptide production [30, 40]. **Table 3** shows the aminograms of different residues of several fish species, some raw and others that

Amino Acid	Red tilapia		Mackerel fish			Yamú viscera		
	RTVH	FRTVH	WT	WM	DM	PI	DH9	DH28
Histidine	4.06	1.99	4.5	3.8	5.2	6.629	5.069	5.222
Isoleucine	2.53	2.44	5.5	6	5.6	4.919	4.073	5.221
Leucine	7.99	8.14	9.4	10	8.8	5.445	5.254	5.267
Lysine	7.68	9.91	7.9	8	7.6	3.54	3.33	2.437
Methionine	1.32	0.14	2.7	4.6	2.8	0.944	1.018	0.656
Arginine	3.97	4.44	7.6	5.9	7.1	—	—	—
Valine	5.27	4.43	7.8	6.8	8.5	1.108	1.901	0.874
Phenylalanine	0.91	1.17	4.2	4	3.1	2.407	1.898	2.406
Threonine	8.04	6.06	5.7	5.5	5.5	1.228	1.927	1.898
Tryptophan	—	—	—	—	—	—	—	—
Ac Aspartic	2.31	4.39	11.8	12.2	11.4	1.837	1.799	2.135
Ac Glutamic	6.6	5.84	15.8	18	15.6	4.329	4.797	5.045
Asparagine	2.31	4.39	—	—	—	1.054	0.984	0.627
Serine	4.26	3.94	4.5	5.2	4.1	3.436	3.72	3.398
Glycine	21.38	17.62	6	6.2	4.6	1.516	1.882	1.508
Alanine	3.8	3.82	7.7	7.3	7.3	2.498	3.18	2.89
Tyrosine	3.11	4.23	3.5	3.9	3.4	17.54	17.66	13.824
Cystine	—	—	—	—	—	3.237	4.302	3.706
Hydroxyproline	—	—	—	—	—	—	—	—
Proline	—	—	1.5	4.6	1.5	—	—	—
Glutamine	6.6	5.84	—	—	—	—	—	—
Total	92.14	88.79	106.1	112	102.1	61.66	62.79	57.114
Reference	[11]		[41]			[14]		

RTVH: Red Tilapia Viscera Hydrolysates.
FRTVH: Fraction <3 kDa of Red Tilapia Viscera Hydrolysates.
PI: Yamú Protein Viscera Isolate.
DH9: Yamú Protein Viscera Hydrolysates with 9% of Degree of hydrolysis.
DH28: Yamú Protein Viscera Hydrolysates with 28% of Degree of hydrolysis.

Table 3.
Amino acids content of fish y-products.

have undergone hydrolysis processes [14], atomization drying [40] and membrane fractionation [11].

4. Enzymatic hydrolysis of fish by-product proteins

4.1 Protein hydrolysis

Protein hydrolysis occurs when a peptide bond is broken by water action, in the presence of a catalyst that may be an enzyme or a chemical agent [42]. Low-cost chemical processes can be by acid or alkaline hydrolysis, but they are non-specific, not reproducible and lead to amino acid denaturation. On the other hand, enzymatic hydrolysis is more expensive but does not deteriorate amino acids [43].

Once the native protein is broken, fragments of the native protein (oligomers) form, which can be a substrate for the subsequent hydrolysis process, so it is a multi-substrate reaction [44], especially in mediums where no pure protein is available, but mixtures of innumerable proteins, such as in fish residues and in general in other agro-industrial waste. Due to the hydrolysis process, the molecular characteristics of the proteins change, because the average molecular weight of the protein fragments present decreases, this increases the surface load, causes the release of hydrophobic groups, and changes functional properties, among other effects [45].

4.2 Enzymatic hydrolysis of protein

This process consists of decomposing proteins into smaller fragments, whose catalysts are enzymes called proteases [11]. This is a set of simultaneous link break reactions, consisting of serial stages, with different species loaded in equilibrium, giving fragments of decreasing size as follows [46]:

$$proteous \rightarrow proteins \rightarrow peptones \rightarrow peptides \rightarrow amino\ acid$$

The catalytic process that occurs is divided into three steps. First, the enzyme (E) should approach the substrate (S) and bind to form the enzyme-substrate complex (ES). Second, the rupture of the peptide bond results in the release of a peptide. Third, the remaining peptide is separated from the enzyme after a nucleophilic attack from a water molecule [11]. Each of these reactions has its speed as described in Eq. (1) [47]. This process can be repeated on any of the peptides formed [46].

$$E + S \underset{K_1}{\Leftrightarrow} ES \underset{K_2}{\rightarrow} EP + H - P' + H_2O \underset{K_3}{\rightarrow} E + P - OH + H - P' \qquad (1)$$

E: Enzyme, S: Substrate, P and P´: Resulting peptides, kx: Constant reaction rate.

This procedure has advantages over chemical hydrolysis as they have high selectivity and low contamination. It is a specific process that is carried out under moderate pH and temperature conditions, which makes it easy to control [30]. The product obtained is called protein hydrolyzate and it consists of peptides generally between 2 and 20 amino acids [48]. However, there are also disadvantages such as the high enzyme costs and long processing times [49].

Critical operating conditions in protein enzymatic hydrolysis include temperature, pH, enzyme type and concentration, substrate and concentration, cofactors, coenzymes, hydrolysis time [50], agitation speed [51], and presence of inhibitors, like fat in fish by-products [11].

On the other hand, variations that enzyme activity may suffer during the reaction should be controlled, such as denaturation, aggregation, or enzyme inactivation, which can be produced by temperature effects, pH shear stress or other substances that interfere with catalysis [12].

4.2.1 Enzymatic hydrolysis kinetics

During the reaction, the enzyme attacks the peptide bond as follows [52, 53]:

Opening of the peptide bond

$$-CHR' - CO - NH - CHR'' - +H_2O\ \overrightarrow{enzyme} - CHR' - COOH + NH_2 - CHR'' \tag{2}$$

Proton exchange

$$-CHR' - COOH - NH2 - CHR' \rightarrow {}' - CHR' - COO^- + N^+H_3 - CHR'' \tag{3}$$

Tritation of amino group

$$N^+H_3 - CHR'' + OH^- \rightleftharpoons NH_2 - CHR'' + H_2O \tag{4}$$

Under neutral or alkaline conditions, the dissociation of the amino group becomes significant, so a decrease in pH may occur due to the accumulation of the protons released, which makes it necessary to add a base to keep pH constant and prevent the enzyme from being affected in its activity [30]. The analysis of the equations above concludes that the amount of hydrolyzed protein is proportional to the amount of base required to neutralize the pH of the reaction medium [30].

4.2.2 Follow-up of hydrolysis reaction

The hydrolysis reaction progress is established by the Hydrolysis Degree (HD), expressed as a fraction or percentage of the number of broken peptide bonds at any given time (h) for the total peptide bonds in the intact protein (htot) (Eq. 5) [54]. Both can be expressed as protein meq/g or as protein mmol/g [30].

$$GH(\%) = \frac{h}{h_{tot}}.100 \tag{5}$$

Methods used to determine Hydrolysis Degree (HD) include the pH-stat method [52], O-phthaldialdehyde (OPA) [54], Trinitrobenesulfonic acid (TNBS) [55], formalin titration, and soluble nitrogen in trichloroacetic acid (TCA) [56]. The fundamental difference between these methods is in the principle that each one is based to measure the number of broken bonds (h) at any given time of the reaction, because htot is usually determined from the analysis of the total amino acid content in the intact protein [57].

4.2.2.1 pH-stat method

This method is based on the fact that in peptide bond hydrolysis, a carboxyl group and an amino group are released. In an aqueous solution, these groups will be more or less ionized depending on pH [55]. At neutral or alkaline pH, carboxyl groups are fully ionized and proton exchange occurs between the carboxyl group and the amino group. At alkaline pH, amino groups will also be partially or fully ionized depending on the pH and amino acid in question, since the pK of the free amino acids N-terminal amino group ranges from 9 to 10.8. The following equations show, in general, the chemical species involved in protein enzymatic hydrolysis [58].

$$P_1 - CO - NH - P_2 + H_2O \xrightarrow{protease} P_1 - COOH + NH_2 - P_2 \tag{6}$$

$$P_1 - COOH \longrightarrow P_1 - COO^- + \boldsymbol{H}^+ \quad (7)$$

$$NH_2 - P_2 \rightleftarrows NH_3^+ - P_2 \quad (8)$$

The resulting free protons cause a pH decrease of the reaction mixture, and a base addition is required to keep pH constant. The amount of base required is directly related to the amount of hydrolyzed peptide bonds, and it can be used to estimate HD. Unfortunately, the relationship between HD and base consumption is not simple and depends on several variables, including pK of the α-amino group released, the temperature of the reaction mixture, and length of the peptide chain [52]. The relationship between the spent base volume and HD has been described by Adler-Nissen, 1986 [55] in Eq. (9).

$$GH(\%) = \frac{BN_B}{M_p \alpha h_{Tot}} .100 \quad (9)$$

where B is the base volume consumed in L to keep pH constant, MP is the protein mass in kg, NB is the base concentration, and α is the dissociation degree of the amino groups released in the reaction. α and pK are calculated with Eqs. (10) and (11), respectively, where T is the temperature (K) [59].

$$\alpha = \frac{10^{pH-pK}}{\left(1 + 10^{pH-pK}\right)} \quad (10)$$

$$pK = 7.8 + \frac{298 - T}{298 * T} * 2400 \quad (11)$$

4.2.2.2 O-phthaldialdehyde method (OPA) and Trinitrobencenesulfonic acid method (TNBS)

Both methods are spectrophotometric, based on the determination of the α-amino groups released, by derivatization with trinitro-bencenesulfonic acid or ortho-phthaldialdehyde, respectively [56]. They are detected in the ultraviolet–visible range for the TNBS method, or fluorescent for the OPA. The absorbance value obtained is then converted into quantitative values using a standard curve prepared with a free amino acid, usually leucine, calculating HD as the percentage proportion of the amino acid released in the hydrolyzed regarding the amino acid amount of the whole protein [54, 55]. In **Figures 1** and **2**, reactions of an amino group with TNBS and OPA, respectively, take place [56].

However, in these methods, derivatization reagents exhibit different reactivity to some amino acids, affecting measurement accuracy. For example, in the case of the OPA method, it will not be accurate when applied on proline- and cysteine-rich hydrolyzates [57].

4.2.3 Proteases most important characteristics

Proteases are the enzymes responsible for catalyzing the hydrolysis reaction of protein-peptide bonds, also known as peptidases [62]. Although, they can be obtained from plants, animals or microorganisms, most commercially viable proteases are obtained from this latter [63], especially Bacillus species, such as Bacillus licheniformis, Bacillus subtilis, and Aspergillus fungal species such as Aspergillus

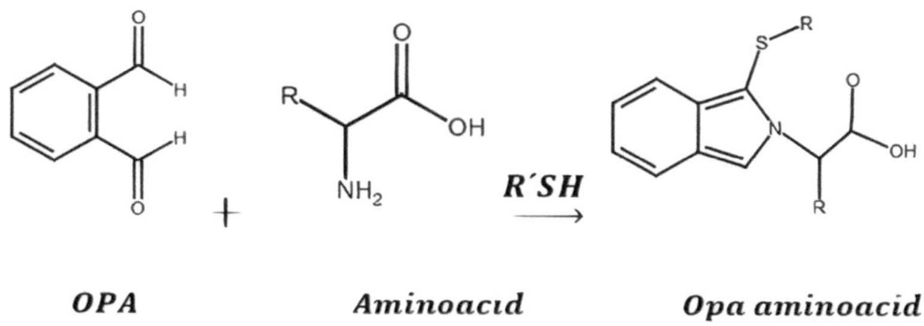

Figure 1.
Reaction of OPA with amino acids. Source [60].

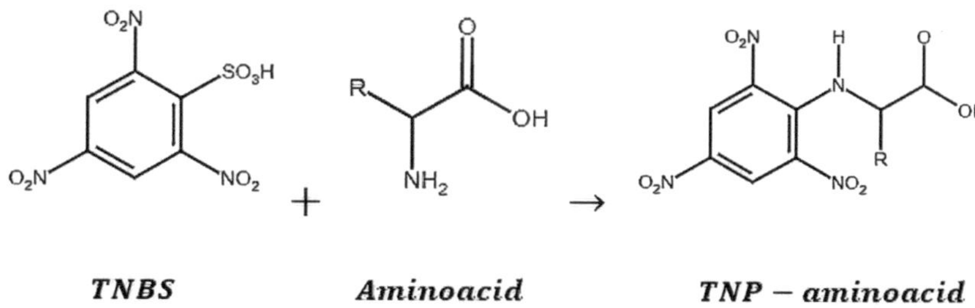

Figure 2.
Reaction of TNBS with amino acids. Source [61].

niger, A. flavus, Ammophilus fumigatus, and A. oryzae [64]. Some of the commercial proteases that have been used to obtain hydrolyzates from fish residues include trypsin, chymotrypsin, pepsin, Alcalase® 2.4 L, Flavourzyme® 500 L, E Properase, pronase, collagenases, bromelain and papain [50].

Proteases belong to the hydrolases group, they constitute a large and complex group of enzymes that differ from each other in their specificity due to substrate, their selectivity, the nature of their active sites, their catalytic mechanism, their stability profile, their pH, and optimum temperature. For these reasons, proteases cannot be classified under the general enzyme nomenclature system, but are classified according to their catalytic action, the nature of their active site, and their optimal pH value [63]. From the point of view of functional groups that have their active site, proteases can be classified into four main groups as follows [62]: Serine Proteases, Aspartic Proteases, Cysteine Proteases, Metalloproteases. On the other hand, when considering its catalytic mode of action, i.e., the excision site of the polypeptide chain, proteases are classified into exopeptidases and endopeptidases [65]. While, based on their optimal pH range, proteases can be classified into alkaline, neutral and acidic.

5. Production of bioactive and techno-functional peptides of fishery by-products

According to the HD achieved, the hydrolyzate obtained will potentially have biological activities or techno-functional properties. HD less than 10% result in improved techno-functional properties, such as emulsification, foaming capacity and greater solubility, whereas a higher HD tend to deliver hydrolyzates with greater potential as bioactive peptide sources [66].

5.1 Bioactive peptides

A bioactive peptide is a sequence of amino acids that is encrypted in the intact protein, in which it remains inactive, but once released, it can interact with certain receptors and regulate the physiological functions of the organism [67]. This may express some kind of effect on metabolic behavior, either human or animal [65]. These peptides can be released from the protein by gastrointestinal digestion, enzyme hydrolysis, or fermentation [68].

Among the most widely studied biological activities, are antihypertensive [69] Antioxidant [11] Antimicrobial [70], antithrombotic [71], anticancer [11] metal chelating agent, anticoagulants, among others [72].

One of the methods currently applied for obtaining bioactive peptides is enzymatic hydrolysis using commercial enzymes, which represents a reproducible, scalable, and industrial-application-capable method [73]. In this technology, biological activities of the peptides obtained may be affected by the operating conditions applied to isolate proteins, hydrolysis degree, protease type, peptide structure, the amino acids sequence, concentration, and the molecular weight of the peptides obtained [74].

The relationship between the peptide's biological activity and their molecular weight has been widely documented [73], so the search for conditions that maximize HD has been one of the priorities in many studies [75] Peptide fractions with molecular weights between 1 and 4 kDa are of the greatest interest for nutritional and/or pharmaceutical uses in particular [75].

5.1.1 Antioxidant peptides

Free radicals and reactive oxygen species ROS [76], can cause DNA, protein, or lipid damage, resulting in human body damage from neurodegenerative, inflammatory, cardiovascular, diabetes, and cancer diseases [76]. This type of effect can be counteracted by substances with antioxidant capacity, which have different mechanisms of action depending on the free radical reduction form, among which are SET (single electron transfer), and HAT (hydrogen atom transfer) [77]. Based on these mechanisms, some methods to evaluate the antioxidant capacity of different substances have been designed. SET-based methods detect the antioxidant's ability to transfer a chemical species such as metals, carbonyls and electrons, the most commonly used methods of this type are ABTS and FRAP. In the case of HAT methods, the antioxidant ability to inactivate a free radical is measured through the donation of a hydrogen atom, in which one of the most commonly used methods is ORAC [77].

On the other hand, some metals such as iron and copper, which are of importance at the physiological level, may also participate in the formation of reactive oxygen species [78], as in the case of hydroxyl radicals (OH), that are formed by the Fenton reaction and can cause damage to different types of tissues (Canabady-Rochelle et al., 2018). In this sense, metal chelation can counteract the formation of metal-catalyzed radicals in some way, which has somewhat been considered as a form of antioxidant activity [79].

Thus, peptide antioxidant activity is related to metal chelating activity and electron donation activity, which facilitates interaction with free radicals and cuts the reaction chain in which they participate [80]. In addition, the presence of hydrophobic sequences in peptides can interact with lipid molecules, eliminating the donation of protons to result in lipid radicals [81]. Thus, the imidazole group in histidine residues participates in hydrogen atom transfer, electron transfer, active oxygen extinction and capture of hydroxyl radicals [82].

The antioxidant capacity in these hydrolyzates has been attributed to the presence at the N-terminal end of peptide sequences of non-polar hydrophobic amino acids, such as phenylalanine, alanine and proline, and hydrophilic amino acids such as tyrosine, histidine and valine [6]. Thus, capturing the activity of hydrogen peroxide, the chelating activity of Fe2+, and reducing the power of Abalone (Haliotis discus hannai) hydrolyzates was related to hydrophobic amino acids in their peptides [83]. The capturing capacity of radicals has also been attributed to the presence of aromatic residues [84]. While tryptophan and tyrosine have been attributed antioxidant activity mediated by their phenolic and indolic groups, capable of donating hydrogen atoms [85]. The **Table 4** lists several sequences of antioxidant peptides, from different kinds of fish by-products.

5.1.2 Antihypertensive peptides

Hypertension is one of the most important cardiovascular risk factors worldwide, since high blood pressure currently affects about 20% of adults around the world [97]. In these blood pressure-increasing processes, the angiotensin I converter enzyme (ACE) plays a crucial role. This enzyme, a dipeptidyl carboxypeptidase (EC. 3.4.15.1), promotes the conversion of angiotensin I to a powerful angiotensin II vasoconstrictor, and inactivates the bradequinine vasodilator, which is a depressant of the renin-angiotensin system action [97]. Angiotensin II is also involved in the release of the steroid Na-retaining, which also tends to increase blood pressure [97]. For these reasons, a first step in the search for potentially useful substances to control high blood pressure is the ability test to inhibit ACE. In this sense, the search for peptides that can reach therapeutic tests as drugs for blood pressure control should initially be evaluated as ACE inhibitors [97]. The **Table 5** lists several sequences of antihypertensive peptides, from different kind of fish by-products.

Source	Sequence	Reference
Amur sturgeon (Acipenser schrenckii) skin	PAGT	[86]
COD (Gadus macrocephalus) gelatin	TCSP, TGGGNV	[87]
Hoki (Johnius belengerii) Skin	HGPLGPL	[88]
Tilapia (Oreochromis niloticus) skin	EGL, TGDET	[89]
Mackerel (Magalaspis cordvla) viscera	ACFL	[90]
Biuefin leatherjacket (Navodon septentrionalis) Head	WEGPLK, GPP, GVPLT	[91]
Salmon Pectoral fin	FLNEFLHV	[92]
Black Pomfret (Parastromateus niger) Viscera	AMTGLEA	[93]
Skate (Raja porosa) Cartilage	F1MGPY, GPACDY, 1VAGPQ	[94]
Grass carp (Ctenopharyngodon Idella) skin	PYSFK, GFGPQLVGGRP	[95]
Squid (Ommastrephes bartrami) Viscera	WVAPLK	[96]
Salmon (Salmo sp.) Fin	FLNEFLHV	[92]

Table 4.
Amino acid sequence of antioxidant peptides from fish by-products.

5.1.3 Anti-carcinogenic peptides

Cancer (malignant tumor), one of the most common diseases in the world [106], consists of abnormal and uncontrolled growth of cells, with proliferation and spread in surrounding tissues [11]. Thus, inhibition of deregulated cell proliferation is one of the strategies for treating this type of disease [107]. Among the broad list of substances that have been evaluated for this purpose are luteinizing hormone-releasing hormone and Atrial natriuretic peptide, for the treatment of prostate and colorectal cancer, respectively [106].

Various fish-derived proteins have been reported as sources of anticancer peptides [11, 108], as in the case with the antiproliferative activity of protein hydrolyzates of 18 fish species against breast cancer cell lines [109]. In **Table 6**, different fishery sources that have been active against some types of cancer are shown.

There are three ways in which antiproliferative peptides act on cancer cells, apoptosis, necrosis, and cell cycle disturbances [11]. These mechanisms of action change according to structural characteristics such as molecular weight and amino acid composition. Thus, smaller peptides have greater molecular mobility and diffusivity, so they can interact better with the components of cancer cells. This activity has been attributed to amino acid sequences between 3 and 25 residues, with the predominance of hydrophobic amino acids, and one or more residues of Lys, Pro, Arg, Ser, Glu, THR Leu, Gly, Ala and Tyr. Because hydrophobic amino acids improve interactions

Source	Sequence	Reference
COD (Gadus macrocephalus) gelatin	TCSP, TGGGNV	[87]
Pollack (Theragra chalcogramma) Skin	GPL, GPM	[98]
Salmon (Oncorhynchus keta) Skin	GLPLNLP	[99]
Sea Bream Scale	GY, VY, GF, VIY	[100]
Sardinella (Sardinella aurita) Head/viscera	FRGLMHY	[101]
Snakehead fish (Channidae sp) Muscle	LYPPP, YSMYPP	[102]
Small-spotted catshark (Scyliorhinus canicula) Muscle, viscera, skin, and frame	VAMPF	[103]
Lizard fish (Saurida elongate) Muscle	RVCLP	[104]
Lizardfish (Synodus macrops) Scale Gelatin	AGPPGSDGQPGAK	[105]

Table 5.
Amino acid sequence of ACE inhibitor peptides from fish by-products.

Source	Type of cancer	Reference
Ruditapes philippinarum hydrolysates	prostate, breast, and lung cancer	[110]
Squid gelatin hydrolysates	CMF-7, U87	[111]
Oyster protein and anchovy hydrolysates	Colon and prostate cancer	[112]
Blood clam muscle	Prostate cancer	[91]
Krab subproducts	Prostate cancer	[113]

Table 6.
Use of peptides from fish by-products in cancer treatment.

between peptides and the outer surface of the bilayer of the tumor cell membrane, due to their phospholipid content and thus, they exert selective cytotoxic activity on these cells to healthy cells [107].

In addition to the amino acid sequence, the anti-cancer peptide's function is influenced by net load, amphipathicity, hydrophobicity, structural membrane folding (including secondary structure, dynamics and orientation), oligomerization, and peptide concentration [11]. The cationic amphibious structure predisposes them to interact with the cell membrane anion surfaces [114]. The α helix is a main structural characteristic of this peptide type, with lateral chains of hydrophilic and hydrophobic amino acids, forming clear hydrophilic and hydrophobic surfaces. On the other hand, they concentrate on the N-terminal and the C-terminal to form different hydrophilic and hydrophobic domains. Anti-cancer peptides with a β sheet structure are generally stabilized by disulfide bonds, and these sheets are in β antiparallel formation. Meanwhile [11]. The net charge and positive charge number also influence these peptides activity, since their association with the cancer cell membrane occurs through electrostatic interactions due to its cationic condition and the anion lipopolysaccharide on the external membrane that causes its disturbance [115].

5.1.4 Anticoagulant peptides

Blood clotting is a crucial process for human health, excessive clotting that leads to blocked blood vessels causes strokes, heart attacks, and pulmonary embolism [11]. This makes anticoagulant compounds vital to preserving life quality in modern times. The anticoagulant is a compound that will stop blood clotting by binding to one or more coagulation factors, preventing it from binding to the membrane phospholipids [11]. Heparin is currently the anticoagulant most commonly used, but heparin has several disadvantages, including thrombocytopenia and non-specific plasma binding. In addition, it can cause platelet dysfunction and aggregation [116]. Therefore, there is a marked interest in the search for new anticoagulant compounds with minor collateral risks for the medical treatment of thromboembolic events [11].

Anticoagulant activity is less investigated than other biological activities, and specifically, peptides with this activity isolated from fish-based by-products have not been reported [11]. This way, an oligopeptide from the blue mussel, with a molecular mass of approximately 2,5 kDa has been isolated, showing anticoagulant activity by the prolongation of both thrombin time and activated partial thromboplastin time, by interaction specifically with blood clotting factors IX, X, and II. Nasri et al. [71], in 2012 isolated four anticoagulant peptides from protein hydrolyzates of goby muscle proteins, in which they found that they had Arg in the C-terminal position. Thus, concluding that small peptides with an amino acid charged at the C-end are considered potential thrombin inhibitors and/or other factors involved in the coagulation process [71]. Anticoagulant peptides from yellow-sole fish skeleton have also been isolated [117].

5.1.5 Antimicrobial peptides

The excessive use of conventional antimicrobial products has caused the emergence of resistant strains, which poses a health threat. Therefore, the development of antimicrobials using mechanisms other than traditional antibiotics is needed [11]. In this context, antimicrobial peptides effectively promote toxicity against invading pathogenic microorganisms, and also modulate the immune response in superior organisms [118]. These peptides are produced in all kingdoms, from bacteria to fungi and plants to mammals. Their unique intrinsic properties make them attractive therapeutic agents, since they show high biological activities associated with low toxicity and high specificity, as well as potentially useful as ingredients of functional or health-promoting foods [119]. These peptides generally contain less than 50 amino acid residues, with a molecular weight less than 10 kDa [120]. Despite their structural diversity, they have common physico-chemical characteristics; they are positively charged (+2 to +9) under physiological conditions due to the presence of lysine, arginine and histidine residues; and contain a substantial portion (50% or more) of hydrophobic residues [118]. These peptides commonly adopt an amphipathic conformation in which positively charged and hydrophobic groups are segregated into opposite faces of a α-helix, a β-leaf, or some other tertiary structure. This gives them the ability to cross the phospholipid membrane. The spectrum of different chemical properties of the amino acid side chain provides a variety of peptide sequences to show a cationic amphibious helical peptide [121]. Having a positive net charge allows them to interact with the anionic phospholipids of the bacterial membrane or other pathogens, and their amphipathicity, i.e., presence of apolar regions (with hydrophobic amino acids), and positive loads regions (cationic amino acids, Arg, Lys or His), facilitates them that, after initial interaction, the polar regions interact with the polar chains of the phospholipids, achieving the insertion of the peptide into the microbial membrane [122]. They are also flexible, which allows their internalization toward the bacterial cytoplasm, and leads to cell death due to ion and metabolic substances loss [123].

The most common mechanisms of action recognized in peptide antimicrobial activity include (i) the barrel model, in which a water-filled channel and an ion channel protein are formed by the interaction of peptides, acting as pores that disrupt the structure of the cell membrane; (ii) toroidal pore, in which less organized pore structures are formed; (iii) carpet models, in which the destabilization of the cell membrane in mycellar structures is caused by the accumulation of peptides above the limit concentration; (iv) molecular electroporation, following the concept that molecular electroporation can be achieved not only by electrical fields externally applied, but also by highly charged molecules that bind to the membrane surface; (v) sinking raft model, product of the induction of the membrane curvature by adsorbed peptides, which is relieved by its aggregation in the bilayer, allowing the aggregate to be translocated into the lumen of the gallbladder by a sinking raft process; and after membrane permeation, intracellular targets activation or blocking occurs [11]. These peptides not only generate toxic effects on microorganisms, but also exert important effects on the host, including immunomodulation, angiogenesis induction, wound healing and gene expression modulation. These effects may complement each other during the control of infectious and inflammatory diseases, and may be highly desirable when considered an optimal combination of an antimicrobial compound and regeneration booster [118]. In recent decades, barbel muscle antimicrobial peptides have been obtained by enzymatic hydrolysis of proteins from aquatic organisms [124]. Mustelus viscera [125], sea cucumber by-products [126], and different fish species [120], among others.

5.2 Commercial peptides obtained from fish sources

Thanks to their potential to produce bioactive compounds, fish parts and their residues have been used to obtain different types of functional inputs that have reached the market in different countries (**Table 7**). It should be noted, however, that few countries in which these products are being marketed. Given that fish, production extends to a much larger number of countries and that waste from that industry is proportional to production, it is clear that there is a latent possibility of expanding the market for products derived from fish sources.

Commercial name	Source	Functionality	Country
Custom Collagen®	Tilapia	Liver and kidney	US
Hydroiyzed Fish Collagen Typc 1	Tilapia	Skin, tendons, and arteries	UK
Amizate®	Atlantic salmon	Muscle anabolism	North America
Protizen®		Stress, weight disorder, sleep trouble	UK
Levenorm®	Sarda	Antihypertensive	Canada
MOLVAL®	Molva	Cholesterol, stress, and cardiovascular health	UK
Norland Hydrolyzed Fish Collagen	Cod	Hair, skin and nails	US
PeptACE®	Sarda	Vascular function and blood pressure	Japan and US
Stabilium®200	Molva dypterygia	Stress, memory, and cognitive function	UK
Seacure®	Hake	Gastrointestinal and bowel function	Canada and US
Seagest™	White fish	Intestinal lining and health	US
Valtyron®	Sardine	Blood pressure	
Vasotensin®	Tuna and verdel	Vascular function and blood pressure	Japan and US
Nutripeptin®	Cod	Weight and blood glucose	US and UK
Liquamen®	Molva	Oxidative stress, glycemic index, and stress	UK

Table 7.
Commercial products obtained by enzymatic Hydroiysis of fish protein by-products [37, 127].

Author details

Jose Edgar Zapata Montoya* and Angie Franco Sanchez
Nutrition and Food Technology Group, Universidad de Antioquia, Colombia

*Address all correspondence to: edgar.zapata@udea.edu.co

References

[1] Ishak NH, Sarbon NM. A review of protein hydrolysates and bioactive peptides deriving from wastes generated by fish processing. Food and Bioprocess Technology. 2018;**11**:2-16

[2] FAO. The State of World Fisheries and Aquaculture: Meeting the Sustainable Development Goals. Rome: FAO; 2018

[3] Ghaly AE, Ramakrishnan VV, Brooks MS, Budge SM, Dave D. Fish processing wastes as a potential source of proteins, amino acids and oils: A critical review. J Microb Biochem Technol. 2013;**5**: 107-129. DOI: 10.4172/1948-5948.1000110

[4] AMEC. Management of wastes from Atlantic seafood processing operations. Dartmouth, Nova Scotia, Canada: AMEC Earth and Environment Limited; 2003

[5] Gaviria YS, Figueroa OA, Zapata JE. Aplicación de la metodología de huella ecológica como indicador de sostenibilidad en el uso de ensilaje de pescado en dietas para alimentación de aves. Información tecnológica. 2021;**32** (5):199–208. DOI: 10.4067/s0718-07642021000500199

[6] Villamil O, Vaquiro H, Solanilla JF. Fish viscera protein hydrolysates: Production, potential applications and functional and bioactive properties. Food Chemistry. 2017;**224**:160-171. DOI: 10.1016/j.foodchem.2016.12.057

[7] Tchoukanova N, Gonzalez M, Poirier S. Best Management Practices: Marine Products Processing. New Brunswick: The Fisheries and Marine Products Division of the Coastal Zones Research Institute Inc; 2012

[8] Suarez LM, Montes JR, Zapata JE. Optimization acid content in silage víscera red tilapia (Oreochromis spp.) with Life cycle assessment of feeds derived. Technol Inf. 2018;**29**(6):83-94

[9] Liu Y, Li X, Chen Z, Yu J, Wang F, Wang J. Characterization of structural and functional properties of fish protein hydrolysates from surimi processing by-products. Food Chemistry. 2014;**151**: 459-465

[10] Zapata JE, Moya M, Figueroa OA. Hidrólisis Enzimática de la Proteína de Vísceras de Trucha Arco Íris (Oncorhynchus mykiss): Efecto del tipo de Enzima, Temperatura, pH y Velocidad de Agitación. Información Tecnológica. 2019;**30**(6):63-72

[11] Gomez LJ, Gomez NA, Zapata JE, Lopez-García G, Cilla A, Alegría A. In-vitro antioxidant capacity and cytoprotective/cytotoxic effects upon Caco-2 cells of red tilapia (Oreochromis spp.) viscera hydrolysates. Food Research International. 2019;**120**:52-61. DOI: 10.1016/j.foodres.2019.02.029

[12] Esteban MB, García AJ, Ramos P, Márquez MC. Evaluation of fruitvegetable and fish wastes as alternative feedstuffs in pig diets. Waste Management. 2007;**27**:193-200

[13] Ghaedian R, Coupland JN, Decker EA, McClemets JD. Ultrasonic determination of fish composition. Journal of Food Engineering. 1998;**35**: 323-337

[14] Carranza-Saavedra D, Zapata-Montoya JE, Váquiro-Herrera HA, Solanilla-Duque JF. Study of biological activities and physicochemical properties of Yamú (Brycon siebenthalae) viscera hydrolysates in sodium alginate-based edible coating solutions. International Journal of Food

Engineering. 2021;**17**:677-691. DOI: 10.1515/ijfe-2021-0036

[15] Leonarduzzi E, Rodrigues KA, Macchi GJ. Proximate composition and energy density in relation to Argentine hake females (Merluccius hubbsi) morphometrics and condition indices. Fisheries Research. 2014;**160**:33-40

[16] Gaviria YS, Figueroa OA, Zapata JE. Efecto de la inclusión de ensilado químico de vísceras de tilapia roja (Oreochromis spp.) en dietas para pollos de engorde sobre los parámetros productivos y sanguíneos. Información tecnológica. 2021;**32**:79-88. DOI: 10.4067/S0718-07642021000300079

[17] Chuesiang P, Sanguandeekul R. Protein hydrolysate from tilapia frame: Antioxidant and angiotensin I converting enzyme inhibitor properties. International Journal of Food Science and Technology. 2015;**50**(6):1436-1444

[18] Bhaskar N, Mahendrakar NS. Protein hydrolysate from visceral waste proteins of Catla (Catla catla): Optimization of hydrolysis conditions for a commercial neutral protease. Bioresour Technology. 2008;**99**(10):4105-4111. DOI: 10.1016/j.biortech.2007.09.006

[19] Pires C, Teixeira B, Cardoso C, Mendes R, Nunes ML, Batista I. Cape hake protein hydrolysates prepared from alkaline solubilised proteins pre-treated with citric acid and calcium ions: Functional properties and ACE inhibitory activity. Process Biochemistry. 2015;**50**(6):1006-1015

[20] Quintero J, Zapata JE. Optimización de la extracción del colágeno soluble en Ácido de subproductos de tilapia roja (oreochromis spp) mediante un diseño de superficie de respuesta. Información tecnológica. 2017;**28**: 109-120. DOI: 10.4067/s0718-07642017000100011

[21] Sousa SC, Vazquez JA, Pérez-Martín RI, Carvalho AP, Gomes AM. Valorization of byproducts from commercial fish species: Extraction and chemical properties of skin gelatins. Molecules. 2017;**22**(1545):1-12

[22] Sampedro LJG, Grimaldos NAG, Pereañez JA, Montoya JEZ. Lipids as competitive inhibitors of subtilisin carlsberg in the enzymatic hydrolysis of proteins in red tilapia (oreochromis sp.) viscera: Insights from kinetic models and a molecular docking study. Brazilian Journal of Chemical Engineering. 2019; **36**:647-655. DOI: 10.1590/0104-6632.20190362s20180346

[23] Taheri A, Anvar SAA, Ahari H, Fogliano V. Comparison the functional properties of protein Hydrolysates from poultry byproducts and rainbow trout (Onchorhynchus mykiss) viscera. Iranian Journal of Fisheries Sciences. 2013;**12**(1):154-169

[24] Opheim M, Šližytė R, Sterten H, Provan F, Larssen E, Kjos NP. Hydrolysis of Atlantic salmon (Salmo salar) rest raw materials- effect of raw material and processing on composition, nutritional value, and potential bioactive peptides in the hydrolysates. Process Biochemistry. 2015;**50**(8):1247-1257

[25] Roslan J, Yunos KFM, Abdullah N, Kamal SMM. Characterization of fish protein hydrolysate from tilapia (Oreochromis Niloticus) by-product. Agriculture and Agricultural Science Procedia. 2014;**2**:312-319. DOI: 10.1016/j.aaspro.2014.11.044

[26] Giraldo-Rios DE, Rios LA, Zapata-Montoya JE. Kinetic modeling of the alkaline deproteinization of Nile-tilapia skin for the production of collagen.

Heliyon. 2020;**6**:1-8. DOI: 10.1016/j.heliyon.2020.e03854

[27] Gencbay G, Turhan S. Proximate composition and nutritional profile of the black sea anchovy (Engraulis encrasicholus) whole fish, fillets, and by products. J Aquat Food Prod T. 2016; **25**(6):864-874

[28] Arias L, Gómez LJ, Zapata JE. Efecto de Temperatura-Tiempo sobre los lípidos extraídos de vísceras de tilapia roja (oreochromis sp.) utilizando un proceso de Calentamiento-Congelación. Información tecnológica. 2017;**28**: 131-142. DOI: 10.4067/s0718-07642017000500014

[29] Soufi-Kechaou E, Derouiniot-Chaplin M, Ben Amar R, Jaouen P, Berge JP. Recovery of valuable marine compounds from cuttlefish by-product hydrolysates: Combination of enzyme bioreactor and membrane technologies. C. R. Chimie. 2017;**20**(9–10):975-985

[30] Sierra-Lopera LM, Zapata-Montoya JE. Optimization of enzymatic hydrolysis of red tilapia scales (Oreochromis sp.) to obtain bioactive peptides. Biotechnology Reports. 2021;**30**:e00611. DOI: 10.1016/j.btre.2021.e00611

[31] Morkore T, Vallet JL, Cardinal M, Guillen MCG, Montero P, et al. Fat content and fillet shape of Atlantic salmon: Relevance for processing yield and quality of raw and smoked products. Journal of Food Science. 2001;**66**:1348-1354

[32] FAO/NACA/WHO J. The State of World Fisheries and Aquaculture [internet]. 2016. USA. Available from: www.fao.org/publications [Accessed: 29 Jan 2021]

[33] Friedman K. Nutritional value of proteins from different food sources: A review. Journal of Agricultural and Food Chemistry. 1996;**44**:6-29

[34] Dunajski E. Texture of fish muscle. Journal of Texture Studies. 1979;**10**: 301-318

[35] Chen J, Li L, Yi R, Xu N, Gao R, Hong B. Extraction and characterization of acid-soluble collagen from scales and skin of tilapia (Oreochromis niloticus). LWT - Food Science and Technology. 2016;**66**:453-459

[36] Zhang Q, Wang Q, Lv S, Lu J, Jiang S, Regenstein JM, et al. Comparison of collagen and gelatin extracted from the skins of Nile tilapia (Oreochromis niloticus) and channel catfish (Ictalurus punctatus). Food Bioscience. 2016;**13**: 41-48

[37] Chalamaiah M, Dinesh Kumar B, Hemalatha R, Jyothirmayi T. Fish protein hydrolysates: Proximate composition, amino acid composition, antioxidant activities and applications: A review. Food Chemistry. 2012;**135**: 3020-3038

[38] Spinelli J, Dassow JA. Fish proteins: Their modification and potential uses in the food industry. In: Chemistry and Biochemistry of Marine Food Products. Westport, CT, USA: AVI Publishing Company; 1982. pp. 13-25

[39] Kristinsson HG, Rasco BA. Fish protein hydrolysates: Production, biochemical, and functional properties. Critical Reviews in Food Science and Nutrition. 2000;**40**:43-81

[40] Sepúlveda CT, Zapata JE, Martínez-Álvarez O, Alemán A, Montero MP, Gómez-Guillén MC. The preferential use of a soy-rapeseed lecithin blend for the liposomal encapsulation of a tilapia viscera hydrolysate. LWT. 2021;**139**: 110530. DOI: 10.1016/j.lwt.2020.110530

[41] Brooks MS, Ghaly AE, Ramakrishnan VV, Budge SM, Dave D. Fish processing wastes as a potential source of proteins, amino acids and oils: A critical review. Journal of Microbial & Biochemical Technology. 2013;**5**(4): 107-129. DOI: 10.4172/1948-5948.1000110

[42] Sierra M. Revaloración de escamas y esqueletos de tilapia roja (oreochromis spp.) para la obtención de péptidos bioactivos. Medellin: Universidad de Antioquia; 2021

[43] Tavano OL. Protein hydrolysis using proteases: An important tool for food biotechnology. Journal of Molecular Catalysis B: Enzymatic. 2013;**90**:1-11. DOI: 10.1016/j.molcatb.2013.01.011

[44] Trusek-Holownia A, Lech M, Noworyta A. Protein enzymatic hydrolysis integrated with ultrafiltration: Thermolysin application in obtaining peptides. Chemical Engineering Journal. 2016;**305**:61-68. DOI: 10.1016/j.cej.2016.05.087

[45] Samaranayaka AGP, Li-Chan ECY. Food-derived peptidic antioxidants: A review of their production, assessment, and potential applications. Journal of Functional Foods. 2011;**3**(4):229-254. DOI: 10.1016/j.jff.2011.05.006

[46] Benítez R, Ibarz A, Pagan J. Hidrolizados de proteína: procesos y aplicaciones Resumen. Acta Bioquímica Clínica Latinoamericana. 2008;**42**(2): 227-237

[47] Mc Auley MT. Computer Modelling for Nutritionists. Chester: Springer; 2019. DOI: 10.1007/978-3-319-39994-2

[48] Zamora-Sillero J, Gharsallaoui A, Prentice C. Peptides from fish by-product protein hydrolysates and its functional properties: An overview. Marine Biotechnology. 2018;**20**(2): 118-130. DOI: 10.1007/s10126-018-9799-3

[49] He S, Franco C, Zhang W. Functions, applications and production of protein hydrolysates from fish processing co-products (FPCP). Food Research International. 2013;**50**(1): 289-297. DOI: 10.1016/j.foodres.2012.10.031

[50] Wang C, Chang T, Shi L, Yang H, Cui M, Tambalu L. Seafood processing by-products: Collagen and gelatin. In: Kim S-K, editor. Seafood Processing By-Products. 1st ed. New York: Springer; 2014. pp. 207-242. DOI: 10.1007/978-1-4614-9590-1_12

[51] Gómez LJ, Zapata JE. Efecto del Nivel de Grasa y Velocidad de Agitación en la Hidrolisis Enzimática de Vísceras de Tilapia Roja (Orechromis sp.). Información tecnológica. 2017;**28**(4): 47-56. DOI: 10.4067/s0718-07642017000400007

[52] Adler-Nissen J. Enzymic hydrolysis of food proteins. London: Elsevier Applied Science Publishers; 1986

[53] Márquez MC, Vázquez MA. Modeling of enzymatic protein hydrolysis. Process Biochemistry. 1999; **35**:111-117

[54] Nielsen PM, Petersen D, Dambmann C. Improved method for determining food protein degree of hydrolysis. Journal of Food Science. 2001;**66**(5):642-646. DOI: 10.1111/j.1365-2621.2001.tb04614.x

[55] Adler-Nissen J. Determination of the degree of hydrolysis of food protein hydrolysates by trinitrobenzenesulfonic acid. Journal of Agricultural and Food Chemistry. 1979;**27**(6):1256-1262

[56] Rutherfurd SM. Methodology for determining degree of hydrolysis of proteins in hydrolysates: A review. Journal of Aoac International. 2010; **93**(5):1515-1522. DOI: 10.1093/jaoac/93.5.1515

[57] Wubshet SG, Lindberg D, Veiseth-Kent E, Kristoffersen KA, Böcker U, Washburn KE, et al. Bioanalytical aspects in enzymatic protein hydrolysis of by products. In: Proteins: Sustainable Source, Processing and Applications. Cambridge: Academic Press; 2019. DOI: 10.1016/b978-0-12-816695-6.00008-8

[58] Mat DJL, Cattenoz T, Souchon I, Michon C, Le Feunteun S. Monitoring protein hydrolysis by pepsin using pH-stat: In vitro gastric digestions in static and dynamic pH conditions. Food Chemistry. 2018;**239**:268-275. DOI: 10.1016/j.foodchem.2017.06.115

[59] Valencia P, Espinoza K, Ceballos A, Pinto M, Almonacid S. Novel modeling methodology for the characterization of enzymatic hydrolysis of proteins. Process Biochemistry. 2015;**50**(4): 589-597. DOI: 10.1016/j.procbio.2014.12.028

[60] Dzwolak W, Ziajka S. Enzymatic hydrolysis of milk proteins under alkaline and acidic conditions. Journal of Food Science. 1999;**64**:393-395

[61] Satake K, Okuyama T, Ohashi M, Shinoda T. Journal of Biochemistry. 1960;**47**:654-660

[62] Rao MB, Tanksale AM, Ghatge MS, Deshpande VV. Molecular and biotechnological aspects of microbial proteases. Microbiology and Molecular Biology Reviews. 1998;**62**(3):597-635

[63] Vazquez JA, Medina A, Duran AI, Nogueira M, Fernandez-Compas A, Perez Martin RI, et al. Production of valuable compounds and bioactive metabolites from by-products of fish discards using chemical processing, enzymatic hydrolysis, and bacterial fermentation. Marine Drugs. 2019;**17**(3): 139. DOI: 10.3390/md17030139

[64] Singh P, Kumar S. Microbial enzyme in food biotechnology. In: Kuddus M, editor. Enzymes in Food Biotechnology. Cambridge: Academic Press; 2019. pp. 19-28. DOI: 10.1016/B978-0-12-813280-7.00002-5

[65] Martínez-Medina GA, Barragán AP, Ruiz HA, Ilyina A, Martínez Hernández JL, Rodríguez-Jasso RM, et al. Fungal proteases and production of bioactive peptides for the food industry. In: Kuddus M, editor. Enzymes in Food Biotechnology. Cambridge: Academic Press; 2019. pp. 221-246. DOI: 10.1016/B978-0-12-813280-7.00014-1

[66] Beaubier S, Framboisier X, Ioannou I, Galet O, Kapel R. Simultaneous quantification of the degree of hydrolysis, protein conversion rate and mean molar weight of peptides released in the course of enzymatic proteolysis. Journal of Chromatography B: Analytical Technologies in the Biomedical and Life Sciences. 2019;**1105**: 1-9. DOI: 10.1016/j.jchromb.2018.12.005

[67] Tkaczewska J, Jamroz E, Kulawik P, Morawska M, Szczurowska K. Evaluation of the potential use of a carp (Cyprinus carpio) skin gelatine hydrolysate as an antioxidant component. Food & Function. 2019;**10**(2):1038-1048. DOI: 10.1039/c8fo02492h

[68] Sabbione AC, Nardo AE, Añón MC, Scilingo A. Amaranth peptides with antithrombotic activity released by simulated gastrointestinal digestion. Journal of Functional Foods. 2016;**20**: 204-214. DOI: 10.1016/j.jff.2015.10.015

[69] Ketnawa S, Suwal S, Huang J, Liceaga AM. Selective separation and characterisation of dual ACE and DPP-IV inhibitory peptides from rainbow trout (Oncorhynchus mykiss) protein hydrolysates. International Journal of Food Science & Technology. 2018;**54**(4):1062-1073. DOI: 10.1111/ijfs.13939

[70] Aissaoui N, Chobert JM, Haertle T, Marzouki MN, Abidi F. Purification and biochemical characterization of a neutral serine protease from Trichoderma harzianum. Use in antibacterial peptide production from a fish by-product hydrolysate. Applied Biochemistry and Biotechnology. 2017;**182**(2):831-845. DOI: 10.1007/s12010-016-2365-4

[71] Nasri R, Amor IB, Bougatef A, Nedjar-Arroume N, Dhulster P, Gargouri J, et al. Anticoagulant activities of goby muscle protein hydrolysates. Food Chemistry. 2012;**133**(3):835-841. DOI: 10.1016/j.foodchem.2012.01.101

[72] Agyei D, Ongkudon CM, Wei CY, Chan AS, Danquah MK. Bioprocess challenges to the isolation and purification of bioactive peptides. Food and Bioproducts Processing. 2016; **98**:244-256. DOI: 10.1016/j.fbp.2016.02.003

[73] Gómez LJ, Figueroa OA, Zapata YJE. Actividad antioxidante de hidrolizados enzimáticos de plasma bovino obtenidos por efecto de alcalasa® 2.4 L. Información Tecnológica. 2013;**24**(1): 33-42. DOI: 10.4067/S0718-07642013000100005

[74] Sarmadi BH, Ismail A. Antioxidative peptides from food proteins: a review. Peptides. 2010;**31**(10):1949-1956. DOI: 10.1016/j.peptides.2010.06.020

[75] Saidi S, Deratani A, Belleville MP, Amar RB. Production and fractionation of tuna by-product protein hydrolysate by ultrafiltration and nanofiltration: Impact on interesting peptides fractions and nutritional properties. Food Research International. 2014;**65**:453-461. DOI: 10.1016/j.foodres.2014.04.026

[76] Ngo D, Vo T, Ngo D, Wijesekara I, Kim S. Biological activities and potential health benefits of bioactive peptides derived from marine organisms. International Journal of Biological Macromolecules. 2012;**51**:378-383

[77] Santos-Sánchez NF, Salas-Coronado R, Villanueva-Cañongo C, Hernández Carlos B. Antioxidant compounds and their antioxidant mechanism. In: Antioxidants. London: IntechOpen; 2019

[78] Canabady-Rochelle LLS, Selmeczi K, Collin S, Pasc A, Muhr L, Boschi-Muller S. SPR screening of metal chelating peptides in a hydrolysate for their antioxidant properties. Food Chemistry. 2018;**239**:478-485. DOI: 10.1016/j.foodchem.2017.06.116

[79] Decker EA, Welch B. Role of ferritin as a lipid oxidation catalyst in muscle food. Journal of Agricultural and Food Chemistry. 1990;**38**(3):674-677. DOI: 10.1021/jf00093a019

[80] You L, Zhao M, Regenstein JM, Ren J. In vitro antioxidant activity and in vivo anti-fatigue effect of loach (Misgurnus anguillicaudatus) peptides prepared by papain digestion. Food Chemistry. 2011;**124**(1):188-194. DOI: 10.1016/j.foodchem.2010.06.007

[81] Je JY, Qian ZJ, Byun HG, Kim SK. Purification and characterization of an antioxidant peptide obtained from tuna backbone protein by enzymatic hydrolysis. Process Biochemistry. 2007; **42**(5):840-846. DOI: 10.1016/j.procbio.2007.02.006

[82] Aluko RE. Functional Foods and Nutraceuticals. Biotechnology: New Ideas, New Developments (A Textbook of Modern Technology). New York, NY: Springer New York; 2012. DOI: 10.1007/978-1-4614-3480-1

[83] Je JY, Park SY, Hwang JY, Ahn CB. Amino acid composition and in vitro antioxidant and cytoprotective activity of abalone viscera hydrolysate. Journal of Functional Foods. 2015;**16**:94-103. DOI: 10.1016/j.jff.2015.04.023

[84] Ghanbari R. Review on the bioactive peptides from marine sources: Indication for health effects. International Journal of Peptide Research and Therapeutics. 2019;**25**(3):1187-1199. DOI: 10.1007/s10989-018-9766-x

[85] Pihlanto A. Antioxidative peptides derived from milk proteins. International Dairy Journal. 2006;**16**(11): 1306-1314. DOI: 10.1016/j.idairyj.2006.06.005

[86] Ngo DH, Ryu B, Vo TS, Himaya SWA, Wijesekara I, Kim SK. Free radical scavenging and angiotensin-I converting enzyme inhibitory peptides from Pacific cod (Gadus macrocephalus) skin gelatin. International Journal of Biological Macromolecules. 2011;**49**: 1110-1116

[87] Nikoo M, Benjakul S, Ehsani A, Li J, Wu F, Yang N, et al. Antioxidant and cryoprotective effects of a tetrapeptide isolated from Amur sturgeon skin gelatin. Journal of Functional Foods. 2014;**7**:609-620

[88] Mendis E, Rajapakse N, Kim SK. Antioxidant properties of a radicals scavenging peptide purified from enzymatically prepared fish skin gelatin hydrolysate. Journal of Agriculture and Food Chemistry. 2005;**53**:581-587

[89] Zhang YF, Duan X, Zhuang YL. Purification and characterization of novel antioxidant peptides from enzymatic hydrolysates of tilapia (Oreochromis niloticus) skin gelatin. Peptides. 2012;**38**:13-21

[90] Kumar S, Sutherland F, Rosso R, Teh AW, Lee G, Heck PM, et al. Effects of chronic omega-3 polyunsaturated fatty acid supplementation on human atrial electrophysiology. Heart Rhythm. 2011;**8**:562-568

[91] Chi CF, Wang B, Wang YM, Zhang B, Deng SG. Isolation and characterization of three antioxidant peptides from protein hydrolysate of bluefin leatherjacket (Navodon septentrionalis) heads. Journal of Functional Foods. 2015;**12**:1-10

[92] Ahn CB, Kim JG, Je JY. Purification and antioxidant properties of octapeptide from salmon byproduct protein hydrolysate by gastrointestinal digestion. Food Chemistry. 2014;**147**:78-83

[93] Jaiganesh R, Nazeer RA, Sampath Kumar NS. Purification and identification of antioxidant peptide from black pomfret, Parastromateus niger (Bloch, 1975) viscera protein hydrolysate. Food Science and Biotechnology. 2011;**20**:1087-1094

[94] Pan X, Zhao Y, Hu F, Wang B. Preparation and identification of antioxidant peptides from protein hydrolysate of skate (Raja porosa) cartilage. Journal of Functional Foods. 2016;**25**:220-230

[95] Cai L, Wu X, Zhang Y, Li X, Ma S, Li J. Purification and characterization of three antioxidant peptides from protein hydrolysate of grass carp (Ctenopharyngodon idella) skin. Journal of Functional Foods. 2015;**16**:234-242. DOI: 10.1016/j.jff.2015.04.042

[96] Song R, Zhang K, Wei R. In vitro antioxidative activities of squid (Ommastrephes bartrami) viscera autolysates and identification of active peptides. Process Biochemistry. 2016; **51**(10):1674-1682

[97] Wijesekara I, Kim SK. Angiotensin-I-converting enzyme (ACE) inhibitors from marine resources: Prospects in the pharmaceutical industry. Marine Drugs. 2010;**8**(4):1080-1093. DOI: 10.3390/md8041080

[98] Byun HG, Kim SK. Purification and characterization of angiotensin I converting enzyme (ACE) inhibitory peptides from Alaska pollack (Theragra chalcogramma) skin. Process Biochemistry. 2001;**36**:1155-1162

[99] Lee JK, Jeon JK, Byun HG. Antihypertensive effect of novel angiotensin I converting enzyme inhibitory peptide from chum salmon (Oncorhynchus keta) skin in spontaneously hypertensive rats. Journal of Functional Foods. 2014;**7**:381-389

[100] Fahmi A, Morimura S, Guob HC, Shigematsu T, Kida K, Uemura Y. Production of angiotensin I converting enzyme inhibitory peptides from sea bream scale. Process Biochemistry. 2004;**39**:1195-1200

[101] Bougatef A, Nedjar-Arroume N, Ravallec-Plé R, Leroy Y, Guillochon D, Barkia A, et al. Angiotensin I-converting enzyme (ACE) inhibitory activities of sardinelle (Sardinella aurita) by-products protein hydrolysates obtained by treatment with microbial and visceral fish serine proteases. Food Chemistry. 2008;**111**:350-356

[102] Ghassem M, Babji AS, Said M, Mahmoodani F, Arihara K. Angiotensin I-converting enzyme inhibitory peptides from snakehead fish sarcoplasmic protein hydrolysate. Journal of Food Biochemistry. 2014; **38**(2):14-19

[103] García-Moreno PJ, Espejo-Carpio FJ, Guadix A, Guadix EM. Production and identification of angiotensin I-converting enzyme (ACE) inhibitory peptides from Mediterranean fish discards. Journal of Functional Foods. 2015;**18**:95-105

[104] Wu X, Cai L, Zhang Y, Mi H, Cheng X, Li J. Compositions and antioxidant properties of protein hydrolysates from the skins of four carp species. International Journal of Food Science & Technology. 2015;**50**(12): 2589-2597. DOI: 10.1111/ijfs.12927

[105] Chen J, Liu Y, Wang G, Sun S, Liu R, Hong B, et al. Processing optimization and characterization of angiotensin-I-converting enzyme inhibitory peptides from lizardfish (Synodus macrops) scale gelatin. Marine Drugs. 2018;**16**(7):228. DOI: 10.3390/md16070228

[106] Baig MH, Ahmad K, Saeed M, Alharbi AM, Barreto GE, Ashraf GM, et al. Peptide based therapeutics and their use for the treatment of neurodegenerative and other diseases. Biomedicine & Pharmacotherapy. 2018; **103**:574-581. DOI: 10.1016/j.biopha.2018.04.025

[107] Chalamaiah M, Yu W, Wu J. Immunomodulatory and anticancer protein hydrolysates (peptides) from food proteins: A review. Food Chemistry. 2018;**245**:205-222. DOI: 10.1016/j.foodchem.2017.10.087

[108] Correia-da-Silva M, Sousa E, Pinto MMM, Kijjoa A. Anticancer and cancer preventive compounds from edible marine organisms. Seminars in

Cancer Biology. 2017;**46**:55-64. DOI: 10.1016/j.semcancer.2017.03.011

[109] Picot L, Bordenave S, Didelot S, Fruitier-Arnaudin I, Sannier F, Thorkelsson G, et al. Antiproliferative activity of fish protein hydrolysates on human breast cancer cell lines. Process Biochemistry. 2006;**41**:1217-1222. DOI: 10.1016/j.procbio.2005.11.024

[110] Kim EK, Kim YS, Hwang JW, Lee JS, Moon SH, Jeon BT, et al. Purification and characterization of a novel anticancer peptide derived from Ruditapes philippinarum. Process Biochemistry. 2013;**48**(7):1086-1090. DOI: 10.1016/j.procbio.2013.05.004

[111] Alemán A, Pérez-santín E, Bordenave-juchereau S, Arnaudin I, Gómez-guillén MC, Montero P. Squid gelatin hydrolysates with antihypertensive, anticancer and antioxidant activity. Food Research International. 2011;**44**:1044-1051. DOI: 10.1016/j.foodres.2011.03.010

[112] Song R, Wei R, Luo H, Yang Z. Isolation and identification of an antiproliferative peptide derived from heated products of peptic hydrolysates of halffin anchovy (Setipinna taty). Journal of Functional Foods. 2014;**10**:104-111

[113] Doyen A, Beaulieu L, Saucier L, Pouliot Y, Bazinet L. Demonstration of in vitro anticancer properties of peptide fractions from a snow crab by-products hydrolysate after separation by electrodialysis with ultrafiltration membranes. Separation and Purification Technology. 2011;**78**(3):321-329. DOI: 10.1016/j.seppur.2011.01.037

[114] Gianfranceschi GL, Gianfranceschi G, Quassinti L, Bramucci M. Biochemical requirements of bioactive peptides for nutraceutical efficacy. Journal of Functional Foods. 2018;**47**:252-263. DOI: 10.1016/j.jff.2018.05.034

[115] Schweizer F. Cationic amphiphilic peptides with cancer-selective toxicity. European Journal of Pharmacology. 2009;**625**(1):190-194. DOI: 10.1016/j.ejphar.2009.08.043

[116] Jin NZ, Gopinath SCB. Potential blood clotting factors and anticoagulants. Biomedicine and Pharmacotherapy. 2016;**84**:356-365. DOI: 10.1016/j.biopha.2016.09.057

[117] Rajapakse N, Jung WK, Mendis E, Moon SH, Kim SK. A novel anticoagulant purified from fish protein hydrolysate inhibits factor XIIa and platelet aggregation. Life Sciences. 2005;**76**(22): 2607-2619. DOI: 10.1016/j.lfs.2004.12.010

[118] Alencar-Silva T, Braga MC, Santana GOS, Saldanha-Araujo F, Pogue R, Dias SC, et al. Breaking the frontiers of cosmetology with antimicrobial peptides. Biotechnology Advances. 2018;**36**(8):2019-2031. DOI: 10.1016/j.biotechadv.2018.08.005

[119] Sila A, Hedhili K, Przybylski R, Ellouz-Chaabouni S, Dhulster P, Bougatef A, et al. Antibacterial activity of new peptides from barbel protein hydrolysates and mode of action via a membrane damage mechanism against Listeria monocytogenes. Journal of Functional Foods. 2014;**11**:322-329. DOI: 10.1016/j.jff.2014.10.006

[120] Najafian L, Babji AS. A review of fish-derived antioxidant and antimicrobial peptides: Their production, assessment, and applications. Peptides. 2012;**33**(1):178-185. DOI: 10.1016/j.peptides.2011.11.013

[121] Deslouches B, Phadke SM, Lazarevic V, Cascio M, Islam K, Montelaro RC, et al. De novo generation

of cationic antimicrobial peptides: Influence of length and tryptophan substitution on antimicrobial activity. Antimicrobial Agents and Chemotherapy. 2005;**49**(1): 316-322. DOI: 10.1128/AAC.49.1. 316-322.2005

[122] Vogel HJ, Schibli DJ, Jing W, Lohmeier-Vogel EM, Epand RF, Epand RM. Towards a structure-function analysis of bovine lactoferricin and related tryptophanand arginine-containing peptides. Biochemistry and Cell Biology. 2002;**80**(1):49-63

[123] Villarruel R, Huizar R, Corrales M, Sánchez T, Islas A. Péptidos naturales antimicrobianos: Escudo esencial de la respuesta inmune. Investigación en Salud. 2004;**6**(3):170-179

[124] Sila A, Nedjar-Arroume N, Hedhili K, Chataigné G, Balti R, Nasri M, et al. Antibacterial peptides from barbel muscle protein hydrolysates: Activity against some pathogenic bacteria. LWT-Food Science and Technology. 2014; **55**(1):183-188. DOI: 10.1016/j. lwt.2013.07.021

[125] Abdelhedi O, Jridi M, Jemil I, Mora L, Toldrá F, Aristoy MC, et al. Combined biocatalytic conversion of smooth hound viscera: Protein hydrolysates elaboration and assessment of their antioxidant, anti-ACE and antibacterial activities. Food Research International. 2016;**86**: 9-23. DOI: 10.1016/j.foodres.2016. 05.013

[126] Tripoteau L, Bedoux G, Gagnon J, Bourgougnon N. In vitro antiviral activities of enzymatic hydrolysates extracted from byproducts of the Atlantic holothurian Cucumaria frondosa. Process Biochemistry. 2015;**50**: 867-875. DOI: 10.1016/j.procbio. 2015.02.012

[127] Sierra Lopera LM, Sepúlveda Rincón CT, Vásquez Mazo P, Figueroa Moreno OA, Zapata Montoya JE. Byproducts of aquaculture processes: Development and prospective uses. Review. Revista Vitae. 2018;**25**(3): 128-140. DOI: 10.17533/udea.vitae. v25n3a03

Section 2

Ligation

Chapter 6

Review of the Structural Basis of Human E2 Conjugating Enzymes in Complexed with RING E3 Ligases

Erin Meghan Gladu, Iman Sayed and Michael Anthony Massiah

Abstract

Protein ubiquitination is a post-translational modification that controls essential biological processes through its regulation of protein concentration, function, and cellular location. RING E3 ligases are a critical component of a three-enzyme cascade that facilitates the ubiquitination of proteins. RING-type E3 ligases represent one class of E3 ligases that function by binding the substrate protein and ubiquitin-conjugating enzymes (E2s). Proteins exhibiting RING-type E3 ligase activities do so via a domain that adopts a ββα-RING fold and coordinates two zinc ions. To date, structural studies show that the RING domain interacts with the catalytic domain of the E2 enzyme. The catalytic domain is approximately 150 amino acids and adopts a canonical structure consisting of four α-helices and 3–4 β-strands. Structural analyses of RING–E2 complexes reveal that RING domains interact on a similar surface of the E2 enzyme. We postulate that the mechanism of interaction between an E2 enzyme and its cognate RING E3 domain may contribute to the extent of substrate modification. In this review, we compare the primary and secondary structures of human E2 enzymes and examine their quaternary structure with RING domains. Our analyses reveal the interactions appear to be relatively conserved with similar types of amino acids involved.

Keywords: ubiquitination, ubiquitin, really interesting new gene, E2–E3 binding, protein degradation; E2 conjugating enzyme, zinc-binding proteins, protein–protein, E2 recognition

1. Introduction

Protein ubiquitination is a highly conserved process in eukaryotic cells that plays key roles in cellular functions [1–4]. Depending on the type and extent of ubiquitin (Ub) modification, the cellular fate of the protein can be defined. Proteins that are composed with a chain of Ub (polyubiquitination) are usually captured by accessory proteins, including deubiquitinating enzymes (dubs) on the proteasome [5]. The Ub are cleaved and recycled while the target protein is degraded. In other instances, proteins with a single Ub (monoubiquitination), diubiquitination or less extensively modified can have their cellular location or function altered [6–10]. As a result, ubiquitination can regulate most, if not all, signaling processes through the temporal and spatial regulation of proteins. Dysregulation in the ubiquitination pathway is associated with several diseases, including cancers, genetic defects and neurodegenerative diseases [11–16].

Ubiquitination is a highly coordinated event that is conserved in plants and animals as well as prokaryotes. The process of protein recycling in bacterial cells is called pupylation [17–20]. In eukaryotic cells, there are several mechanisms by which a protein is covalently modified with the small and stable Ub protein (**Figure 1**). The process involving RING E3 ligases is the most common because of the abundance of proteins with RING domains. Typically, RING-mediated ubiquitination involves three classes of enzymes. The first enzyme is common to all types of protein ubiquitination mechanisms. The ubiquitin-activating enzyme (E1) prepares the C-terminal carboxylate group of Gly-76 for chemistry by first catalyzing the addition of adenine monophosphate (AMP). The AMP group serves as a good leaving group and prepares the C-terminus carboxylate group for nucleophilic attack. The activated Ub ~ AMP first becomes covalently attached to the thiol group of an active site cysteine residue via a thioester bond. To date, there are 8–10 different types of E1 enzymes [21] and of these two are associated with activation of Ub [22]. In step two, the Ub is transferred [17] to an active site cysteine residue on one of four classes of human Ub-conjugating enzymes (E2). The thioester bond between C-terminal carboxylate group and the cysteine residue preserves the bond energy associated with the phosphoester bond with the AMP and the thioester bond with the E1 enzyme. RING-type E3 ligases promote substrate ubiquitination by binding the E2 enzyme and the substrate. In this case, a lysine residue serves as the nucleophile to attack the E2 ~ Ub thioester bond to form a stable isopeptide bond with the Ub. Subsequently, Ub can be attached to other lysine residues on the chain to generate a multi-ubiquitinated protein or Ub can form a Ub-chain on a lysine. This means that Ub can form a chain via one or several of its seven lysine residues. While lysine sidechain is the most common and most stable

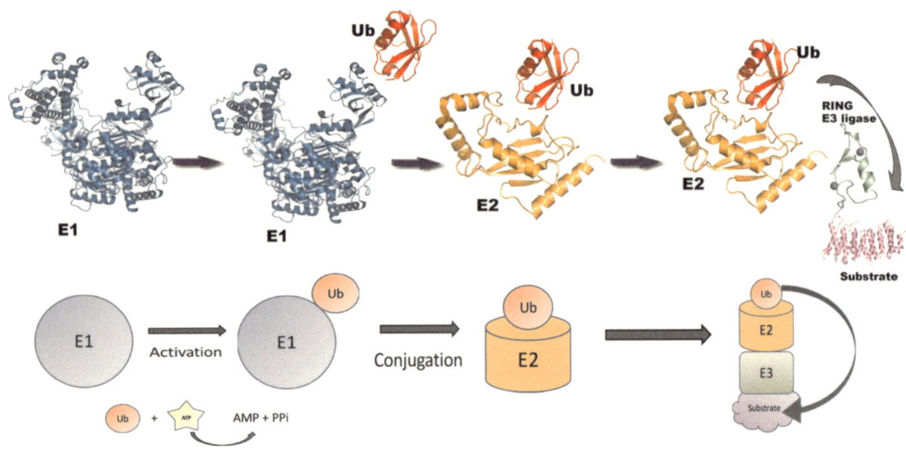

Figure 1.
Schematics of the ubiquitination cascade associated RING E3 ligases. The process begins with the ubiquitin-activating enzyme (known as E1), adding an AMP moiety to the C-terminal carboxylate group before transferring it to an internal cysteine to form an E1 ~ Ub intermediate. The ubiquitin is then transferred to the ubiquitin conjugating enzyme (E2) via a thioester bond involving a cysteine residue and Gly76 of ubiquitin. A RING E3 ligase then interacts with the E2 enzyme, influencing, among several things, the closed conformation between the E2 and ubiquitin. The closed conformation allows for aminolysis.

covalent linkage with Ub, evidence suggests that the Ub can be transferred to the side-chains of cysteines, serine, and threonine of substrates [23, 24]. The lysineless Pexp5 protein was observed to be monoubiquitinated via one of its cysteine residue and this modification affect its translation [25]. Hydroxyester linkages with Ub were observed for MHC-1 heavy chain in the ER-associated degradation pathway [26]. It would appear that both these types of linkages should be transient because they would be very susceptible to hydrolysis.

Furthermore, it is observed with in vitro assays, which are routinely performed to confirm that proteins with RING domain possess E3 ligase activity, that RING E3 ligases also facilitate autoubiquitination [2, 27, 28]. While the function and consequences of self-ubiquitination are poorly understood, it was demonstrated that autoubiquitination of Mdm2 and Nedd4 RING E3 ligases enhance their substrate ubiquitination activity [27, 29]. In contrast, autoubiquitination can result in self-induced degradation [30, 31].

To analyze the sequence, structure and interactions of E2 and RING E3 enzymes, this review focuses on a limited number of E2 and E3's from humans, and for the most part, the analysis presented is consistent with proteins in cells from plants and animals. From the literature, there are 394 reports of E2 enzymes with their cohort E3 enzymes from animal and human origins. These complexes represent the interactions from 23 specific E2 enzymes with 247 different E3 ligases. Of these, the quaternary structures of 33 E2–E3 pairs are reported in the protein database (PDB) for which the structure of the E2 and the E3 enzymes are known individually as well. These proteins are of human origin.

This review focuses on RING-type ubiquitination that includes E2 enzymes representing 4 classes with specific RING-type E3 ligases (**Figure 2**, **Tables 1** and **2**).

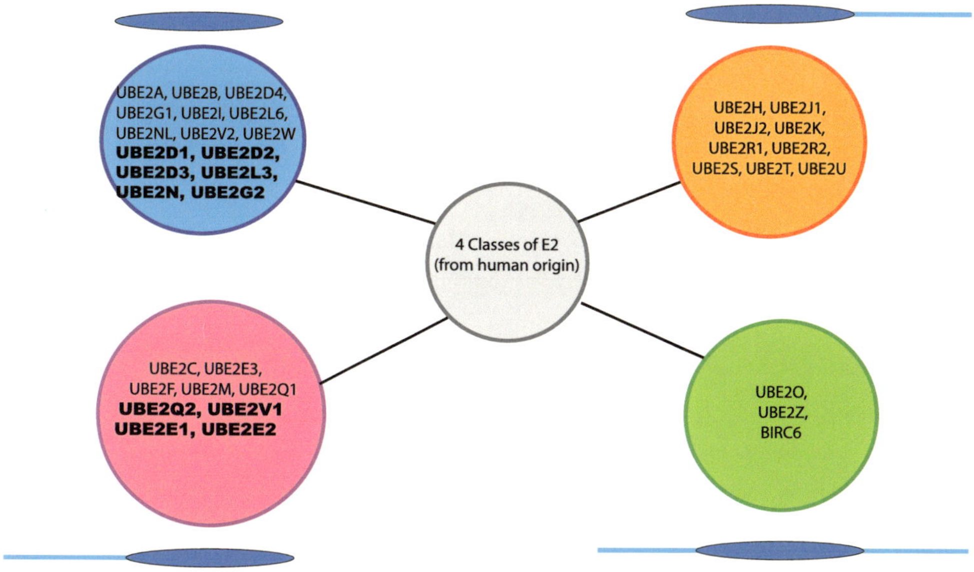

Figure 2.
Classification of 36 current human E2 enzymes. The E2 enzymes are classified based on the differences in number of amino acids for at their termini. Class one E2's do not have any terminal extensions and is essentially approximately 150 amino acids that adopts the catalytic domain, class 2 E2's have >20 amino acids at their N-terminus, class 3 enzymes have > 20 amino acids at their C-terminus, and class 4 enzymes have extensions at their C and N termini. The bolded E2's represent the E2's that for which structural analysis of E2 and E3 interactions were discussed.

Class I Class II Class III Class IV		Ube2A Ube2NL Ube2H	Ube2B Ube2V2 Ube2J1	Ube2d1 Ube2W Ube2J2	Ube2d2 Ube2C Ube2K	Ube2d3 Ube2E1 Ube2R1	Ube2d4 Ube2e2 Ube2R2	Ube2g1 Ube2e3 Ube2S	Ube2g2 Ube2F Ube2T	Ube2i Ube2M Ube2U	Ube2L3 Ube2q1 Ube2O	Ube2L6 Ube2q2 Ube2z	Ube2N Ube2v1 Birc6
		Percent identity matrix of E2 enzymes in human binary reactions											
1	UBE2A	**100 (100)**	95[a] (99)[b]	41 (78)	40 (78)	40 (77)	42 (78)	34 (73)	41 (73)	42 (73)	26 (72)	31 (70)	37 (70)
		33 (68)	22 (56)	31 (62)	42[c] [32]	39 (63)	38 (62)	38 (57)	25 (57)	27 (61)	24 (57)	23 (56)	24 (46)
		32 (62)	24 (52)	29 (59)	29 (57)	39 (65)	39 (67)	31 (61)	32 (60)	33 (58)	26 (57)	30 (58)	30 (62)
2	UBE2B	95 (99)	**100 (100)**	40 (78)	38 (76)	38 (79)	40 (76)	34 (74)	41 (78)	42 (75)	26 (74)	30 (69)	37 (69)
		33 (63)	22 (56)	32 (62)	42 (64)	39 (64)	39 (63)	39 (63)	25 (55)	27 (60)	24 (56)	22 (57)	23 (46)
		32 (64)	25 (53)	29 (58)	29 (57)	39 (64)	39 (67)	32 (61)	33 (60)	32 (57)	27 (57)	30 (61)	29 (63)
3	UBE2D1	41 (78)	40 (78)	**100 (100)**	89 (97)	88 (97)	92 (97)	37 (77)	39 (72)	34 (76)	38 (69)	38(71)	44 (71)
		40 (65)	25 (53)	35 (61)	36 (63)	61 (80)	61 (80)	61 (80)	35 (63)	30 (66)	24 (54)	23 (53)	25 (45)
		34 (62)	31 (61)	24 (54)	41 (69)	38 (62)	35 (63)	35 (66)	43 (68)	31 (61)	22 (54)	33 (63)	31 (59)
4	UBE2D2	40 (78)	38 (76)	89 (97)	**100 (100)**	97 (99)	93 (99)	34 (71)	38 (75)	36 (68)	37 (67)	36 (69)	45 (71)
		41 (64)	23 (52)	33 (63)	35 (78)	64 (78)	64 (78)	65 (79)	34 (63)	31 (64)	24 (55)	24 (53)	25 (41)
		31 (63)	29 (57)	24 (56)	42 (69)	33 (60)	33 (63)	37 (67)	43 (67)	35 (64)	24 (55)	36 (62)	31 (59)
5	UBE2D3	40 (77)	38 (79)	88 (97)	97 (100)	**100 (100)**	92 (99)	34 (71)	38 (76)	36 (68)	36 (67)	35 (69)	45 (72)
		41 (65)	23 (52)	33 (60)	35 (64)	63 (80)	63 (80)	64 (80)	34 (63)	31 (64)	24 (53)	24 (52)	25 (41)
		30 (63)	30 (59)	24 (54)	41 (69)	33 (60)	33 (62)	35 (67)	42 (67)	35 (63)	24 (56)	36 (62)	31 (58)
6	UBE2D4	42 (78)	40 (76)	92 (97)	93 (99)	92 (99)	**100 (100)**	36 (79)	39 (73)	35 (69)	38 (68)	38 (70)	45 (70)
		41 (67)	25 (50)	35 (58)	38 (61)	63 (80)	63 (80)	63 (81)	36 (64)	32 (66)	25 (54)	24 (53)	26 (43)
		32 (62)	29 (59)	24 (56)	43 (67)	36 (60)	35 (63)	37 (65)	41 (66)	32 (62)	23 (55)	35 (63)	32 (59)
7	UBE2G1	34 (73)	34 (74)	37 (77)	34 (71)	34 (71)	36 (79)	**100**	50 (69)	34 (61)	28 (69)	24 (62)	32 (69)
		29 (63)	23 (54)	30 (50)	33 (57)	31 (55)	32 (55)	32 (55)	27 (58)	26 (55)	15 (45)	16 (46)	24 (49)
		27 (52)	23 (54)	24 (54)	27 (57)	50 (69)	54 (73)	26 (62)	29 (58)	29 (59)	22 (55)	25 (59)	25 (57)

Class I Class II Class III Class IV		Percent identity matrix of E2 enzymes in human binary reactions											
		Ube2A Ube2NL Ube2H	Ube2B Ube2V2 Ube2J1	Ube2d1 Ube2W Ube2J2	Ube2d2 Ube2C Ube2K	Ube2d3 Ube2E1 Ube2R1	Ube2d4 Ube2e2 Ube2R2	Ube2g1 Ube2e3 Ube2S	Ube2g2 Ube2F Ube2T	Ube2i Ube2M Ube2U	Ube2L3 Ube2q1 Ube2O	Ube2L6 Ube2q2 Ube2z	Ube2N Ube2v1 Birc6
8	UBE2G2	41 (73)	41 (78)	39 (72)	38 (75)	38 (76)	39 (73)	50 (73)	100	37 (75)	26 (68)	23 (65)	30 (66)
		28 (60)	29 (48)	27 (57)	33 (56)	31 (61)	32 (61)	32 (60)	21 (56)	25 (55)	18 (53)	18 (53)	18 (47)
		27 (54)	27 (55)	27 (52)	26 (57)	48 (71)	48 (73)	30 (62)	27 (55)	30 (56)	24 (57)	29 (59)	28 (58)
9	UBE2I	42 (73)	42 (75)	34 (76)	36 (68)	36 (68)	35 (69)	34 (61)	37 (65)	100	27 (61)	27 (63)	31 (62)
		31 (61)	23 (50)	27 (58)	36 (57)	35 (59)	36 (64)	36 (63)	26 (55)	31 (58)	23 (47)	23 (49)	23 (48)
		26 (56)	22 (51)	22 (56)	27 (58)	36 (60)	35 (58)	30 (61)	30 (61)	28 (52)	21 (56)	30 (55)	30 (58)
10	UBE2L3	26 (72)	26 (74)	38 (69)	37 (67)	36 (67)	38 (68)	28 (69)	26 (68)	27 (61)	100	55 (80)	30 (62)
		28 (63)	24 (53)	20 (61)	26 (56)	35 (60)	35 (60)	35 (60)	29 (57)	32 (58)	18 (50)	19 (50)	25 (45)
		28 (56)	25 (52)	21 (54)	29 (51)	29 (62)	29 (63)	24 (51)	34 (58)	22 (56)	19 (53)	21 (53)	26 (53)
11	UBE2L6	31 (70)	30 (69)	38 (71)	36 (69)	35 (69)	38 (70)	24 (62)	23 (65)	27 (65)	55 (80)	100	29 (63)
		27 (63)	21 (52)	24 (57)	26 (56)	33 (57)	32 (57)	31 (56)	27 (52)	29 (59)	18 (50)	19 (46)	24 (49)
		24 (55)	25 (52)	23 (56)	27 (58)	27 (60)	26 (61)	25 (55)	33 (59)	25 (57)	20 (48)	23 (59)	24 (58)
12	UBE2N	37 (70)	37 (71)	44 (71)	45 (71)	45 (72)	45 (70)	32 (69)	30 (66)	31 (62)	20 (62)	29 (63)	100
		92 (97)	21 (49)	27 (57)	38 (64)	45 (64)	44 (64)	44 (64)	27 (61)	26 (62)	19 (50)	19 (53)	22 (46)
		29 (53)	25 (58)	27 (59)	44 (71)	32 (65)	30 (69)	36 (65)	44 (71)	29 (59)	20 (58)	30 (64)	25 (62)
13	UBE2NL	33 (68)	33 (69)	40 (65)	41 (64)	41 (65)	41 (69)	29 (63)	28 (60)	31 (61)	28 (63)	27 (63)	92 (97)
		100	17 (52)	24 (56)	35 (61)	42 (61)	41 (61)	40 (61)	24 (56)	23 (61)	16 (51)	17 (52)	18 (45)
		25 (50)	23 (53)	25 (55)	43 (68)	30 (59)	28 (63)	34 (63)	41 (69)	25 (52)	17 (52)	28 (62)	22 (54)
14	UBE2V2	22 (56)	22 (56)	25 (53)	23 (52)	23 (52)	25 (50)	23 (54)	19 (48)	23 (50)	24 (53)	21 (52)	21 (49)
		17 (52)	100	22 (50)	23 (57)	23 (57)	23 (55)	23 (54)	18 (54)	16 (53)	18 (58)	19 (57)	88 (94)
		21 (44)	23 (52)	18 (51)	26 (54)	22 (54)	21 (54)	19 (48)	26 (46)	18 (53)	15 (51)	20 (57)	20 (48)

Percent identity matrix of E2 enzymes in human binary reactions

Class I Class II Class III Class IV		Ube2A Ube2NL Ube2H	Ube2B Ube2V2 Ube2J1	Ube2d1 Ube2W Ube2J2	Ube2d2 Ube2C Ube2K	Ube2d3 Ube2E1 Ube2R1	Ube2d4 Ube2e2 Ube2R2	Ube2g1 Ube2e3 Ube2S	Ube2g2 Ube2F Ube2T	Ube2i Ube2M Ube2U	Ube2L3 Ube2q1 Ube2O	Ube2L6 Ube2q2 Ube2z	Ube2N Ube2v1 Birc6
15	UBE2W	31 (62)	32 (62)	35 (61)	33 (59)	33 (60)	35 (58)	30 (60)	27 (57)	27 (58)	20 (61)	24 (57)	27 (57)
		24 (56)	22 (50)	100	27 (54)	34 (57)	35 (55)	36 (54)	20 (49)	20 (53)	23 (50)	23 (52)	24 (44)
		19 (44)	23 (52)	22 (51)	27 (54)	31 (54)	33 (54)	23 (48)	28 (46)	24 (53)	17 (51)	26 (57)	22 (48)
16	UBE2C	42 (65)	42 (64)	36 (63)	35 (63)	35 (61)	38 (64)	33 (57)	33 (56)	36 (57)	26 (56)	26 (56)	38 (64)
		35 (61)	23 (57)	27 (54)	100	32 (57)	33 (61)	34 (55)	25 (60)	26 (64)	20 (56)	19 (57)	25 (57)
		29 (56)	22 (50)	20 (51)	31 (57)	33 (57)	33 (57)	28 (54)	34 (55)	30 (54)	20 (59)	21 (55)	24 (56)
17	UBE2E1	39 (63)	39 (64)	61 (80)	64 (78)	63 (80)	63 (80)	31 (55)	31 (61)	35 (59)	35 (60)	33 (57)	45 (64)
		42 (61)	23 (57)	34 (58)	32 (57)	100	88 (89)	84 (88)	27 (59)	30 (65)	23 (63)	21 (59)	24 (57)
		28 (50)	30 (46)	25 (51)	40 (54)	29 (44)	28 (47)	31 (52)	41 (53)	25 (44)	18 (53)	26 (56)	23 (56)
18	UBE2E2	38 (62)	39 (63)	61 (80)	64 (78)	63 (80)	63 (80)	32 (55)	32 (61)	36 (64)	35 (60)	32 (57)	44 (64)
		41 (61)	23 (55)	35 (58)	33 (61)	88 (89)	100	87 (93)	28 (62)	30 (64)	21 (61)	20 (60)	24 (54)
		28 (50)	29 (43)	25 (48)	41 (52)	29 (42)	28 (44)	31 (49)	41 (52)	23 (41)	19 (52)	26 (57)	24 (53)
19	UBE2E3	38 (57)	39 (63)	61 (80)	65 (79)	64 (80)	63 (81)	32 (55)	32 (60)	36 (63)	35 (60)	31 (56)	44 (64)
		40 (61)	23 (54)	36 (58)	34 (55)	84 (88)	87 (93)	100	29 (63)	29 (67)	20 (62)	19 (60)	24 (53)
		28 (50)	28 (42)	25 (46)	40 (52)	29 (42)	28 (42)	32 (48)	41 (52)	25 (40)	18 (50)	25 (56)	23 (58)
20	UBE2F	25 (57)	25 (55)	35 (63)	34 (63)	34 (63)	36 (64)	27 (58)	21 (56)	26 (55)	29 (57)	27 (52)	27 (61)
		24 (56)	18 (54)	20 (49)	25 (60)	27 (59)	28 (62)	29 (63)	100	38 (69)	15 (52)	15 (51)	17 (52)
		27 (46)	21 (38)	15 (43)	32 (55)	27 (51)	23 (53)	28 (49)	26 (44)	23 (45)	17 (56)	24 (54)	20 (47)
21	UBE2M	27 (61)	27 (60)	30 (66)	31 (64)	31 (64)	32 (66)	26 (55)	25 (55)	31 (58)	32 (58)	29 (59)	26 (62)
		23 (51)	16 (53)	20 (53)	26 (64)	30 (65)	30 (64)	29 (67)	38 (69)	100	18 (59)	18 (58)	16 (60)
		22 (46)	15 (43)	16 (48)	24 (51)	23 (56)	25 (54)	24 (52)	27 (50)	24 (52)	20 (48)	25 (51)	24 (55)

Class I Class II Class III Class IV		Percent identity matrix of E2 enzymes in human binary reactions											
		Ube2A Ube2NL Ube2H	Ube2B Ube2V2 Ube2J1	Ube2d1 Ube2W Ube2J2	Ube2d2 Ube2C Ube2K	Ube2d3 Ube2E1 Ube2R1	Ube2d4 Ube2e2 Ube2R2	Ube2g1 Ube2e3 Ube2S	Ube2g2 Ube2F Ube2T	Ube2i Ube2M Ube2U	Ube2L3 Ube2q1 Ube2O	Ube2L6 Ube2q2 Ube2z	Ube2N Ube2v1 Birc6
22	UBE2Q1	24 (57)	24 (56)	24 (54)	24 (55)	24 (53)	25 (54)	15 (45)	18 (53)	23 (47)	18 (50)	18 (50)	19 (50)
		16 (51)	18 (58)	23 (50)	20 (56)	23 (63)	21 (61)	20 (62)	15 (52)	18 (59)	100	74 (75)	19 (59)
		13 (46)	18 (28)	14 (33)	18 (46)	16 (37)	16 (34)	15 (37)	20 (39)	17 (29)	18 (50)	17 (40)	18 (45)
23	UBE2Q2	23 (56)	22 (57)	23 (53)	24 (53)	24 (52)	24 (53)	16 (46)	18 (53)	23 (49)	19 (50)	19 (46)	19 (53)
		17 (52)	18 (57)	23 (52	19 (57)	21 (59)	20 (60)	19 (60)	15 (51)	18 (58)	74 (75)	100	20 (56)
		14 (44)	18 (29)	17 (33)	19 (46)	16 (36)	16 (33)	16 (37)	20 (40)	16 (27)	19 (46)	17 (31)	18 (48)
24	UBE2V1	24 (46)	23 (46)	25 (45)	25 (41)	25 (41)	26 (43)	24 (49)	18 (47)	23 (48)	25 (45)	24 (49)	22 (46)
		18 (45)	88 (94)	24 (44)	25 (57)	24 (57)	24 (54)	24 (53)	17 (52)	16 (60)	19 (59)	20 (56)	100
		18 (44)	23 (48)	18 (43)	27 (50)	21 (55)	22 (57)	21 (47)	26 (47)	19 (45)	16 (54)	21 (59)	19 (50)
25	UBE2H	32 (62)	32 (64)	34 (62)	31 (63)	30 (63)	32 (62)	27 (52)	27 (54)	26 (56)	28 (56)	24 (55)	29 (53)
		25 (50)	21 (44)	19 (49)	29 (56)	28 (50)	28 (50)	28 (50)	27 (46)	22 (46)	13 (46)	14 (44)	18 (44)
		100	20 (61)	19 (59)	29 (67)	30 (65)	29 (67)	26 (65)	27 (61)	20 (59)	22 (54)	18 (50)	23 (48)
26	UBE2J1	24 (52)	25 (53)	31 (61)	29 (57)	30 (59)	29 (59)	23 (54)	27 (55)	22 (51)	25 (52)	25 (52)	25 (58)
		23 (53)	23 (52)	23 (50)	22 (50)	30 (46)	29 (43)	28 (42)	21 (38)	15 (43)	18 (28)	18 (29)	23 (48)
		20 (61)	100	31 (67)	21 (54)	23 (56)	25 (58)	21 (64)	20 (54)	16 (49)	20 (47)	21 (44)	19 (43)
27	UBE2J2	29 (59)	29 (58)	24 (54)	24 (56)	24 (54)	24 (56)	24 (54)	27 (52)	22 (56)	21 (54)	23 (56)	27 (59)
		25 (55)	18 (51)	22 (54)	20 (51)	25 (51)	25 (48)	25 (46)	15 (43)	16 (48)	14 (33)	17 (33)	18 (43)
		19 (59)	31 (67)	100	19 (57)	23 (50)	24 (53)	19 (53)	20 (63)	18 (50)	16 (57)	23 (48)	18 (44)
28	UBE2K	29 (57)	29 (57)	41 (69)	42 (69)	41 (69)	43 (67)	27 (57)	26 (57)	27 (58)	29 (51)	27 (58)	44 (71)
		43 (68)	26 (54)	27 (58)	31 (57)	40 (54)	41 (52)	40 (52)	32 (55)	24 (51)	18 (46)	19 (46)	27 (50)
		29 (67)	21 (54)	19 (57)	100	27 (62)	27 (64)	29 (61)	36 (62)	24 (62)	21 (52)	21 (55)	22 (53)

Percent identity matrix of E2 enzymes in human binary reactions

Class I Class II Class III Class IV		Ube2A Ube2NL Ube2H	Ube2B Ube2V2 Ube2J1	Ube2d1 Ube2W Ube2J2	Ube2d2 Ube2C Ube2K	Ube2d3 Ube2E1 Ube2R1	Ube2d4 Ube2e2 Ube2R2	Ube2g1 Ube2e3 Ube2S	Ube2g2 Ube2F Ube2T	Ube2i Ube2M Ube2U	Ube2L3 Ube2q1 Ube2O	Ube2L6 Ube2q2 Ube2z	Ube2N Ube2v1 Birc6
29	UBE2R1	39 (65)	39 (64)	38 (62)	33 (60)	33 (60)	36 (60)	50 (69)	48 (71)	36 (60)	29 (62)	27 (60)	32 (65)
		30 (59)	22 (54)	31 (53)	33 (57)	29 (44)	29 (42)	29 (42)	27 (51)	23 (56)	16 (37)	16 (36)	21 (55)
		30 (65)	23 (56)	23 (58)	27 (62)	100	82 (94)	31 (59)	30 (61)	25 (56)	23 (52)	25 (53)	27 (52)
30	UBE2R2	39 (67)	39 (67)	35 (63)	33 (63)	33 (62)	35 (63)	54 (73)	48 (73)	35 (58)	29 (63)	26 (61)	30 (69)
		28 (63)	21 (54)	33 (55)	33 (57)	28 (47)	28 (44)	28 (42)	23 (53)	25 (54)	16 (34)	16 (33)	22 (57)
		29 (67)	25 (58)	24 (53)	27 (64)	82 (94)	100	27 (63)	29 (60)	24 (58)	21 (53)	23 (49)	24 (55)
31	UBE2S	31 (61)	32 (61)	35 (66)	37 (67)	35 (67)	37 (65)	26 (62)	30 (62)	30 (61)	24 (51)	25 (55)	36 (65)
		34 (63)	19 (48)	23 (58)	28 (54)	31 (52)	31 (49)	32 (48)	28 (49)	24 (52)	15 (37)	16 (37)	21 (47)
		26 (65)	21 (64)	19 (53)	29 (61)	31 (59)	29 (63)	100	30 (62)	22 (56)	25 (61)	23 (55)	22 (53)
32	UBE2T	32 (60)	33 (60)	43 (68)	43 (67)	42 (67)	41 (66)	29 (58)	27 (55)	30 (61)	34 (58)	33 (59)	44 (71)
		41 (69)	26 (46)	28 (53)	34 (55)	41 (53)	41 (52)	41 (52)	26 (44)	27 (50)	20 (39)	20 (40)	26 (47)
		27 (61)	20 (54)	20 (63)	36 (64)	30 (61)	29 (60)	30 (62)	100	21 (59)	25 (54)	28 (56)	27 (53)
33	UBE2U	33 (58)	32 (57)	31 (61)	35 (64)	35 (63)	32 (62)	29 (59)	30 (56)	28 (52)	22 (56)	25 (57)	29 (59)
		25 (52)	18 (53)	24 (54)	30 (54)	25 (44)	25 (41)	25 (40)	23 (45)	24 (52)	17 (29)	16 (27)	19 (45)
		20 (59)	16 (49)	18 (50)	24 (62)	25 (56)	24 (58)	22 (56)	21 (59)	100	16 (45)	22 (41)	18 (44)
34	UBE2O	26 (57)	27 (57)	22 (54)	24 (55)	24 (56)	23 (55)	22 (55)	24 (57)	21 (56)	19 (53)	20 (48)	20 (58)
		17 (52)	15 (51)	17 (51)	20 (59)	18 (53)	19 (52)	18 (50)	17 (56)	20 (48)	18 (50)	19 (46)	16 (54)
		22 (54)	20 (47)	16 (57)	21 (52)	23 (52)	21 (53)	25 (61)	25 (54)	16 (45)	100	25 (27)	21 (47)
35	UBE2Z	30 (58)	30 (61)	33 (63)	36 (62)	36 (62)	35 (63)	25 (59)	29 (59)	30 (55)	21 (53)	23 (59)	30 (64)
		28 (62)	20 (57)	26 (57)	21 (55)	26 (56)	26 (57)	25 (56)	24 (54)	25 (51)	17 (40)	17 (31)	21 (59)
		18 (50)	21 (44)	23 (48)	21 (55)	25 (53)	23 (49)	23 (56)	28 (56)	22 (41)	25 (27)	100	26 (63)

Class I Class II Class III Class IV		Percent identity matrix of E2 enzymes in human binary reactions											
		Ube2A Ube2NL Ube2H	Ube2B Ube2V2 Ube2J1	Ube2d1 Ube2W Ube2J2	Ube2d2 Ube2C Ube2K	Ube2d3 Ube2E1 Ube2R1	Ube2d4 Ube2e2 Ube2R2	Ube2g1 Ube2e3 Ube2S	Ube2g2 Ube2F Ube2T	Ube2i Ube2M Ube2U	Ube2L3 Ube2q1 Ube2O	Ube2L6 Ube2q2 Ube2z	Ube2N Ube2v1 Birc6
36	BIRC6	30 (62)	29 (63)	31 (59)	31 (59)	31 (58)	32 (59)	25 (57)	28 (58)	30 (58)	26 (53)	24 (58)	25 (62)
		22 (54)	20 (48)	22 (53)	24 (56)	23 (56)	24 (53)	23 (58)	20 (47)	24 (55)	18 (45)	18 (48)	19 (50)
		23 (48)	19 (43)	18 (44)	22 (53)	27 (52)	24 (55)	22 (50)	27 (53)	18 (44)	21 (47)	26 (63)	100

[a]Pairwise percent identity between E2 enzymes between the classes. Comparison made using ClustalW.
[b]Pairwise percent homology, in parenthesis, between E2 enzymes between the classes. Comparison made using ClustalW.
[c]The names and %identity and %homology are color coded to corresponds to class 1 (blue), class2 (red), class3 (orange), and class4 (green).

Table 1.
Matrix of pairwise sequence comparisons of 36 human E2 conjugating enzymes representing four classes.

For Class 1, there are 20 structures between Ube2D1–3 and RING E3 domains, class 2 involves Ube2e1–2 with 2 structures bound to RING E3 domains, class 3 involves Ube2g2 with 2 structures with RING E3 domains, class 4 consists of 2 structures with Ube2l3 bound with RING E3 domains, and class 5 consists of seven structures of Ube2n bound with RING E3 domains. This review discusses the structures of RING domains and the E2 catalytic domains, their interactions, and evaluating patterns of key amino acids and the role they may play in how E2 and RING E3 domains interact and facilitate substrate ubiquitination.

2. Ub E2 conjugating enzymes are categorized into four classes

To date, there are between 40 and 75 different Ub E2 enzymes identified in plants and animals [33, 34]. In mammals and humans, there are currently 36 known Ub E2 enzymes [35]. It is expected that this number should increase as more research in this central area continues. All E2 enzymes consist of a catalytic domain of approximately 150-amino acids. Given this commonality, the human E2 enzymes are categorized into four classes based on the number of amino acids that precede or follow the catalytic domain (**Figure 2**). Class 1 E2 enzymes are essentially just the catalytic domain and are approximately 145–160 amino acids. Class 1 has the most members, consisting of 15 of the 36 E2 enzymes. The most studied, and possibly the most promiscuous when it comes to binding different RING E3 domains, are the three isoforms, Ube2D1–3. Class 2 E2 enzymes have a considerably larger number of residues (20–45 amino acids) N-terminus to the catalytic domain and include the enzymes Ube2e1, Ube2e2, Ube2e3, and Ube2f. Class 3 E2 enzymes have between 10 and 50 amino acids at the C-terminus and includes Ube2r1, Ube2r2, Ube2s, Ube2t, and Ube2u, which represent 5 of the 9 E2 members [35]. Both classes 2 and 3 consist of nine E2 enzymes. Class 4 E2s have both N- and C-terminal extensions and consists of three members, Ube2Z, Ube2O and Birc6. While the exact role the N- or C-terminal residues have on impacting specificity with specific E3 ligases is unclear, they bind is unclear but they have been shown to contribute to substrate binding and the linkages of Ub chain elongation [36–38].

3. Primary sequence alignments show high identity and homology

To understand how the E2 enzymes of the four classes are related and whether the classifications using N- and C-terminal extensions are appropriate criteria, the sequences of the catalytic domains are compared (**Table 1**). The matrix shows the percent identity and homology (in parenthesis) between pairs of E2 enzymes within and between classes. Within class 1, the identities of the primary sequences range between 25% and 45% and the similarity/homology between 70% and 78%. The Ube2A enzyme has sequence identities between 40% and 42% with Ube2D1–4, Ube2g2, and Ube2i. The Ube2A and Ube2B enzymes may be isoforms because they are 95% identical and 99% homologous. The Ube2D [1–4] family has the highest percent identities and homology among each other with values of 88–92% and 97%, respectively. Interestingly, Ube2v2 shows the lowest percent identity among its class with an average identity of 21 $\pm$ 4%. The E2 with the next lowest identity is Ube2L3 with an average value of 26%; however Ube2D1–4 shows ~38% identity with Ube2L3. The Ube2L3 E2 enzyme only reacts with cysteine, indicating it would interact with the other classes of E3 enzymes; only two HECT-type E3 ligases that interact with Ube2L3 have been identified. In general, the remaining E2 show an average identity of 42 ± 8%. When comparing the class 1 and 2 enzymes, Ube2V2 not only has the lowest percent identity within class 1, but also with almost all of the E2 enzymes with an average value of 20 ± 3%. In contrast, it has 88% identity with Ube2V1. Many of the class 1 E2s show considerably higher identity and homology with Ube2C and Ube2E1–3 of ~32% and 70%, respectively. Ube2D1-4 show higher identities with these four enzymes, which is ~63%. The identities between class 1 E2s with Ube2F, Ube2M, Ube2Q1–2, and Ube2V1 are considerably lower. The sequences between classes 1 and 3 have identities between 30 and 45% with homologies in the 55–70% range. Again, Ube2D1–4 sequences show the highest sequence identity with class 3 sequences. Class 1 shows high sequence identities with the Class 4 E2 enzymes. Specifically, Ube2A, Ube2B, Ube2D1, Ube2D2, Ube2D3, Ube2D4, and Ube2N show sequence identities and homolgies of approximately 30% and 60%, respectively, with the class 4 Ube2Z. Ube2O shows the lowest sequence identity with the class 1 sequences.

The class 2 enzymes appear to fall in two groups based on their sequence identities. Six of the nine show identities between 25 and 35, while conversely showing identities ranging between 15 and 25% with Ube2Q1, Ube2Q2, Ube2V1. Ube2Q1 and Ube2Q2 are 75% identical but show 20% identities with Ube2V1. Similarly to Ube2D1–4, the Ube2E1, Ube2E2, and Ube2E3 enzymes have the highest identities at >85% among each other. Class 3 E2 enzymes show identities in the 16–30% range, with an average of 26 ± 11% among each other. The similarities in sequence are in the 60–70% range. Incidentally, this range and average values are lower than those values when comparing with class 1 sequences. The enzymes Ube2R1 and Ube2R2 are 82% identical. Given that the other E2 enzymes in this class are not as related sequentially with each other, Ube2K and Ube2T stand out with showing the next highest identity of 36%. These enzymes show the lowest sequence identities with class 4 enzymes, with an average percentage of 22 ± 3%. Lastly, the class 4 enzymes show the average percent identities of 24 ± 2%. Based on sequences, the Ube2Z enzyme has 25% and 26% identities with BIRC6 and Ube2O, respectively, while BIRC6 and Ube2O have 21% with each other. As noted above, the three class 4 enzymes have the highest sequence identities with class 1 enzymes than with the other two classes.

Identities compare identical amino acids in the same location between proteins. Sequence homology, which includes identical and amino acids with similar properties, provides another perspective on how these enzymes preserve their structure and function. For class 1 enzymes, the Ube2d1, Ube2d2, Ube2d3, and Ube2d4 are the most related and commonly used in ubiquitination assays, sharing 97–100% homology with each other. In fact, these four enzymes will most likely interact with the same RING E3 ligase if one is shown to be a cohort. Similarly, Ube2A and Ube2B share 99% homology and Ube2NL share 97% homology with Ube2n. Ube2V2 has the lowest percent homologies when compared with other E2's within class 1, and those values range between 48-56%. In summary, class 1 enzymes have an average pairwise homology of 56% with each other. In class 2, Ube2E1, Ube2E2, Ube2E3 share 88-93% homologies with another, and the average pairwise homology among members is 54%. For Class 3, the E2's share greater than 50% homology with one another (average of 54%), with the exception of Ube2U and Ube2J1, which share 49% homology. Ube2R1 and Ube2R2 share 94% homology. For Class 4, Ube2O and Ube2Z share only 27% homology with each other.

Comparing the homologies between classes, class 1 Ube2V2 shares 94% homology with class 2 Ube2V1, and class 1 Ube2D1-4 share 78-81% homologies with class 2

Ube2E1, Ube2E2, and Ube2E3. 93% of class 1 E2 enzymes shared the lowest percent homologies with Ube2V2 (class 2), ranging between 41 to 49%. Most of class 1 E2's shows the least percent homology with class 3 Ube2J1 and Ube2J2, ranging from 52 to 56%. Class 1 Ube2V2 and Ube2W share 44% homology with Ube2H, the lowest homology between these two classes. Class 2 Ube2Q1 and Ube2Q2 share 27–29% homologies with class 3 Ube2U. 93% of class 1 E2 enzymes have the lowest percent homology with class 4 Ube2O, compared with Ube2Z and Birc6, however, the values are greater than 48% homology.

Based on these sequence alignments, it appears that proteins with sequence identities as low as 16% and sequence homologies at 50% with similar function will adopt a very similar tertiary structure. It would be important to analyze the role of the amino acids that are not conserved among those proteins to determine their role in catalytic rates, substrate binding, mechanism of Ub binding, and specificity for RING domains. Furthermore, given that class 1 enzymes show the highest sequence identity among each other than those of the other classes, it would be interesting to determine how residues on the N- and C-termini may compensate for the lower sequence identities within the catalytic region.

4. The tertiary structure of the catalytic subunit of E2's is conserved

The structure of the catalytic subunit of E2 enzymes consists of four-antiparallel β strands that forms one surface (**Figure 3**). On one end of the surface is the C-terminal helix-turn-helix region and on the other side is the N-terminal helix. Across the inner surface of the β-sheet sits the fourth helix, along with a helical turn located adjacent to

A. B.

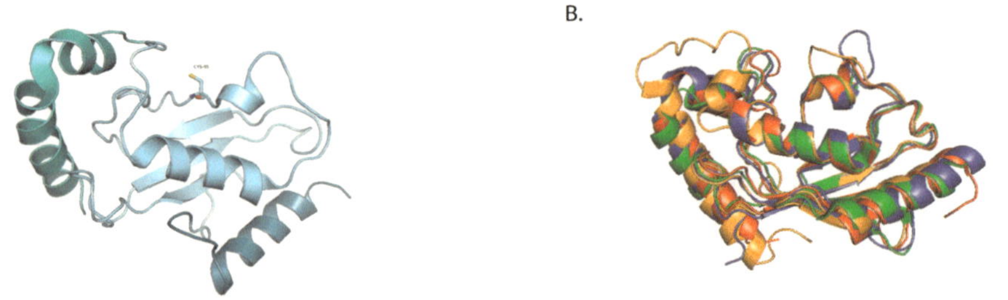

C.

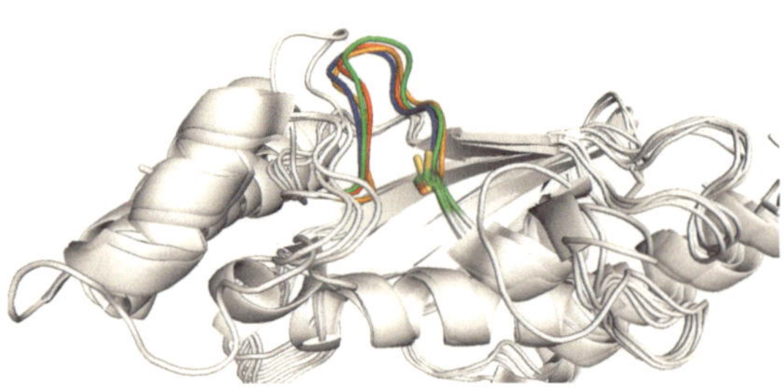

D.

Structural Similarity Root-Mean Squared Deviation (RMSD)				
	Ube2A	Ube2C	Ube2T	Ube2Z
Ube2A	—	0.85	1.17	1.26
Ube2C	0.85	—	1.14	1.20
Ube2T	1.17	1.34	—	1.3
Ube2Z	1.26	1.20	1.34	—

Figure 3.
Ribbon representation of E2 structures. A. Tertiary structure of Ube2D2 class1 E2 enzyme showing the structure of the catalytic domain. The catalytic cysteine that forms the thioester bond with Ub is shown. B. Overlay of the catalytic subunits of an E2 enzyme from each of the four classes. Class1 (Ube2A, 6cyo) is colored red, class2 (Ube2C, 1ik7) is in green, class3 (Ube2T,1yh2) is in blue and class4 (Ube2Z, 5a4p) is in yellow-orange. C. Same superposition as B but with the loop containing the catalytic cysteine highlighted. Ube2C is solved with a serine instead of the cysteine and it adopts a very similar orientation. D. Table of the pairwise RMSD values for the superposition of the backbone atoms between structures.

the catalytic cysteine. This helical turn is important for interaction with the RING E3 domain and activation for nucleophilic attack of the thioester linkage with Ub [37]. There are several loops connecting the β-strands that the quaternary structures of E2 enzymes in complex with RING domains show which are important for contacts with the RING E3 domains.

Not surprisingly, structures of the catalytic domains of E2 enzymes belonging to the four classes show remarkable similarity (**Figure 3B**). For the most part, all four proteins have the same number of amino acids that contribute to the four β-strands. The number of amino acids associated with the helices are very similar. The N-terminal and central helices as well as the 1-turn helix appears to have the same number of amino acids in their composition. Helix 1 is formed by ~15 amino acids of the first residues that form the start of the secondary and tertiary structure. The last 25 amino acids of the catalytic core adopt a 4-turn α-helix, followed by what appears to be a reverse turn and a 2-turn terminal α-helix.

Interestingly, the significant difference between an overlay of four structures belonging to each class is observed within the C-terminal helix-turn-helix region. The lengths of these helices differ, with the helices of Ube2Z (class4) being the longest with two additional 2-helical turns. There are also greater variations in the position from of the C-terminal helices, which is surprising given the fact that its location is on the opposite surface to where the RING domains typically bind E2 enzymes. The residues that precedes the C-terminal helix for Ube2Z did not superimpose well with the other structures. The relative positions of the loops are very similar among the four structures, however, Ube2Z has two additional loops that are not present in the other classes [39]. One of these loops is absent in all Class1–3 E2 enzymes and only present in class 4 E2 BIRC6.

To measure the overall similarity and consistency in the structures, the backbone Cα,C,N atoms are superimposed. The root mean square deviation (RMSD) values of the superposition of the backbone atoms between the four structures range from 0.8 to 1.3 Å, confirming the similarity in structures. The lower the RMSD between pairs of structures, the more closely aligned the structures. RMSD values ranging from 0.8–1.3 Å indicate that while there are small differences in the structures, overall they are very similar. The RMSD values between Ube2A (class1) and Ube2C (class2) is 0.8 Å, and with Ube2T (class3) and Ube2Z (class4) are 1.1 and 1.2 Å, respectively. The RMSD values of the superposition of Ube2Z with Ube2A and Ube2T are 1.26 Å and 1.34A. The similarity in structures indicates a common mechanism of function.

The catalytic cysteine that forms the thioester bond with the C-terminal carboxylate group of Gly76 is located on a 11-amino acid structured loop that precedes the single helical turn. The sequences of this loop show a 74% sequence homology with a consensus sequence of H-P-N-h-D/Y-x-x-G-p-I/V-C-L, where h, p and x indicate a hydrophobic, a polar residue, and any residue, respectively. The proline residue is important because it introduces a bend in the loop that positions the backbone carbonyl of the adjacent Asn (N) residue to form a hydrogen bond with the backbone NH of the Cys residue, which is located between two aliphatic hydrophobic residues. The side-chain amine (NH_2) hydrogens form two hydrogen bonds with the backbone carbonyl groups of a conserved N-x-x-S motif on the structured loop that precedes the C-terminal helix-turn-helix region. Residues D/Y-x-x-G are involved in forming a tight type-I turn in the loop. The positions of the loops between the four structures superimposed well (<0.6 Å). The conserved sequence and structure indicate their importance in chemistry associated with the covalent binding of an Ub molecule. The structure shown for Ube2C E2 enzyme is that of a mutant that consists of a serine instead of a cysteine (**Figure 3C**). The structure of the serine side-chain is positioned very similar to those of the cysteine residues of the other structures. Interestingly, the C/S mutant Ube2C was able to form ester linkages with the Ub in vitro [40]. This observation suggests that the native Ube2NL (Ube2N-linked) should function as an E2 ligase given that it has a serine in place of the catalytic cysteine.

Furthermore, sequence alignment of all 36 E2 enzymes revealed several motifs, conserved residues, and conserved properties in certain parts of their sequences (**Figure 4A**). Starting from the N-terminus, there are conserved I-h-P-G and P-a-Q/E-GG motifs, where 'h' represents a hydrophobic residue and 'a' indicates an aromatic residue, at around position 37. Near the 60th position, there is a PF motif (Y-P-F) followed by a conserved proline three amino acids downstream. There are also several additional conserved regions that include K/R-I-Y/a, H-P-N, I-C-L-D-I-L, a conserved tryptophan around position 93, followed by S-P-A-L and S-L-L motifs, and a hydrophobic residue at positions near 105 and 109.

In **Figure 4B**, the conserved motifs are mapped to their locations on the 3-dimensional structure of an E2 enzyme with respect to the active cysteine site. The ICLDIL motif consists of the active site cysteine and encompass the single helical turn, and the HPN and SPAL motifs are spatially located to the left and top right, respectively, to this cysteine. The LLS motif is located at the center of the α-helix that lays across the β-sheet. The I-h-P-G and P-a-Q/E-GG motifs, in which h and a represent hydrophobic and aromatic residues, respectively, are located on the loops preceding the C-terminal helices on the opposite surface of the cysteine. The conserved

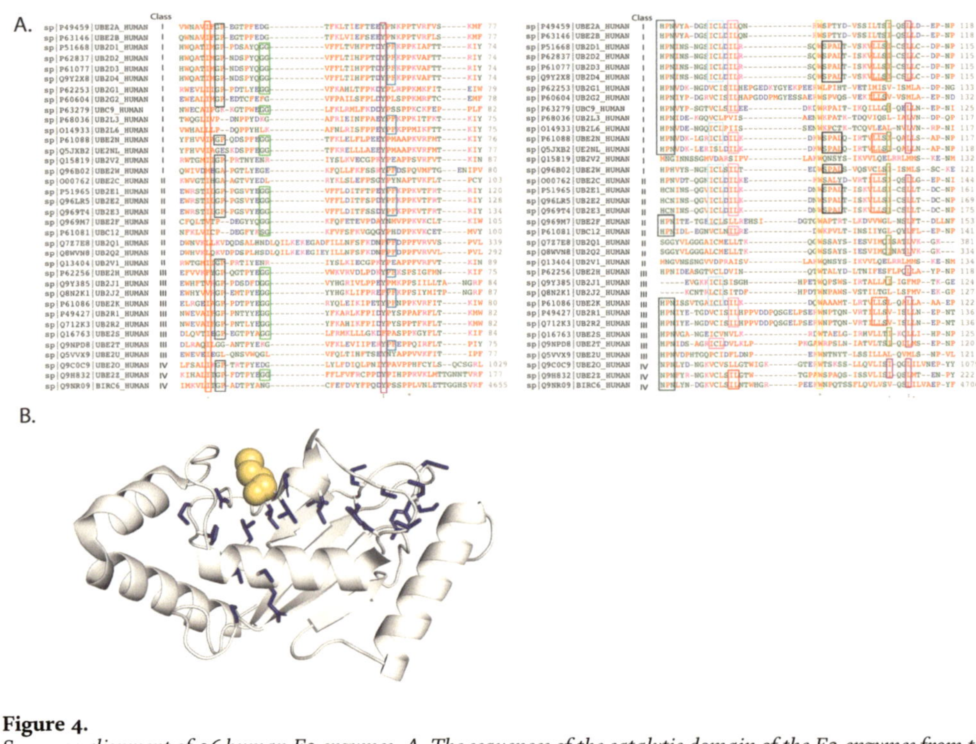

Figure 4.
Sequence alignment of 36 human E2 enzymes. A. The sequences of the catalytic domain of the E2 enzymes from the 4 classes are aligned. motifs that are conserved among the enzymes are highlighted; these residues mostly surround the catalytic cysteine and maybe important for allosteric communication between the E2 and RING domains. The alignment is performed using ClustalW [41, 42]. B. Location of the sidechains of several conserved sequence motifs on the E2 catalytic domain, noted in the text and identified from the sequence alignment. The catalytic cysteine residue is shown as yellow spheres.

tryptophan and proline residues at positions near 93 and 65, respectively, are spatially located in the region above the N-terminal α-helix, adjacent to that SPAL motif. The side-chains of two conserved hydrophobic residues located on the α-helix point towards the cysteine while another on a β-strand appears to make hydrophobic contacts with those on the α-helix. Based on the alignment of these residues along the pathway between with the RING domain binds the E2 enzyme and the catalytic side, it appears that some of these conserved residues are involved in RING domain binding (on the right), while most may be important for allosteric effects to activate the reactivity of the cysteine and its thioester bond with the Ub (**Figure 4B**).

5. Proteins with RING domains exhibit E3 ligase activities

E3 ligases include the HECT (homologous to the E6AP carboxyl terminus)-type, SCF (Skp1–Cullin–F-box-protein), and RING (really interesting new gene) protein families [43, 44]. The RING-type represents the largest class with several hundred members in animals, including humans [45], and which includes the RBR (RING-inbetween-RING), U-box and B-box families [46–52]. Unlike the prototypical RING zinc-finger domain, U-box domains adopt the same ββα-RING fold but are not cysteine and histidine rich and do not bind any zinc ions. It is believed that RING E3 ligases specify themselves and specific proteins for ubiquitination. Some RING E3 ligases can target several substrates [16, 53, 54].

Proteins with a RING domain are typically hypothesized to possess E3 ligase activity. Commonly, RING proteins are demonstrated to have E3 ligase activity by performing in vitro autoubiquitination assays. In these assays, the reaction mixture consists of the E1 and E2, enzymes, Ub, the RING E3 protein, and ATP. Often, the isolated RING domain is used to confirm that it is the RING domain that confer E3 ligase activity to the RING containing protein [3, 28]. Western blot analysis of the reaction mixture is probed with either an antibody specific for the E3 protein or domain, but more often specific for a modified Ub, such as with biotin, His_6- or HA tag [3, 7, 28]. RING E3 ligase activity is confirmed by the presence of mono-ubiquitinated Ub (mono-Ub), di-Ub, and/or poly-Ub; the polyubiquitinated products with various amounts of Ub appear as a smear of high-molecular weight bands [28].

While most RING E3 ligases possess a single RING domain, the TRIM (tripartite motif) family possesses two or three domains with RING folds, with the RING domain found at the N-terminus [46, 55–60]. RING domains are 50 to 90-amino acid regions that typically consist of at least eight cysteine and histidine residues uniquely spaced along the primary sequence (**Figure 5**). RING domains are classified as zinc-fingers. As with most zinc-finger domains, the zinc ion (Zn^{2+}) is tetrahedrally coordinated. RING domains bind two zinc ions in a unique cross-braced arrangement (**Figure 5A**) [40, 57]. The cross-braced mechanism involves the first and third pair of zinc-ligands binding one zinc ion, and the second and fourth pair of zinc-binding residues coordinating the other zinc ion. The consensus sequences for RING domains ($C\text{-}X_{(2)}\text{-}C\text{-}X_{(9–39)}\text{-}C\text{-}X_{(1–3)}\text{-}H\text{-}X_{(2–3)}\text{-}C/H\text{-}X_{(2)}C\text{-}X_{(4–48)}C\text{-}X_{(2)}C$) [61] indicate that one zinc ion is normally bound by the sulfur atom of three cysteines and the imidazole nitrogen of a histidine residue while the other is usually bound by the sulfur of four cysteine residues. Other common consensus are C3H2C3, C4HC3 and C8. The number of amino acids between the Cys/His residues varies considerably in two regions, designated loops 1 and 2; the lengths are between 3 and 39 and 4–48 amino acids, respectively.

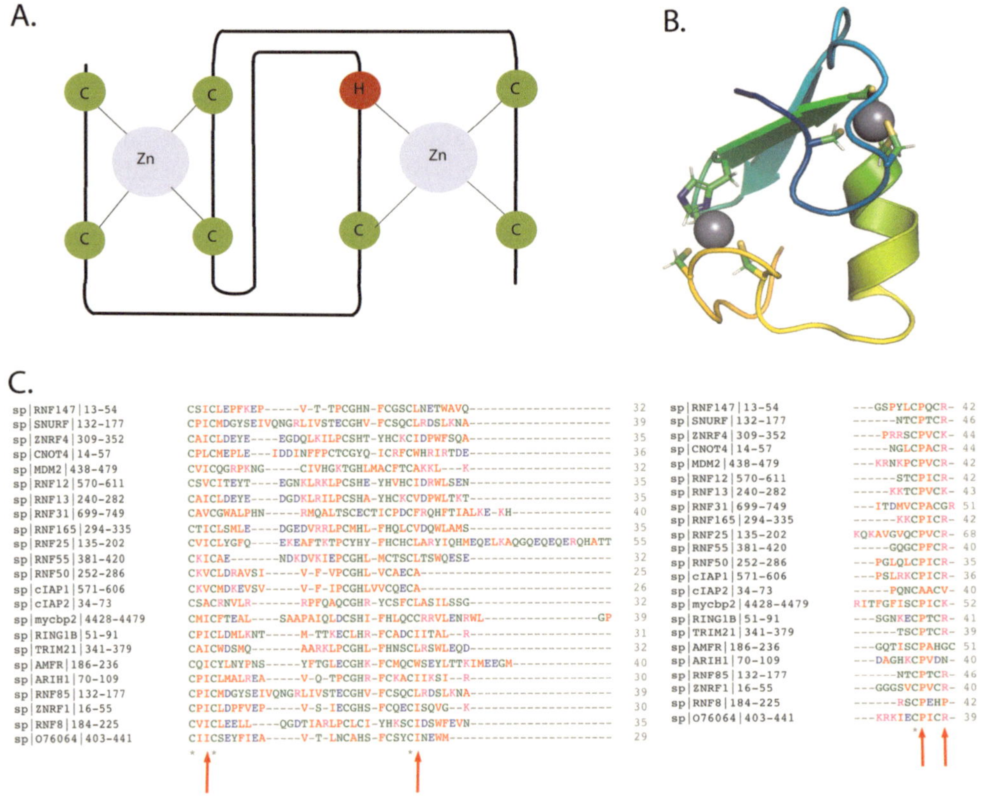

Figure 5.
*Sequences and structures of RING domains. A. Cross-braced arrangement of the zinc-binding mechanism in which the first and third pair of zinc-ligands bind one zinc ion, and the second and fourth pair binds the other zinc. B. Prototypical structure of the RING ββα-fold adopted by RING domains exhibiting E3 ligase activity. The spheres present the two zinc ions bound by key cysteine and histidine residues. It should be noted that some RING domains do not have well-defined β-strands in their loop1. C. Sequence alignment of several RING domains for which the structures of E2–E3 complexes are known (**Table** 2).*

Despite having eight covalent bonds to two zinc ions to stabilize the tertiary structure, RING domains can be susceptible to unfolding or aggregation when produced at the concentration to perform structural studies. Mutation of any of its eight zinc-binding ligands to a non-zinc binding amino acid, for example cysteine to serine, results in loss of binding to both zinc ions and unfolding of the domain [62].

As noted, RING domains can vary in the lengths of their sequences. In addition, with the exception of the conserved cysteine and histidine residues, RING domains do not share high sequence homology, which is normally in the 20% range. Despite these differences, RING domains adopt an overall beta-beta-alpha (ββα) canonical RING fold (**Figure 5B**). RING fold is characterized by two large loops (L1, L2) in which L1 precedes the 2- to 3-turn α-helix, and L2 follows the helix. Loop 1 is typically more structured and consists of two short β-strands separated by a type-2 turn. Loop 1 consists of the first and second pairs of zinc-binding residues, while the first helical turn of the α-helix has another pair. Loop 2 consists of the last pair of zinc-binding residues and tends to exhibit more dynamic properties than L1. The relative locations of the two zinc ions are similar for all RING domains.

Despite the low sequence homologies among RING domains, there are several amino acids that are fairly well conserved at specific locations in the sequence. There is a conserved hydrophobic residue (I/V/L) located between the first pair of cysteine residues and an acidic residue located two residues after the cysteine (**Figure 5C**). In addition, there are an additional 2–3 acidic (D/E) residues located between the first and second pair on L1. Just preceding and following the third pair of ligands are hydrophobic residues, in which the preceding residue is usually aromatic. For the last pair of zinc-binding residues on L2, there is a conserved proline and a hydrophobic residue located between them, while an arginine immediately follows this pair. The similarity in structures of RING E3 domains suggests a common mechanism of action.

While the RING domain confer E3 ligase activity, this function is generally in concert with it binding the substrate in order for the substrate to be ubiquitinated [63–65]. For instance, loss of interaction between the RING domain and its substrate results in loss of substrate ubiquitination [16, 65]. However, since the RING domain is only part of a larger protein, often, other domains or regions of the protein play key roles in substrate recognition and binding [16, 66, 67]. For example, mutations of the B-box domains in the MID1 protein that disrupted binding to a substrate protein prevents ubiquitination of the substrate despite MID1 possessing wild-type level ligase activity [16, 32]. Unfortunately, while the number of RING E3 ligases are prevalent, their substrates are not always known or fully characterized; so how RING E3 ligases recognize their substrates is still a work in progress.

6. Mapping E2-RING domain interactions

Unlike the other classes of E3 ligases (HECT, RBR, SCF) that first transfer the ubiquitin from the E2 enzyme to themselves, RING E3 ligases facilitate the concerted Ub transfer to the substrate protein by interacting with both the E2 and substrate, placing the substrate in close proximity to the E2 $\sim$ Ub thioester bond. Thus, the interactions between RING E3 ligase, its cognate E2s and its substrate protein are essential for Ub transfer [63–65]. The inability of the RING domain to interact with either the E2 or substrate will result in loss of ubiquitination of either the RING protein (autoubiquitination) or the substrate (ubiquitination) [16, 32]. The ability of Ub to transfer from the E2 enzyme to the substrate is based on the reactivity towards the amino group of a lysine (aminolysis) or cysteine (transthiolation). Aside from lysine, its surrounding residues also play a crucial role in the determination of whether the protein will be ubiquitinated. The level of ubiquitination is determined by how the E2 enzyme accommodates the preferred lysine [8, 68].

7. RING domains are positioned far from the active site

Given that the interactions of RING E3 ligases and E2 enzymes impact auto- and substrate-ubiquitination, the structures of several RING domains and their cognate E2 enzymes are examined [69–74].

Evaluation of the literature and PDB reveal that four E2 members are highly studied: Ube2D1-3, Ube2N, Ube2G2, and Ube2L3 (class 1) (**Table 2**). The Ube2D1–3 family appears to be the most promiscuous; these three enzymes are usually the first to be employed to test whether a protein with a RING domain possess E3 ligase activities. The complexes of Ube2D1 with 12 RING E3 domains (MDM2, RNF12,13,25,31,165, CNOT4, c-CBL, BIRC2,7) have been characterized [75–83]. Sequence analysis of these RING domains reveal low pairwise sequence identities. The

E2's	Interacting E3's	E2-E3 structures	E3 structures
Ube2d1	RNF147	5FER	6FLM
2CP4	SNURF	4AP4	2EA6
	ZNRF4	3OJ4	
	CNOT4	5AIE	2CPI
Ube2d2	CNOT4	1UR6	2CPI
3L1Y	MDM2	6SQS	5AFG
	RNF12	6W7Z	
	RNF13	5ZBU	5ZC4
	RNF31	5EDV	4LJQ
	RNF165	5ULK	5D0I
	RNF165	5D0K	5D0I
	RNF25	5D1K	
	RNF55	4A49	2JUJ
	RNF50	4AUQ	3GT9
	cLAP1	6HPR	3M1D
	cLAP2	3EB6	2UVL
Ube2d3	mycbp2	6T7F	5O6C
1X23	UBE4B	3L1Z	3KRE
	RING1B	4S3O	2CKL
	RING1B	3RPG	2CKL
Ube2e1	TRIM21	6FGA	5OLM
3BZH			
Ube2e2	RNF12	6W9A	
1Y6L			
Ube2g2	AMFR	2LXP	4G30, 2LXH(NMR)
2CYX	AMFR	3H8K	4G30, 2LXH(NMR)
Ube2l3	ARIH1	5TTE	4KC9
6XXU	ARIH1	5UDH	4KC9
Ube2n	RNF85	5VNZ	2JMD
1JBB	RNF85	3HCU	2JMD
	SNURF	5AIT	2EA6
	TRIM21	6S53	5OLM
	ZNRF1	5YWR	
	RNF8	4WHV	4AYC
	RNF8	4ORH	4AYC

Table 2.
PDB IDs corresponding E3, cognate E2, and E2-E3 structures.

highest sequence identity is between RNF165 and RNF12 at 25%, while the remaining RING domains have identities of 15 ± 4%. The structures of Ube2D1 were solved in complexes with ZNRF7, CNOT4, TRIM25, and RNF4. NOT4 and ZNRF4 have the lowest sequence identity of 13%, while the highest pairwise identity is between TRIM25 and RNF4 at 30%. The average is ~20%. The Ube2D3 was solved in complexes with Ube4B, Mycb2 and RING2. Structural studies of Ube2N with ZNRF1, TRAF6, TRIM21, RNF4 and 8 RING E3 ligases are also reported [84–88]. The average pairwise identity among these RING domains is 23 ± 6%, while the highest is 36% between RNF4 and TRIM21; the lowest is between RNF8 and ZNRF1 at 16%.

Interestingly, despite the low sequence identities between the RING domains, the structures of the E2-RING complexes reveal a common mode of interaction (**Figure 6**). All the RING domains are located on a similar site of the E2 enzyme and oriented in a similar manner. The face of the RING domain that interacts with the E2 enzymes involves residues of the structured loop1 and its two β-strands and loop2, and incorporates both zinc ions. The zinc ions do not provide any stabilizing interaction with the E2 amino acids but may contribute an electronic effect that adds to the allosteric effects. The RING domains interact with the surface opposite to the β-sheet of the E2 structure, and contacts residues at the N-terminal end of helix 1 the loops connecting the helix traversing the β-sheet, the helical turn adjacent to the catalytic

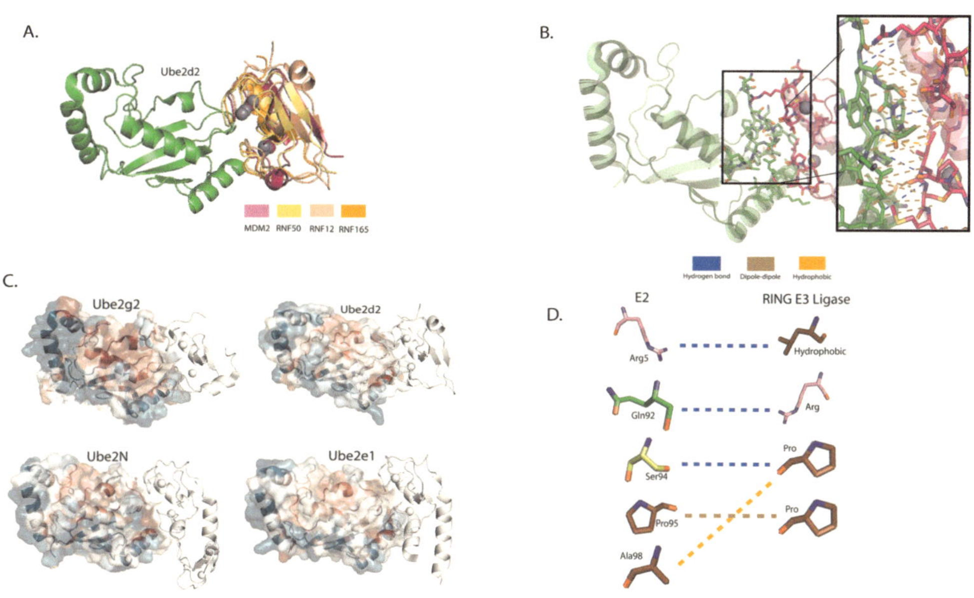

Figure 6.
Interactions of RING E3 domains with Ube2D2. A. Structures of Ube2D2 bound to MDM2 (purple), RNF12 (salmon), RNF165 (orange), and RNF50 (yellow). This mode of binding is prototypical for all known E2–RING interactions. The closest residue of the RING domain from the catalytic cysteine is ~15 a. the binding interface averages ~550 ± 30 Å [2]. B. Structure of an E2–RING (pdb: 4ap5) with key pairs of residues highlighted at the interface. Some of these residues are identified in 'D'. C. the tertiary structures of E2–RING pairs from each of the four E2 classes were analyzed by the program ConSurf to identify the locations of conserved residues. Residues/regions on the protein that are more conserved are colored bright pink/red and regions with variable residues are colored blue. Bright pink/red regions are located at the interface with the RING domains for all four types of Ub E2 enzymes. D. Summary of the types of interactions associated at the interface. There is usually one salt-bridge, four hydrogen bonds and ~ 40 non-covalent contacts.

cysteine, and a loop connecting two of the β-strands. The area of the interface averages to ~550 Å [2] for the various structures, as calculated by PDBsum [89, 90].

Interestingly, the location of the RING domains is found to be at least ~15 Å from the thioester linkage catalytic site (**Figure 6B**), which would suggest that RING domains are not directly involved in the chemistry of the Ub transfer. In fact, the distal location indicates that RING domains exert an allosteric effect as the key factor in influencing Ub transfer to the substrate. Based on the sequence alignment, and the structures of the complexes, it appears that E2 members of each family have uniquely conserved residues involved with their interactions with different RING domains and these residues can contribute to the long range of structural communication. Analyzing the structures and sequences of these complexes with the program ConSurf [91, 92], confirmed conserved residues at the interface. The structures of Ube2D2, Ube2N, Ube2G2 and Ube2L3 all show a large patch of conserved residues at the interface where the RING domains interact (**Figure 6B**). Most of these residues are found on the N-terminal half of helix 1, the central helix that traverse the β-sheet, on the loops connecting the β-strands and on the helical turn. Residues that are less conserved are located away from this E2-RING interface (**Figure 6C**).

Among the class 1 enzymes, for which the most structural information is available, there are four key sets of interactions with RING domains. The amino group of an arginine near position 5 (helix 1) forms a hydrogen bond with the carbonyl group of a hydrophobic residue (valine, leucine, or isoleucine) on the RING domain. The hydrophobic property of the hydrocarbon chain contributes to the arginine interaction with the hydrophobic residue. For many of E2s, a glutamine (Gln) or asparagine (at position 92 of Ube2D2), located on the helical turn, forms a hydrogen bond with an arginine side-chain found on loop2 of the RING domains. This interaction appears to be essential for the activation of the thioester bond for nucleophilic attack. Mutation of this glutamine dramatically affects Ub transfer [63, 64]. Interestingly, other members of the E2 (ex: Ube2G2, Ube2N) have a basic residue (R/K) at this position, indicating that these E2 enzyme may interact with a RING domain that instead of having an arginine on loop2 may have a complementary Asn/Gln residue. Such complementarity may provide some clue about specificity. For example, Ube2D2 with a Gln may not bind the same RING domain as Ube2G2, which has a Lys in place of the Gln. It would seem logical then that the cohort RING domain will have a Gln or Asn to form a hydrogen bond with the E2 enzyme.

The class 1 E2's have several fairly conserved proline residues, but the proline (position 95 of Ube2D2) makes hydrophobic contacts with a proline or isoleucine residue on the RING domain. Multiple sequence alignment of the RING domains confirmed these conserved residues (**Figure 6C**). The hydrophobic residue that interacts with Arg5 of Ube2D2 and the arginine that interacts with Gln92 are conserved in RING E3 domains. Adjacent to Pro95 (Ube2D2) is Ser94 that forms a hydrogen bond between its sidechain OH and backbone carbonyl group of the proline/isoleucine of the RING domain that interact with Pro95 (**Figure 6B, D**).

The Ube2N family is also found to have four conserved interactions with three RING domains. The Serine and Proline interaction is conserved, however in the Ube2N family it corresponds to Ser96. The arginine interaction with a hydrophobic residue is also conserved for this family, however it is Arg7 instead of Arg5. An interaction between Lys10 and Leucine is also conserved in almost all catalog entries, with hydrocarbons in their side chains interacting. Lastly, Arg6 is found to be involved in a conserved interaction with a negatively charged residue (Asp or Glu). It is possible that the slight differences in positions of the various RING domains with the Ube2D2 enzyme may reflect the RING domains adjusting to make these conserved interactions with the E2.

Although the Ube2G2 and Ube2L3 families had significantly less cataloged entries, a few conserved interactions with their RING families are noted. Despite Ube2G2 being solved in a complex with only one RING domain (RNF45), the serine and proline interaction is conserved, and in Ube2G2's case it is Ser111 interacting. An interaction between Ser67 of Ube2G2 and a Gln of RING domain is conserved, as well as an electrostatic interaction between Glu108 and an arginine of RNF45. Ube2L3 has conserved interactions between its Pro95 and the RING domain proline or isoleucine, and between its Arg5 and an isoleucine of RING E3 ligase. Notably absent from conserved residues in Ube2G2 is the interaction of serine and proline. Ube2L3 is also missing the conserved interaction of Arg5 and a hydrophobic residue.

The conserved interactions among the E2s and RING domains, noted above, are confirmed by analyzing the various structures using PDBsum. The analyses reveal an average of one salt-bridge, four hydrogen bonds and ~ 40 non-covalent interactions. The surface area of the interface is relatively small, usually less than 600 Å [2]. These observations would indicate a rather labile interaction between RING and E2s. NMR studies probing the interactions of RING and E2 enzymes confirm fast exchange in binding by the observation of very small chemical shift changes in the protein NH NMR signals when these proteins are titrated with respect to each other [69, 93–95]. NMR, isothermal calorimetry, and SPR binding studies of 23 pairs of RING E3 and E2 proteins exhibit dissociation constants (K_d) are in the sub-millimolar (60–200 μM) range [64].

Despite this common binding mechanism, it is not exactly clear how specificity is established between different E2 enzymes and their cognate RING E3 ligases. It is possible that although the catalytic domain of these proteins have a high level of sequence and structural similarity, minor differences of amino acids may dictate binding specificity to different RING E3 ligases. Interestingly, yeast two-hybrid screening studies reveal that some E2s interact specifically with one RING protein, while others can interact with over a hundred different RING E3s [36]. For instance, Ube2U has 52 interactors, UBE2D1–4 have 29–35 interactors, and UBE2N has 28 [36]. These observations of E2s promiscuity for many RING E3s is not unsurprising given that human cells have only a few dozen E2s and hundreds of RING E3s that must associate to promote the ubiquitination of protein substrate. There are also RING E3 ligase that can interact with multiple E2s [28, 36]. However, there are still RING E3s that only bind specific E2s [36]. It is possible that a RING domain interacts with several E2s and that some of these interactions may dictate different levels of substrate ubiquitination, i.e. mono- vs. di- vs. polyubiquitination [28].

8. RING domains as activators for aminolysis of E2 ~ Ub linkage

As noted, the distal binding of RING domain from the catalytic cysteine exerts allosteric effects on the reactivity of the active site [71, 72]. Structures of a RING domain interacting with an E2 enzyme with an Ub reveal that the RING domain binds in the same position compared with when it binds the free E2 enzyme [48, 71, 80, 96, 97]. NMR and computational studies have shown that the covalently attached Ub is highly mobile making transient interactions with the E2 enzyme, but in the presence of the bound RING domain the Ub exhibits more interactions with the E2 enzyme [35, 71]. This confirmation is referred to as the closed conformation. In the absence of the RING domain, the Ub is substantially more flexible. Thus, the RING domain promotes a more closed E2 ~ Ub conformation (more E2:Ub contacts). Mutations that destabilize the closed conformation of the E2 and Ub are shown to disrupt Ub transfer activity [71, 98]. The closed E2 ~ Ub conformation is important for activation of the thioester bond for nucleophilic attack from a lysine side chain amino group [64]. Examination of the several E2–RING structures reveals that the central helix moves slightly outwards and the N-terminal helix becomes longer by a helical turn for several E2 enzymes. Some of the β-strands also move positions ever so slightly. Despite the small interface, the binding of the RING is sufficient to induce electronic and conformational changes as part of its allosteric effect.

While this review focuses on the interaction of E2 enzymes and monomeric RING domains, there is now studies that suggest that RING homo- and hetero-dimers and multimers are important for increasing the rate of aminolysis and Ub transfer [15, 76, 85, 97, 99–106]. BRCA1 is shown to have enhanced ligase activity when in a RING-RING complex with BARD1 [101, 105, 107], and MID1 RING domain exhibits increased activity in complex with the B-box domain [16, 28, 53]. The enhanced effects of RING dimerization is observed in vivo and in vitro and cannot be rationalized structurally [63, 64, 65, 107, 108]. The structures of RING dimers reveal that they are symmetrical and that the interface involves the surface opposite to the one involved in the interaction with E2 enzymes. How this interaction enhances the allosteric and electronic effects that RING has on the reactivity of the thioester linkage is not clear. Structures of RING dimers with E2 enzymes are also dimeric structures, in which each RING binds its own E2 enzyme [80, 106, 109]. It is possible that the increased activity for some RING proteins may be due to an increased apparent concentration effect due to their dimerization or multimerization.

9. Conclusion

Ubiquitination is an essential process that serves to regulate many cellular processes, most notably in regulating the cellular concentrations of proteins (homeostasis) through cellular degradation. The pathway to covalently attach a Ub to a substrate protein is highly coordinated. Errors in this pathway have significant consequences to cell function and contribute to the pathogenesis of several human illnesses including cancers, genetic disorders, and brain disorders [13, 110–114]. Many of these defects are associated with a dysfunctional RING protein that obviously leads to an increase in concentration of their target protein.

Though the eukaryotic system have created redundancies in how it labels proteins for proteasomal degradation, RING E3 ligases are overwhelmingly the most prevalent with currently over 700 members in humans. It is expected that this number will increase to also include proteins with domains that have ββα-RING folds. For instance, the U-box domain is shown to have the same RING fold despite not being cysteine and histidine rich and not binding any zinc ions [47, 115, 116]. The U-box domain is shown to interact on the same interface and manner on Ube2D2 as the zinc-binding RING domains [71, 72], suggesting that function preserves structure. Furthermore, the MID1 B-box1 and B-box2 domains are shown to bind two zinc ions and adopt a similar RING fold as monomers and a RING dimers in tandem, despite having less than 25% sequence homology with RING domains [46, 58, 59]. The B-box domains are similar to RING domains in the manner in which they bind the zinc ions but the sizes of their L1 and L2 regions are on the smaller ranges of RING domains.

In contrast, there are some zinc-binding domains that adopt the same ββα-RING fold but function as E3 ligase enhancers or E4 ligases. The BARD1 RING domain dimerizes with BRCA1 to enhance the E3 ligase activity of BRCA1 [101, 105, 107]; similarly, MdmX dimerizes with Mdm2 to target p53 [68, 76]. It is unclear why these E4 RING domains do not facilitate ubiquitination but it is possible that their interaction with the E2 enzyme is considerably different than that of RING E3 domains. There are currently no structures of an E4 enhancer with an E2, possibly because the interaction is so weak. The structures of BRCA1:BARD1 and Mdm2:MdmX are essentially very similar to homo-RING dimers suggesting that the mechanism of enhanced activities observed for RING homo-dimers may be the same for RING hetero-dimers.

It has long been postulated that the RING E3 ligases target a specific protein for ubiquitination and therefore how a large number of proteins interact with a little over three dozen E2 enzymes is not fully understood. As noted, RING domains share very low sequence homology with each other yet they all adopt a very similar structure, indicating that their mechanism of function is very similar. To try and provide insights into how RING domains are recognized, the sequences and structures of RING domains and E2 enzymes are evaluated. And while this review does not answer all the questions about the interaction, many of which are being investigated, it does provide some clues. As noted, E2 enzymes have a central catalytic domain that adopt a common structure. Despite variations in sequence identities and similarities, it appears that the catalytic domain preserves conserved residues at the interface that bind RING domains and residues that will transmit the allosteric effect of the bound RING domain (**Figures 3** and **4**). Similarly, the RING domain with far less sequence homologies also maintain key amino acids in specific locations to interact with the E2 catalytic domain in a similar location.

RING domains are relatively small (50–70 amino acids) in the context of their roles as enzymes and in facilitating the ubiquitination of substrate proteins. It would seem quite daunting for such a small protein domain to bind large proteins (E2, substrate) while maintaining their role to influence Ub transfer [16, 28, 53]. Despite the importance of substrate ubiquitination via RING E3 ligases, the mechanism is still not completely understood. Furthermore, it appears that the mechanism of interaction helps to determine which lysine residue on a substrate gets ubiquitinated. It should be noted that the RING domain exists as part of a larger protein and that the binding of substrates is not always relegated to the RING domain; more often substrates are proteins that bind other domains within the RING-containing E3 ligase protein. Mutations in the B-box1 domain of the MID1 RING protein, which binds alpha4 and the catalytic subunit of protein phosphatase 2A (PP2Ac), prevented polyubiquitination of these proteins despite MID1 maintaining full autoubiquitination activity [53]. Similarly, the SH2 domain of the c-CBL E3 ligase functions to bind the substrate while the RING domain interacts with the E2 enzyme and facilitates ubiquitination [45, 65]. In fact, it is possible to create chimeras with a RING domain followed with a domain that specifically binds any protein. Such substrate trapping strategy can be useful in regulating proteins whose upregulation is associated with cellular dysfunction and human diseases.

Finally, while there has been considerable research in this exciting area of protein ubiquitination, specifically in understanding how E2 and E3 proteins interact, there are still many questions left unanswered, such as how does specific E2-RING interaction influences the level and type of ubiquitination, and the lysine that becomes covalently modified.

Authors contribution

EMG and IS performed literature review, data collection (sequences and structures) and performed analysis. They also drafted the manuscript. MM oversaw the whole project, guiding the analysis and writing the manuscript.

Author details

Erin Meghan Gladu†, Iman Sayed† and Michael Anthony Massiah*
Department of Chemistry, George Washington University, Washington DC, USA

*Address all correspondence to: massiah@gwu.edu

† Authors are undergraduate students who contributed equally.

References

[1] Pickart CM, Vella AT. Ubiquitin carrier protein-catalyzed ubiquitin transfer to histones. Mechanism and specificity. The Journal of Biological Chemistry. 1988;**263**:15076-15082

[2] Pickart CM. Mechanisms underlying ubiquitination. Annual Review of Biochemistry. 2001;**70**:503-533

[3] Pickart CM. Ubiquitin biology: An old dog learns an old trick. Nature Cell Biology. 2000;**2**:E139-E141

[4] Callis J. The ubiquitination machinery of the ubiquitin system. Arabidopsis Book. 2014;**12**:e0174

[5] Glickman MH, Ciechanover A. The ubiquitin-proteasome proteolytic pathway: Destruction for the sake of construction. Physiological Reviews. 2002;**82**:373-428

[6] Kaiser P, Flick K, Wittenberg C, Reed SI. Regulation of transcription by ubiquitination without proteolysis: Cdc34/SCF(Met30)-mediated inactivation of the transcription factor Met4. Cell. 2000;**102**:303-314

[7] Mastrandrea LD, You J, Niles EG, Pickart CM. E2/E3-mediated assembly of lysine 29-linked polyubiquitin chains. The Journal of Biological Chemistry. 1999;**274**:27299-27306

[8] Sadowski M, Suryadinata R, Tan AR, Roesley SN, Sarcevic B. Protein monoubiquitination and polyubiquitination generate structural diversity to control distinct biological processes. IUBMB Life. 2012;**64**:136-142

[9] Deng L, Wang C, Spencer E, Yang L, Braun A, You J, et al. Activation of the IkappaB kinase complex by TRAF6 requires a dimeric ubiquitin-conjugating enzyme complex and a unique polyubiquitin chain. Cell. 2000;**103**: 351-361

[10] Hofmann RM, Pickart CM. Noncanonical MMS2-encoded ubiquitin conjugating enzyme functions in assembly of novel polyubiquitin chains for DNA repair. Cell. 1999;**96**:645-653

[11] Ong JY, Torres JZ. E3 Ubiqutin ligase in Cancer and their pharmacological targeting. Rijeka: IntechOpen Book Series; 2018

[12] Borg NA, Dixit VM. Ubiquitin in Cell-cycle regulation and dysregulation in cancer. Annual Review of Cancer Biology. 2017;**1**:59-77

[13] Loman N, Johannsson O, Kristoffersson U, Olsson H, Borg A. Family history of breast and ovarian cancers and BRCA1 and BRCA2 mutations in a population-based series of early-onset breast cancer. Journal of the National Cancer Institute. 2001;**93**: 1215-1223

[14] Zheng Q, Huang T, Zhang L, Zhou Y, Luo H, Xu H, et al. Dysregulation of ubiquitin-proteasome system in neurodegenerative diseases. Frontiers in Aging Neuroscience. 2016;**8**:303

[15] Short KM, Hopwood B, Yi Z, Cox TC. MID1 and MID2 homo- and heterodimerise to tether the rapamycin-sensitive PP2A regulatory subunit, alpha 4, to microtubules: implications for the clinical variability of X-linked Opitz GBBB syndrome and other developmental disorders. BMC Cell Biology. 2002;**3**:1

[16] Du H, Wu K, Didoronkute A, Levy MV, Todi N, Shchelokova A, et al. MID1 catalyzes the ubiquitination of

protein phosphatase 2A and mutations within its Bbox1 domain disrupt polyubiquitination of alpha4 but not of PP2Ac. PLoS One. 2014;**9**:e107428

[17] Darwin KH. Prokaryotic ubiquitin-like protein (Pup), proteasomes and pathogenesis. Nature Reviews. Microbiology. 2009;**7**:485-491

[18] Barandun J, Delley CL, Weber-Ban E. The pupylation pathway and its role in mycobacteria. BMC Biology. 2012;**10**:95

[19] Burns KE, Darwin KH. Pupylation versus ubiquitylation: Tagging for proteasome-dependent degradation. Cellular Microbiology. 2010;**12**:424-431

[20] Striebel F, Imkamp F, Ozcelik D, Weber-Ban E. Pupylation as a signal for proteasomal degradation in bacteria. Biochimica et Biophysica Acta. 2014; **1843**:103-113

[21] Groettrup M, Pelzer C, Schmidtke G, Hofmann K. Activating the ubiquitin family: UBA6 challenges the field. Trends in Biochemical Sciences. 2008;**33**: 230-237

[22] VanDemark AP, Hill CP. Two-stepping with E1. Nature Structural Biology. 2003;**10**:244-246

[23] McDowell GS, Philpott A. Non-canonical ubiquitylation: Mechanisms and consequences. The International Journal of Biochemistry & Cell Biology. 2013;**45**:1833-1842

[24] Vosper JM, McDowell GS, Hindley CJ, Fiore-Heriche CS, Kucerova R, Horan I, et al. Ubiquitylation on canonical and non-canonical sites targets the transcription factor neurogenin for ubiquitin-mediated proteolysis. The Journal of Biological Chemistry. 2009;**284**: 15458-15468

[25] Williams C, van den Berg M, Sprenger RR, Distel B. A conserved cysteine is essential for Pex4p-dependent ubiquitination of the peroxisomal import receptor Pex5p. The Journal of Biological Chemistry. 2007;**282**:22534-22543

[26] Wang X, Herr RA, Chua WJ, Lybarger L, Wiertz EJ, Hansen TH. Ubiquitination of serine, threonine, or lysine residues on the cytoplasmic tail can induce ERAD of MHC-I by viral E3 ligase mK3. The Journal of Cell Biology. 2007;**177**:613-624

[27] Ranaweera RS, Yang X. Auto-ubiquitination of Mdm2 enhances its substrate ubiquitin ligase activity. The Journal of Biological Chemistry. 2013; **288**:18939-18946

[28] Han X, Du H, Massiah MA. Detection and characterization of the in vitro e3 ligase activity of the human MID1 protein. Journal of Molecular Biology. 2011;**407**:505-520

[29] Xie W, Jin S, Wu Y, Xian H, Tian S, Liu DA, et al. Auto-ubiquitination of NEDD4-1 Recruits USP13 to Facilitate Autophagy through Deubiquitinating VPS34. Cell Reports. 2020;**30**(2807–2819):e2804

[30] Noels H, Somers R, Liu H, Ye H, Du MQ, De Wolf-Peeters C, et al. Auto-ubiquitination-induced degradation of MALT1-API2 prevents BCL10 destabilization in t(11;18)(q21;q21)-positive MALT lymphoma. PLoS One. 2009;**4**:e4822

[31] Amemiya Y, Azmi P, Seth A. Autoubiquitination of BCA2 RING E3 ligase regulates its own stability and affects cell migration. Molecular Cancer Research. 2008;**6**:1385-1396

[32] Wright KM, Du H, Massiah MA. Structural and functional observations of

the P151L MID1 mutation reveal alpha4 plays a significant role in X-linked Opitz Syndrome. The FEBS Journal. 2017;**284**: 2183-2193

[33] Liu W, Tang X, Qi X, Fu X, Ghimire S, Ma R, et al. The ubiquitin conjugating enzyme: An important ubiquitin transfer platform in ubiquitin-proteasome system. International Journal of Molecular Sciences. 2020;**21**: 2824

[34] Kraft E, Stone SL, Ma L, Su N, Gao Y, Lau OS, et al. Genome analysis and functional characterization of the E2 and RING-type E3 ligase ubiquitination enzymes of Arabidopsis. Plant Physiology. 2005;**139**:1597-1611

[35] Stewart MD, Ritterhoff T, Klevit RE, Brzovic PS. E2 enzymes: more than just middle men. Cell Research. 2016;**26**: 423-440

[36] van Wijk SJ, de Vries SJ, Kemmeren P, Huang A, Boelens R, Bonvin AM, et al. A comprehensive framework of E2-RING E3 interactions of the human ubiquitin-proteasome system. Molecular Systems Biology. 2009;**5**:295

[37] Ye Y, Rape M. Building ubiquitin chains: E2 enzymes at work. Nature Reviews. Molecular Cell Biology. 2009; **10**:755-764

[38] Haldeman MT, Xia G, Kasperek EM, Pickart CM. Structure and function of ubiquitin conjugating enzyme E2-25K: the tail is a core-dependent activity element. Biochemistry. 1997;**36**: 10526-10537

[39] Schelpe J, Monte D, Dewitte F, Sixma TK, Rucktooa P. Structure of UBE2Z Enzyme Provides Functional Insight into Specificity in the FAT10 Protein Conjugation Machinery. The Journal of Biological Chemistry. 2016; **291**:630-639

[40] Lin Y, Hwang WC, Basavappa R. Structural and functional analysis of the human mitotic-specific ubiquitin-conjugating enzyme, UbcH10. The Journal of Biological Chemistry. 2002; **277**:21913-21921

[41] Li KB. ClustalW-MPI: ClustalW analysis using distributed and parallel computing. Bioinformatics. 2003;**19**: 1585-1586

[42] Thompson JD, Gibson TJ, Higgins DG. Multiple sequence alignment using ClustalW and ClustalX. In: Curr Protoc Bioinformatics Chapter 2. National Library of Medicine. 2002;**2**(2):3

[43] Borden KL, Lally JM, Martin SR, O'Reilly NJ, Etkin LD, Freemont PS. Novel topology of a zinc-binding domain from a protein involved in regulating early Xenopus development. The EMBO Journal. 1995;**14**:5947-5956

[44] Borden KL, Boddy MN, Lally J, O'Reilly NJ, Martin S, Howe K, et al. The solution structure of the RING finger domain from the acute promyelocytic leukaemia proto-oncoprotein PML. The EMBO Journal. 1995;**14**:1532-1541

[45] Deshaies RJ, Joazeiro CA. RING domain E3 ubiquitin ligases. Annual Review of Biochemistry. 2009;**78**: 399-434

[46] Massiah MA, Simmons BN, Short KM, Cox TC. Solution structure of the RBCC/TRIM B-box1 domain of human MID1: B-box with a RING. Journal of Molecular Biology. 2006;**358**: 532-545

[47] Ohi MD, Vander Kooi CW, Rosenberg JA, Chazin WJ, Gould KL.

Structural insights into the U-box, a domain associated with multi-ubiquitination. Nature Structural Biology 4. 2003;**10**:250-255

[48] Dove KK, Olszewski JL, Martino L, Duda DM, Wu XS, Miller DJ, et al. Structural Studies of HHARI/UbcH7 approximately Ub Reveal Unique E2 approximately Ub Conformational Restriction by RBR RING1. Structure. 2017;**25**(890–900):e895

[49] Hatakeyama S, Nakayama KI. U-box proteins as a new family of ubiquitin ligases. Biochemical and Biophysical Research Communications. 2003;**302**:635-645

[50] Patterson C. A new gun in town: The U box is a ubiquitin ligase domain. Science's STKE. 2002:pe4

[51] Dove KK, Klevit RE. RING-Between-RING E3 Ligases: Emerging Themes amid the Variations. Journal of Molecular Biology. 2017;**429**:3363-3375

[52] Reiter KH, Klevit RE. Characterization of RING-between-RING E3 ubiquitin transfer mechanisms. Methods in Molecular Biology. 2018;**1844**:3-17

[53] Du H, Huang Y, Zaghlula M, Walters E, Cox TC, Massiah MA. The MID1 E3 ligase catalyzes the polyubiquitination of Alpha4 (alpha4), a regulatory subunit of protein phosphatase 2A (PP2A): novel insights into MID1-mediated regulation of PP2A. The Journal of Biological Chemistry. 2013;**288**:21341-21350

[54] Schweiger S, Dorn S, Fuchs M, Kohler A, Matthes F, Muller EC, et al. The E3 ubiquitin ligase MID1 catalyzes ubiquitination and cleavage of Fu. The Journal of Biological Chemistry. 2014;**289**:31805-31817

[55] Meroni G. Preface. TRIM/RBCC proteins. Advances in Experimental Medicine and Biology. 2012;**770**:vii-viii

[56] Sardiello M, Cairo S, Fontanella B, Ballabio A, Meroni G. Genomic analysis of the TRIM family reveals two groups of genes with distinct evolutionary properties. BMC Evolutionary Biology. 2008;**8**:225

[57] Short KM, Cox TC. Sub-classification of the rbcc/trim superfamily reveals a novel motif necessary for microtubule binding. The Journal of Biological Chemistry. 2006;**281**:8970-8980

[58] Tao H, Simmons BN, Singireddy S, Jakkidi M, Short KM, Cox TC, et al. Structure of the MID1 tandem B-boxes reveals an interaction reminiscent of intermolecular ring heterodimers. Biochemistry. 2008;**47**:2450-2457

[59] Massiah MA, Matts JA, Short KM, Simmons BN, Singireddy S, Yi Z, et al. Solution structure of the MID1 B-box2 CHC(D/C)C(2)H(2) zinc-binding domain: insights into an evolutionarily conserved RING fold. Journal of Molecular Biology. 2007;**369**:1-10

[60] Borden KL. RING fingers and B-boxes: zinc-binding protein-protein interaction domains. Biochemistry and Cell Biology. 1998;**76**:351-358

[61] Massiah MA, Blake PR, Summers MF. Nucleic Acid Interactive Protein Domains That Require Zinc. New York, NY: Oxford University Press; 1998

[62] Wright KM, Wu K, Babatunde O, Du H, Massiah MA. XLOS-observed mutations of MID1 Bbox1 domain cause domain unfolding. PLoS One. 2014;**9**:e107537

[63] Metzger MB, Pruneda JN, Klevit RE, Weissman AM. RING-type E3 ligases:

master manipulators of E2 ubiquitin-conjugating enzymes and ubiquitination. Biochimica et Biophysica Acta. 2014; **1843**:47-60

[64] Das R, Liang YH, Mariano J, Li J, Huang T, King A, et al. Allosteric regulation of E2:E3 interactions promote a processive ubiquitination machine. The EMBO Journal. 2013;**32**:2504-2516

[65] Joazeiro CA, Weissman AM. RING finger proteins: Mediators of ubiquitin ligase activity. Cell. 2000;**102**:549-552

[66] Ru Y, Wang Q, Liu X, Zhang M, Zhong D, Ye M, et al. The chimeric ubiquitin ligase SH2-U-box inhibits the growth of imatinib-sensitive and resistant CML by targeting the native and T315I-mutant BCR-ABL. Scientific Reports. 2016;**6**:28352

[67] Thien CB, Langdon WY. c-Cbl and Cbl-b ubiquitin ligases: substrate diversity and the negative regulation of signalling responses. The Biochemical Journal. 2005;**391**:153-166

[68] Wang X, Jiang X. Mdm2 and MdmX partner to regulate p53. FEBS Letters. 2012;**586**:1390-1396

[69] Christensen DE, Brzovic PS, Klevit RE. E2-BRCA1 RING interactions dictate synthesis of mono- or specific polyubiquitin chain linkages. Nature Structural & Molecular Biology. 2007;**14**: 941-948

[70] Hodson C, Purkiss A, Miles JA, Walden H. Structure of the human FANCL RING-Ube2T complex reveals determinants of cognate E3-E2 selection. Structure. 2014;**22**:337-344

[71] Pruneda JN, Littlefield PJ, Soss SE, Nordquist KA, Chazin WJ, Brzovic PS, et al. Structure of an E3:E2~Ub complex reveals an allosteric mechanism shared among RING/U-box ligases. Molecular Cell. 2012;**47**:933-942

[72] Nordquist KA, Dimitrova YN, Brzovic PS, Ridenour WB, Munro KA, Soss SE, et al. Structural and functional characterization of the monomeric U-box domain from E4B. Biochemistry. 2010;**49**:347-355

[73] Benirschke RC, Thompson JR, Nomine Y, Wasielewski E, Juranic N, Macura S, et al. Molecular basis for the association of human E4B U box ubiquitin ligase with E2-conjugating enzymes UbcH5c and Ubc4. Structure. 2010;**18**:955-965

[74] Xu Z, Kohli E, Devlin KI, Bold M, Nix JC, Misra S. Interactions between the quality control ubiquitin ligase CHIP and ubiquitin conjugating enzymes. BMC Structural Biology. 2008;**8**:26

[75] Gundogdu M, Walden H. Structural basis of generic versus specific E2-RING E3 interactions in protein ubiquitination. Protein Science. 2019;**28**:1758-1770

[76] Linke K, Mace PD, Smith CA, Vaux DL, Silke J, Day CL. Structure of the MDM2/MDMX RING domain heterodimer reveals dimerization is required for their ubiquitylation in trans. Cell Death and Differentiation. 2008;**15**: 841-848

[77] Middleton AJ, Zhu J, Day CL. The RING domain of RING finger 12 efficiently builds degradative ubiquitin chains. Journal of Molecular Biology. 2020;**432**:3790-3801

[78] Dominguez C, Bonvin AM, Winkler GS, van Schaik FM, Timmers HT, Boelens R. Structural model of the UbcH5B/CNOT4 complex revealed by combining NMR, mutagenesis, and docking approaches. Structure. 2004;**12**:633-644

[79] Zheng N, Wang P, Jeffrey PD, Pavletich NP. Structure of a c-Cbl-UbcH7 complex: RING domain function in ubiquitin-protein ligases. Cell. 2000; **102**:533-539

[80] Dou H, Buetow L, Sibbet GJ, Cameron K, Huang DT. BIRC7-E2 ubiquitin conjugate structure reveals the mechanism of ubiquitin transfer by a RING dimer. Nature Structural & Molecular Biology. 2012;**19**:876-883

[81] Li S, Liang YH, Mariano J, Metzger MB, Stringer DK, Hristova VA, et al. Insights into ubiquitination from the unique clamp-like binding of the RING E3 AO7 to the E2 UbcH5B. The Journal of Biological Chemistry. 2015; **290**:30225-30239

[82] Wright JD, Mace PD, Day CL. Secondary ubiquitin-RING docking enhances Arkadia and Ark2C E3 ligase activity. Nature Structural & Molecular Biology. 2016;**23**:45-52

[83] Patel A, Sibbet GJ, Huang DT. Structural insights into non-covalent ubiquitin activation of the cIAP1-UbcH5B~ubiquitin complex. The Journal of Biological Chemistry. 2019; **294**:1240-1249

[84] Behera AP, Naskar P, Agarwal S, Banka PA, Poddar A, Datta AB. Structural insights into the nanomolar affinity of RING E3 ligase ZNRF1 for Ube2N and its functional implications. The Biochemical Journal. 2018;**475**: 1569-1582

[85] Yin Q, Lin SC, Lamothe B, Lu M, Lo YC, Hura G, et al. E2 interaction and dimerization in the crystal structure of TRAF6. Nature Structural & Molecular Biology. 2009;**16**:658-666

[86] Dickson C, Fletcher AJ, Vaysburd M, Yang JC, Mallery DL, Zeng J, et al. Intracellular antibody signalling is regulated by phosphorylation of the Fc receptor TRIM21. eLife. 2018;**8**: 9396-9410

[87] Branigan E, Plechanovova A, Jaffray EG, Naismith JH, Hay RT. Structural basis for the RING-catalyzed synthesis of K63-linked ubiquitin chains. Nature Structural & Molecular Biology. 2015;**22**:597-602

[88] Hodge CD, Ismail IH, Edwards RA, Hura GL, Xiao AT, Tainer JA, et al. RNF8 E3 ubiquitin ligase stimulates Ubc13 E2 conjugating activity that is essential for DNA double strand break signaling and BRCA1 tumor suppressor recruitment. The Journal of Biological Chemistry. 2016;**291**:9396-9410

[89] Laskowski RA, Jablonska J, Pravda L, Varekova RS, Thornton JM. PDBsum: Structural summaries of PDB entries. Protein Science. 2018;**27**:129-134

[90] Laskowski RA, Hutchinson EG, Michie AD, Wallace AC, Jones ML, Thornton JM. PDBsum: A Web-based database of summaries and analyses of all PDB structures. Trends in Biochemical Sciences. 1997;**22**:488-490

[91] Glaser F, Pupko T, Paz I, Bell RE, Bechor-Shental D, Martz E, et al. ConSurf: Identification of functional regions in proteins by surface-mapping of phylogenetic information. Bioinformatics. 2003;**19**:163-164

[92] Armon A, Graur D, Ben-Tal N. ConSurf: An algorithmic tool for the identification of functional regions in proteins by surface mapping of phylogenetic information. Journal of Molecular Biology. 2001;**307**:447-463

[93] Huang A, de Jong RN, Wienk H, Winkler GS, Timmers HT, Boelens R. E2-c-Cbl recognition is necessary but not

sufficient for ubiquitination activity. Journal of Molecular Biology. 2009;**385**: 507-519

[94] Mercier P, Lewis MJ, Hau DD, Saltibus LF, Xiao W, Spyracopoulos L. Structure, interactions, and dynamics of the RING domain from human TRAF6. Protein Science. 2007;**16**:602-614

[95] Bijlmakers MJ, Teixeira JM, Boer R, Mayzel M, Puig-Sarries P, Karlsson G, et al. A C2HC zinc finger is essential for the RING-E2 interaction of the ubiquitin ligase RNF125. Scientific Reports. 2016;**6**:29232

[96] Buetow L, Gabrielsen M, Anthony NG, Dou H, Patel A, Aitkenhead H, et al. Activation of a primed RING E3-E2-ubiquitin complex by non-covalent ubiquitin. Molecular Cell. 2015;**58**:297-310

[97] Koliopoulos MG, Esposito D, Christodoulou E, Taylor IA, Rittinger K. Functional role of TRIM E3 ligase oligomerization and regulation of catalytic activity. The EMBO Journal. 2016;**35**:1204-1218

[98] Brzovic PS, Lissounov A, Christensen DE, Hoyt DW, Klevit RE. A UbcH5/ubiquitin noncovalent complex is required for processive BRCA1-directed ubiquitination. Molecular Cell. 2006;**21**:873-880

[99] Bell JL, Malyukova A, Holien JK, Koach J, Parker MW, Kavallaris M, et al. TRIM16 acts as an E3 ubiquitin ligase and can heterodimerize with other TRIM family members. PLoS One. 2012;**7**: e37470

[100] Bellon SF, Rodgers KK, Schatz DG, Coleman JE, Steitz TA. Crystal structure of the RAG1 Dimerization domain reveals multiple zinc-binding motifs including a novel zinc binuclear cluster. Nature Structural Biology. 1997;**4**:587

[101] Brzovic PS, Rajagopal P, Hoyt DW, King MC, Klevit RE. Structure of a BRCA1-BARD1 heterodimeric RING-RING complex. Nature Structural Biology. 2001;**8**:833-837

[102] Huang A, Hibbert RG, de Jong RN, Das D, Sixma TK, Boelens R. Symmetry and asymmetry of the RING-RING dimer of Rad18. Journal of Molecular Biology. 2011;**410**:424-435

[103] Liew CW, Sun H, Hunter T, Day CL. RING domain dimerization is essential for RNF4 function. The Biochemical Journal. 2010;**431**:23-29

[104] Mace PD, Linke K, Feltham R, Schumacher FR, Smith CA, Vaux DL, et al. Structures of the cIAP2 RING domain reveal conformational changes associated with ubiquitin-conjugating enzyme (E2) recruitment. The Journal of Biological Chemistry. 2008;**283**: 31633-31640

[105] Mallery DL, Vandenberg CJ, Hiom K. Activation of the E3 ligase function of the BRCA1/BARD1 complex by polyubiquitin chains. The EMBO Journal. 2002;**21**:6755-6762

[106] Plechanovova A, Jaffray EG, McMahon SA, Johnson KA, Navratilova I, Naismith JH, et al. Mechanism of ubiquitylation by dimeric RING ligase RNF4. Nature Structural & Molecular Biology. 2011;**18**:1052-1059

[107] Xia Y, Pao GM, Chen HW, Verma IM, Hunter T. Enhancement of BRCA1 E3 ubiquitin ligase activity through direct interaction with the BARD1 protein. The Journal of Biological Chemistry. 2003;**278**:5255-5263

[108] Fiorentini F, Esposito D, Rittinger K. Does it take two to tango? RING domain self-association and activity in TRIM E3 ubiquitin ligases.

Biochemical Society Transactions. 2020; **48**:2615-2624

[109] Plechanovova A, Jaffray EG, Tatham MH, Naismith JH, Hay RT. Structure of a RING E3 ligase and ubiquitin-loaded E2 primed for catalysis. Nature. 2012;**489**:115-120

[110] Meroni G. TRIM E3 Ubiquitin ligases in rare genetic disorders. Advances in Experimental Medicine and Biology. 2020;**1233**:311-325

[111] Burger A, Amemiya Y, Kitching R, Seth AK. Novel RING E3 ubiquitin ligases in breast cancer. Neoplasia. 2006; **8**:689-695

[112] Humphreys LM, Smith P, Chen Z, Fouad S, D'Angiolella V. The role of E3 ubiquitin ligases in the development and progression of glioblastoma. Cell Death and Differentiation. 2021;**28**:522-537

[113] Lescouzeres L, Bomont P. E3 ubiquitin ligases in neurological diseases: Focus on gigaxonin and autophagy. Frontiers in Physiology. 2020;**11**:1022

[114] De Falco F, Cainarca S, Andolfi G, Ferrentino R, Berti C, Rodriguez Criado G, et al. X-linked Opitz syndrome: novel mutations in the MID1 gene and redefinition of the clinical spectrum. American Journal of Medical Genetics. 2003;**120A**:222-228

[115] Vander Kooi CW, Ohi MD, Rosenberg JA, Oldham ML, Newcomer ME, Gould KL, et al. The Prp19 U-box crystal structure suggests a common dimeric architecture for a class of oligomeric E3 ubiquitin ligases. Biochemistry. 2006;**45**:121-130

[116] Wang Q, Ru Y, Zhong D, Zhang J, Yao L, Li X. Engineered ubiquitin ligase PTB-U-box targets insulin/insulin-like growth factor receptor for degradation and coordinately inhibits cancer malignancy. Oncotarget. 2014;**5**: 4945-4958

Chapter 7

Functional Significance of the E3 Ubiquitin Ligases in Disease and Therapeutics

Julius Tieroyaare Dongdem and Cletus Adiyaga Wezena

Abstract

E3 ubiquitin ligases of which there are >600 putative in humans, constitute a family of highly heterogeneous proteins and protein complexes that are the ultimate enzymes responsible for the recruitment of an ubiquitin loaded E2 ubiquitin-conjugating enzyme, recognise the appropriate protein substrate and directly or indirectly transfer the ubiquitin load onto the substrate. The aftermath of an E3 ligase activity is usually the formation of an isopeptide bond between the free carboxylate group of ubiquitin's C-terminal Gly76 and an ε-amino group of the substrate's Lys, even though non-canonical ubiquitylation on non-amine groups of target proteins have been observed. E3 ligases are grouped into four distinct families: HECT, RING-finger/U-box, RBR and PHD-finger. E3 ubiquitin ligases play critical roles in subcellular signalling cascades in eukaryotes. Dysfunctional E3 ubiquitin ligases therefore tend to inflict dramatic effects on human health and may result in the development of various diseases including Parkinson's, Amyotrophic Lateral Sclerosis, Alzheimer's, cancer, etc. Being regulators of numerous cellular processes, some E3 ubiquitin ligases have become potential targets for therapy. This chapter will present a comprehensive review of up-to-date findings in E3 ligases, their role in the pathology of disease and therapeutic potential for future drug development.

Keywords: classification, disease, E3 ligases, dysfunction, mechanism, therapeutics, ubiquitin

1. Introduction

1.1 E3 ubiquitin ligases in Ubiquitylation

E3 ubiquitin ligases play crucial roles in ubiquitin conjugation to substrates and therefore ubiquitin signalling. Protein ubiquitylation (also referred to as ubiquitination) is a dynamic multifaceted post-translational modification in which ubiquitin is covalently attached to a specific protein target in a three-step enzymatic cascade involving the action of an E1 ubiquitin-activating enzyme, E2 ubiquitin-conjugating enzyme and E3 ubiquitin ligase [1]. An E3 ubiquitin ligase is the ultimate enzyme which directly or indirectly catalyses the transfer and subsequent ligation of an ubiquitin monomer to a specific target protein (i.e., the substrate). The aftermath of ubiquitylation is usually the formation of an isopeptide bond catalysed by the E3 ligase between the free carboxylate group of ubiquitin's *C*-terminal Gly76 and an ε-amino group of the substrate's Lys [2]. However, non-canonical ubiquitylation in which ubiquitin is conjugated to the target protein's *N*-terminal amino group of Met has been observed in more than 22 proteins including ubiquitin itself [3–5]. There is also increasing evidence of ubiquitylation on non-amine groups of target proteins including the thiol groups of Cys [6] and the hydroxyl groups of Thr, Ser and probably Tyr [7].

In the initial ATP-dependent activation step of ubiquitylation, E1 catalyses the acyl-adenylation of ubiquitin's *C*-terminus for conjugation by forming an ubiquitin-adenylate intermediate (**Figure 1**). In the second step, ubiquitin is transferred to the active site Cys residue resulting in the formation of a thioester linkage between the *C*-terminal carboxyl group of ubiquitin and the E1 Cys sulphhydryl (-SH) group with a consequential release of AMP [8]. Ubiquitin is then transferred from the ubiquitin-adenylate intermediate in the subsequent transthiolation reaction to the -SH group of the catalytic Cys of an E2 enzyme [9]. In the final ligation step, E3 binds both the target protein and the ubiquitin-charged E2 and facilitates the transfer of ubiquitin from E2 to the ε-amino group of a Lys in the target protein. E3 enzymes thus function as the substrate recognition modules and consequently determine substrate specificity of ubiquitin conjugation, while the E2s determine the chain type of polyubiquitylation [10, 11]. Human cells possess two E1 enzymes (UBA1, UBA6), approximately 40 E2 enzymes and more than 600 putative E3 enzymes [12–15]. This notwithstanding,

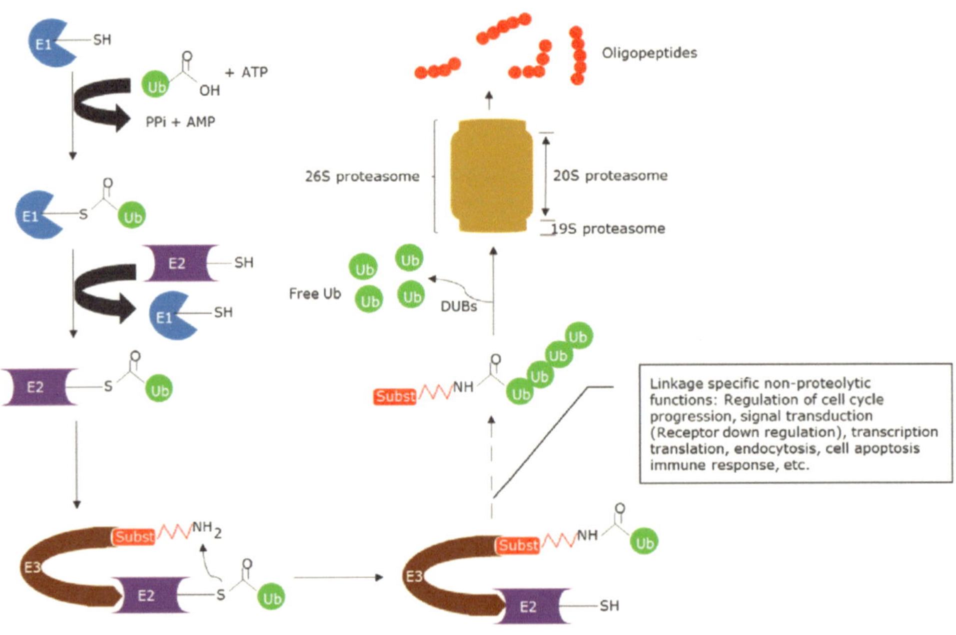

Figure 1.
The ubiquitin-mediated proteasome proteolytic pathway. Ubiquitin modification is an ATP-dependent process carried out by three classes of enzymes, E1, E2 and E3 which specifically target proteins to either change their activity or their location, or in some cases earmark target proteins for proteasome-mediated degradation. E1 forms a thioester bond with ubiquitin which allows subsequent transfer of ubiquitin to E2, followed by the E3 ligation of an isopeptide bond between the carboxyl-terminus of ubiquitin and a Lys residue on the substrate protein. Isopeptide bonds between ubiquitin and ubiquitin or ubiquitin and other target proteins can, however, be hydrolysed by deubiquitylating enzymes (DUBs).

ubiquitin-chain elongation factors (E4 enzymes e.g., the mouse double minute 2 homologue (mdm2)) which extend pre-existing polyubiquitin chains on substrate proteins (e.g., the tumour suppressor p53) have also been reported [16, 17].

1.2 Cellular functions of ubiquitylation in eukaryotes

Ubiquitin's versatility in the regulation of cellular processes is by virtue of its ability to covalently modify other proteins [18]. Post-translational modifications, which are usually covalent and reversible, may alter the properties and therefore, the functions of the modified protein. A major function of ubiquitin is regulation of the degradation of other specific proteins, literally referred to as *'the molecular kiss of death'* [19–21]. The 26S proteasome is responsible for degradation and recycling of unwanted, short-lived, inactive, oxidised, unfolded and/ or misfolded proteins (**Figure 1**) [22, 23]. Proteasomal proteolysis enables the cell to rid itself of these misfolded or damaged proteins and re-adjusts the concentration of essential proteins so that cellular homeostasis is maintained [24]. For a condemned protein to be recognised by the 26S proteasome, a polyubiquitin chain of at least four ubiquitin molecules must be covalently attached to a substrate Lys residue [25]. The proteasome is a barrel-shaped multisubunit protein complex, consisting of two chambers within which proteolysis occurs. The eukaryotic 20S proteasome is the catalytic portion of the 26S proteasome. The 19S regulatory complex mediates substrate recognition and substrate unfolding (**Figure 1**). Exploration of the ubiquitin system in eukaryotes has shown that the chemical modification of proteins by ubiquitin is an incredibly important post-translational event that is crucial to numerous complicated cellular processes beyond the ubiquitin proteasome system. Ubiquitin conjugation also plays a wide variety of roles that are independent of proteasomal degradation [2, 18]. The ubiquitin code modulates cell cycle progression, differentiation, signal transduction, protein–protein interactions, and intracellular protein trafficking. Ubiquitin regulates subcellular localisation of proteins where they control other protein function and cell mechanisms. Transcription, autophagy, inflammatory signalling, modulation of enzymatic activity, DNA repair, heat shock responses, chromatin structure, embryogenesis, cell apoptosis, virus budding, vacuolar protein sorting, inflammatory response and receptor endocytosis are also regulated by ubiquitin-mediated signalling [26, 27]. The cellular environment must constantly maintain homeostatic conditions. The ubiquitin-proteasome system is the major ubiquitin-mediated process recognised as the cellular quality control system [28]. Deubiquitylating enzymes (DUBs) cleave isopeptide bonds releasing free ubiquitin residues from protein substrates. As such DUBs are also implicated in the regulation of cellular events by trimming (poly) ubiquitin conjugates and recycling ubiquitin monomers.

1.2.1 Ubiquitin modifications

The functional consequences of ubiquitylation vary because of recognition by different ubiquitin-binding modules which can distinguish different polyubiquitin modifications. Monoubiquitylation is the conjugation of a single ubiquitin molecule to a single Lys of the target protein. Multimonoubiquitylation occurs when a target protein is tagged with more than one single molecule of ubiquitin. In polyubiquitylation, the target protein is tagged with an ubiquitin chain linked through the *C*-terminal Gly of each ubiquitin unit and a specific internal Lys of the previously attached ubiquitin through sequential rounds of ubiquitylation (**Figure 1**). The presence of *N*-terminal

Met including seven Lys residues per ubiquitin moiety on which polyubiquitylation occurs, empowers ubiquitin with potential to exhibit diverse and highly complicated linkage specific post-translational modification of target proteins [29]. Unlike homogenous chains of ubiquitin which contain a single ubiquitin-ubiquitin linkage type, heterogeneous polyubiquitin chains contain more than one linkage type. In mixed polyubiquitin chains therefore, one linkage type can be extended by a second type, forming a non-branched structure. In branched polyubiquitin chains however, different linkage types form one or more branches (i.e., multiple Lys residues in the same ubiquitin) [29].

Monoubiquitylation has been implicated in the endocytic trafficking of certain cargo proteins, e.g., small GTPases and receptors (e.g., Epidermal Growth Factor Receptor, EGFR) to specific cellular compartments at different stages of the endocytic pathway. Monoubiquitylation has also been implicated in gene expression and DNA repair [30–33]. Multimonoubiquitylation is important for receptor endocytosis [34]. Lys48-linked polyubiquitylation (i.e., polyubiquitin chains linked via Lys48 of the proximal ubiquitin to the next ubiquitin moiety in the chain) has predominantly been linked to targeting proteins for proteasomal degradation [35]. Lys63-linked polyubiquitin chains function as scaffolds to assemble signalling complexes e.g., activation of transcription factor NF-κB involved in inflammatory and immune response, DNA damage tolerance, the endocytic pathway and ribosomal protein synthesis [36]. It has been demonstrated that unanchored (substrate-free) Lys63-linked polyubiquitin chain assembled via UBE2N/UBE2V1 (E2) and TRAF6 (E3) enzymes activate the *NF-κB* pathway by activation of TAK1 which in turn phosphorylates and activates IκB kinase (IKK) [37]. Linear linkage via *N*-terminal Met is also reported to regulate NF-κB signalling [38–40].

2. Types of E3 ubiquitin ligases

Based on considerations of structure, chemistry, and mechanisms by which ubiquitin is transferred to the substrate, four families of E3 ubiquitin ligases are distinguishable. These include homologous to the E6AP carboxyl terminus (HECT), Really Interesting New Gene (RING)-finger/U-box, RING-between RING (RBR) and the recently characterised plant homeodomain (PHD)-finger ligases.

2.1 RING-finger E3 ligases

The RING E3s constitute the largest family of E3 ligases. RING-finger E3 family of ligases include the U-box ligases. Although able to promote the formation of polyubiquitin chains, RING-Finger/U-box E3s lack a catalytic site and hence, do not participate directly in catalysis. RING-type E3s are characterised by the presence of the canonical Cys3HisCys4 amino acid motif (i.e., the RING domain) and consequently mediate the direct transfer of ubiquitin from E2 ~ ubiquitin complex to the substrate (**Figure 2a**). Each canonical Zn finger (Cys3HisCys4) type domain binds two Zn^{2+} ions which are critical to its stability. RING E3 ligases can exist and act as single-subunits e.g., CHIP/Stub1, mdm2, RNF4, RNF114, UBE4B (**Figure 2a**) or may be assembled on a Cullin scaffold to form multiple subunits. Multi-subunit RING E3s contain a substrate receptor (responsible for substrate specificity) at the *C*-terminus, an adaptor(s), a Cullin- and a RING-box at the *N*-terminus (**Figure 2b**). The APC (anaphase-promoting complex) and the SCF (Skp1-Cullin-F-box protein) complex are examples of multi-subunit RING

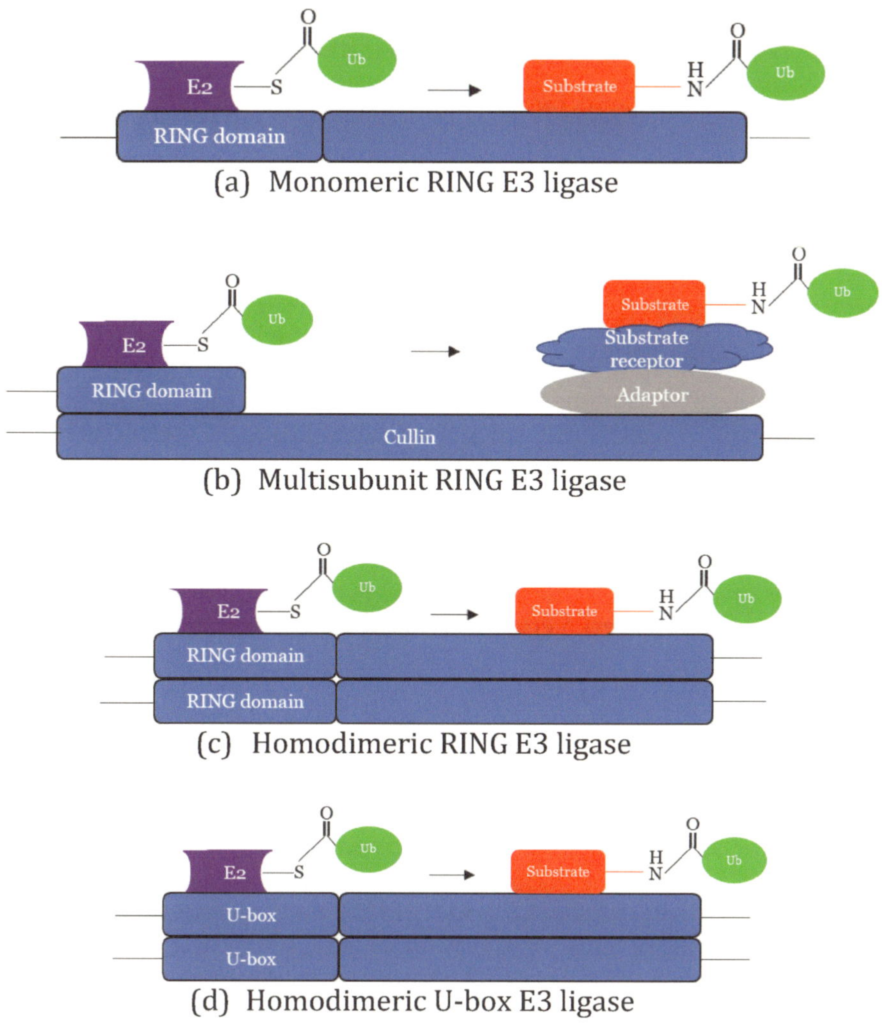

Figure 2.
Illustration of RING/U-box E3 ligase types and their mechanism of ubiquitin transfer to target proteins.

E3s involved in substrate recognition and are the most abundant type of RING E3s [41, 42]. RING-finger E3 ubiquitin ligases regulate several cellular processes, including cell metabolism, cell proliferation, apoptosis, differentiation, and DNA repair making them potential targets for anti-cancer drug development. There is evidence to suggest that RING-finger domains can also allosterically activate the E2 enzymes [43]. RING E3 ligase activity is often regulated by neddylation, phosphorylation and protein–protein interactions with small molecules among others.

2.1.1 U-box E3 ligases

The U-box proteins contain a U-box domain of ~70 amino acids which lacks the characteristic Zn chelating Cys and His residues in RING-finger domain and are characteristically stabilised by a network of H-bonds within each loop, flanked by a central α-helix [44–46]. U-box E3s are more abundant in plants than animals [47].

Both RING and U-box domains are responsible for binding the ubiquitin-charged E2 and stimulating the transfer of ubiquitin to substrate (**Figure 2**). Additionally, RING and U-box E3 ligases can function as monomers (**Figure 2a**), homodimers (**Figure 2c** and **d**), or heterodimers. In a homodimer, each monomer can bind an E2, but apparently not the case in the heterodimeric RINGs.

2.2 PHD-finger E3 ligases

Another RING-related family of E3s are the PHD E3 ligases. Unique sequence and structural signatures that distinguish the PHD-finger from RING fingers have been demonstrated indicating that the PHD-fingers function primarily as E3 ligases that promote protein degradation and constitute a distinct class of E3 ligases. The PHD or leukaemia-associated-protein (LAP) domain resembles the RING finger domain [48–50]. It also has the eight conserved metal binding ligands, Cys4HisCys3 consensus, with similar spacing [51] however, it represents a variant of the RING finger. An example of a PHD domain E3 ligase is Mekk1 kinase. The second *N*-terminal Cys-rich domain of Mekk1 kinase has been shown to exhibit E3 ubiquitin ligase activity toward ERK1/2 and is involved in the down regulation of the MAP kinase cascade [52]. The PHD domain is found in many proteins involved in chromatin-mediated transcriptional regulation [53], however, very little is known about their precise functions.

2.3 HECT E3 ligases

Unlike the RING-finger E3s, the HECT type E3s instead contain a conserved *C*-terminus HECT domain which consists of a larger *N*-terminal bi-lobe architecture encompassing the E2-binding site and a smaller *C*-terminal lobe which comprises the active-site Cys residue (**Figure 3a**). Their reaction cycle consists of three steps; binding to an E2 ~ ubiquitin, transiently loading ubiquitin on themselves via formation of a covalent ubiquitin-thioester linked intermediate with the catalytic Cys, before transferring the ubiquitin molecule onto the target protein [54]. These two lobes are connected by a flexible hinge region, which is critical for juxtaposing the catalytic Cys residues of the E2 and E3 during ubiquitin transfer (**Figure 3a**). A conformational change involving an alteration in the relative orientation of the two lobes is thought to facilitate the transthiolation reaction [54–56]. Whereas the *C*-terminal HECT domain is responsible for E3 catalytic activity, the *N*-terminal portion is highly variable and determines the substrate specificity. HECT domain containing E3 ligases are estimated at 30 in mammals and 28 in humans out of the over 600 E3s [57]. Human HECT E3s can further be categorised into subfamilies based on the mode of protein–protein interactions of their *N*-terminal domain extensions which determine their substrate specificity. These include the Nedd4 family, which are characterised by the presence of tryptophan-tryptophan (WW) motifs, the HERC (HECT and RCC1-like domain) family, which contain one or more regulators of chromosome condensation 1 (RCC1)-like domains (RLDs), and the SI (ngle)-HECT/"other" HECT E3s lacking either WW or RLDs domains but contain various other domains. These subfamilies have been extensively reviewed [58]. HECT domain E3s play several roles. They determine the specificity of ubiquitylation and mediate the trafficking of many receptors. They are also regulators of immune response and several signalling pathways in cell proliferation [59]. Naturally, the intrinsic catalytic activity of HECT E3s is normally folded together in an autoinhibited state by intramolecular interactions between domains which can be released to an active form of the enzyme in response to various signals by unfolding to expose the catalytic Cys.

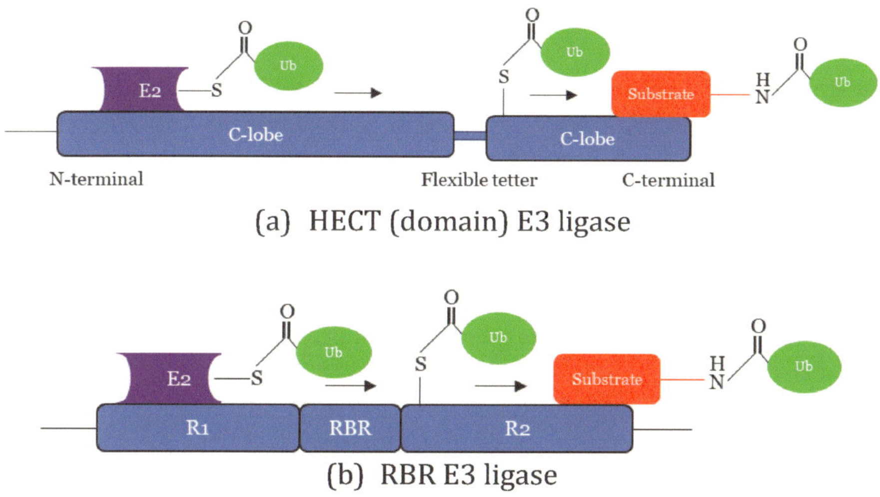

Figure 3.
Mechanism of ubiquitin transfer to target protein by (a) HECT and (b) RBR E3 ubiquitin ligases.

2.4 RBR E3 ligases

RBR E3 ligases employ the characteristic RING-HECT hybrid mechanism [60–65]. Like the HECT E3 ligases, the RBR E3 ligases catalyse the transfer of ubiquitin from E2 to the substrate through a three-step reaction where the RBR E3 first binds the E2, transfers the ubiquitin load to its catalytic Cys and subsequently to the substrate (**Figure 3b**). RBR E3 ubiquitin ligases differ from RING-type E3 mainly because they possess an active site which is absent in other RING-type E3s. However, like RING E3s, the RBR E3 ligases have four RING Zn^{2+} finger domains. Each of these domains coordinates two Zn^{2+} ions through His and Cys residues. They include the canonical Cys3HisCys4-type RING (named RING1) domain as in RING E3s, that binds the E2 enzyme followed by; in-between-RING (IBR) domain and RING2 domain which contains the active site Cys residue (**Figure 3b**). The name RBR was derived from the presence of two predictable RING1 and RING2, separated by an IBR (i.e., the RING1-IBR-RING2 module). Though RING2 domain possesses the catalytic Cys (**Figure 3b**), it does not conform to the canonical RING E3 structure, and it has also been called Rcat (required-for-catalysis) domain. The IBR domain is conserved among RBR E3 family of ligases. Its specific function remains elusive. The IBR domain adopts the same structural fold as RING2 domain, however, it lacks the essential catalytic Cys residue and is sometimes referred to as the BRcat (benign-catalytic) domain [66]. The Zn^{2+} ions bound within RING domains are also reported to be essential for structural stability and proper regulation of its intrinsic enzymatic activity. Their removal from parkin for instance, result in near complete unfolding of the protein [67, 68]. HHARI and Parkin were initially characterised to have the hybrid mechanism [64]. TRIAD1, RNF144A, HOIP, and HOIL-1 L have later been characterised to employ the same RING-HECT hybrid mechanism [69]. RBR RING domains are also normally involved in intramolecular interactions between amino acids of different domains that keep the enzyme in a coiled autoinhibited state. Through different molecular mechanisms such as phosphorylation and protein–protein interactions, the uncoiling of closed-compact or folded autoinhibited states of RBR E3 ligases may be triggered thereby exposing the catalytic sites and increasing the intrinsic E3 ligase activity.

3. Implication of the E3 ubiquitin ligases in disease and therapy

3.1 Role of E3 ligases in disease

Aberrations of ubiquitin signalling are often associated with pathogenesis of several diseases and genetic disorders [58, 70–73]. Errors in ubiquitin signalling processes result in defective autophagy and mitophagy, DNA repair mechanisms, NF-*k*B signalling, etc. [74]. Subsequently, associated diseases including Parkinson's, Amyotrophic Lateral Sclerosis, Alzheimer's, cancer, systemic lupus erythematosus, rheumatoid arthritis, and inflammatory bowel diseases (e.g., Crohn's disease and ulcerative colitis) among several others may ensue [75, 76]. Ubiquitin E3 ligases, most notably RING finger and RING finger-related E3s are fundamental to the specificity of the ubiquitin proteasome system. Many RING finger E3s are implicated in either the suppression or the progression of cancer and cancer chemoresistance [58, 75, 77, 78]. Due to limited space only some key E3-linked diseases have been explained below.

Parkin, a RBR E3 ligase functions in the covalent attachment of ubiquitin to specific substrates e.g., outer mitochondrial membrane proteins – Mfn1, Mfn2, and Miro GTPases [62]. Parkin is involved in protein degradation, collaborating with the ubiquitin-conjugating enzyme, UbcH7 [79, 80]. Even though much of the aetiology of Parkinson's disease remains largely unknown, malfunctioning of PINK1 and/or parkin causes accumulation of damaged mitochondria, which trigger familial parkinsonism. Parkinson's disease (named after Dr. James Parkinson, AD1783–1824) is a neurodegenerative movement disorder caused by the progressive death of dopamine producing neurons in the substantia nigra pars compacta of the mid-brain [81]. The characteristic symptoms of Parkinson's disease include muscle tremor, muscle rigidity, slowness of movement (bradykinesia) and postural instability [82–85]. Parkinson's disease is the second most common neurodegenerative disease after Alzheimer's, affecting predominantly, individuals above 65 years even though earlier onset have been reported [79, 84, 86, 87]. Early-onset Parkinson's disease (EOPD - onset in individuals before 50 years) which account for approximately 5–10% of all cases of Parkinson's disease are attributable to monogenic causes [88, 89]. Mutations in coding regions of *PARK2* gene are often implicated as the most common cause of autosomal-recessive juvenile Parkinsonism (AR-JP) (EOPD), followed by *PINK1* gene variants resulting in a loss of kinase function which apparently, is required for parkin phosphorylation and subsequent activation [84, 90–93]. This renders PINK1 and parkin proteins vital targets for drug development. Greater than 120 pathogenic Parkinson's disease mutations are spread throughout parkin's four domains, indicating the critical functions for each of the individual domains [94].

The multi-functional pathways regulated by RING-type E3 ubiquitin ligases in inflammatory signalling and consequential inflammatory bowel disease have been expansively reviewed by several researchers [73]. Inflammatory bowel disease is characterised by inflammation of the digestive tract and patients present with anomalies in gut microbiota composition e.g., increased levels of harmful bacterial strains, reduced levels of bacterial diversity and protective probiotics. These trigger proinflammatory intestinal pathogenic immune responses which in turn induce intestinal mucosal inflammation. Examples of implicated RING E3 ligases in inflammatory bowel diseases include, TRAF5, TRAF3, TRAF2, UHRF1, RNF183, RNF40, RNF20, RNF170, and RNF186 [73]. For instance, RNF170 E3 ligase ubiquitylates TLR3 for proteasomal degradation. TLR3 is a pattern recognition receptor which recognises pathogen-associated molecular patterns (PAMPs) of lipopolysaccharides, flagellin, and microbial nucleic acids and triggers activation of downstream effectors in innate immune responses. Proteasomal degradation of TL3 therefore suppresses TLR3-mediated innate immunity in macrophages thereby promoting inflammatory diseases [95]. Several sites of the NF-*k*B pathway are regulated by ubiquitylation. NF-*k*B constitutes a family of conserved transcription factors well known to regulate several cell processes particularly inflammatory responses, cell proliferation and apoptosis. NF-*k*B signalling is therefore associated with several diseases such as asthma, arthritis, cancer, etc. [96]. For instance, RNF183 upregulates proinflammatory responses via NF-κB signalling by ubiquitylating IκBα- the inhibitor of c-Rel/p50 heterodimer, for proteasomal degradation which may induce intestinal mucosal inflammation [97]. Receptors that stimulate the NF-*k*B pathway following various stimuli include CD30, CD40, RABK, TNF-α, TCR, and TLRs. In contrast, TRAF2 and TRAF3 catalyse Lys48-linked ubiquitylation of c-Rel and interferon regulatory factor 5 thereby signalling their degradation by the proteasome. This in turn inhibits biosynthesis of proinflammatory cytokines, thereby down regulating immune responses in macrophages [98, 99].

The roles of E3 ubiquitin ligases in the RTK pathway (e.g. EGFR) and MAPK pathway and other oncogenic/tumour suppressive signalling pathways in glioblastoma have been expounded including validation of the potential of E3 ligases as future therapeutic interventions for glioblastoma treatment [72]. Glioblastoma is a malignant brain tumour which is characterised by a mutation in the metabolic enzyme isocitrate dehydrogenase 1 with limited treatment options [100]. In glioblastoma, 67% of cases have mutation in at least 1 RTK and about 20% of classical tumours express a truncated form of EGFR (EGFRvIII). RTK signalling is down regulated by several E3 ligases such as Cbl, Chip and parkin [72]. Ubiquitylation of EGFR by Cbl E3 ligase results in clathrin-mediated internalisation of the receptor and subsequent sorting into lysosomes where the receptor is degraded and therefore reduces EGFR signalling in glioblastoma. LZTR1 is the substrate recognition domain of a Cul3 E3 ubiquitin ligase complex. Mutations in LZTR1 are associated with schwannomatosis and Noonan syndrome in which loss of LZTR1 function drives de-differentiation and proliferation of cells. LZTR1 is also mutated or deleted in about 4% or 20% of glioblastoma cases respectively, where mutations of LZTR1 increase Ras-dependent proliferation of cells coupled with increased resistance to tyrosine kinase inhibitors (glioblastoma chemotherapy) because of enhanced MAPK signalling [101, 102]. The MAPK pathway is a commonly mutated pathway in human cancers. It upregulates cellular phenotypes such as proliferation, differentiation, migration, and invasion.

The BRCA1 RING-type E3 ubiquitin ligase is a human tumour suppressor gene and plays critical roles in DNA repair. Mutation of BRCA1 is associated with the inherited predisposition for breast and ovarian cancers [102]. The E3/E4 ubiquitin ligase, mdm2 is an important negative regulator of the p53 tumour suppressor gene as mentioned earlier. p53 protein regulates the cell cycle, DNA repair and induces cell apoptosis, hence it functions as a tumour suppressor. Mdm2 serves as an E3 ubiquitin ligase of p53. This implies that increased activity of the mdm2 oncogenic protein via augmented mdm2 expression induces tumorigenicity especially those 50% that retain wild-type p53. In addition, inactive mdm2 results in increased cellular levels of p53, which is detrimental to cells and may accelerate the ageing process by excessive apoptosis [103]. It has also been reported that SIAH2 is a RING-finger E3 ligase which contributes to the progression of various malignant tumours, including breast carcinoma, lung and prostate tumours, and oral cancer [104–106]. Furthermore, the E3 ligase, c-Cbl has also been demonstrated to be frequently dysregulated in myelodysplastic myeloproliferative neoplasms and additionally associated with myelodysplastic syndromes, myeloid neoplasms, and primary colorectal cancer [75].

Multiple studies have implicated several E3 ubiquitin ligases in ovarian cancer chemoresistance which hamper improvement of ovarian cancer patient outcome through degradation of various chemoresistance-related substrates in ovarian cancer. Among several studies, [107] demonstrated that NEDD4-2 protein (an E3 ligase that regulates endocytosis and lysosomal degradation of ENaC and other channels) expression is reduced in invasive ovarian epithelial cancer tissues compared with non-cancer ovarian tissue suggesting an important role of NEDD4-2 in the regulation of chemoresistant ovarian cancer [58]. Indeed Nedd4-1 and Nedd4-2 E3 ligases have been demonstrated to mediate numerous pathophysiological processes [108]. Briefly, Nedd4-1 mediates endosomal trafficking of receptor tyrosine kinases, such as EGFR and fibroblast growth factor receptor (FGFR), by ubiquitylating endocytic or vesicle sorting proteins. Nedd4-1 is overexpressed in lung epithelial cells and is associated with lung cancer progression [109]. NEDD4-1 can promote Alzheimer's disease by weaking synaptic strength through ubiquitylation of AMPAR (alpha-amino-3-hydroxy-5-methyl-4-isoxazolepropionic acid receptor) and cytoplasmic internalisation. Nedd4-2 knockout mice reportedly die perinatally due to failure of the pups to breathe resulting from increased ENaC expression, extensive fibrosis, and infiltration of inflammatory cells in the alveolar spaces [110]. This indicates the important regulatory functions of Nedd4-2 in the respiratory system. Ubiquitylation of ENaC by Nedd4-2 E3 ligase through direct binding to its conserved proline-rich PY motif in the *C*-terminals results in sodium current decrease and associated lung oedema [111]. UBR5/EDD is an E3 ligase that mediates the accumulation of ubiquitylated-H2A and -H2AX at DNA damage sites, and G2 checkpoint control. UBR5 is reported to enhance cell survival and cisplatin resistance by regulation of expression of the pro-survival protein myeloid cell leukaemia sequence 1 (Mcl-1) and is an undesirable prognostic factor for patients with serous epithelial ovarian cancer [112]. UBR5 influences ovarian cancer cell cisplatin resistance by mediating MOAP-1 ubiquitylation and degradation through cooperation with Dyrk2 kinase [113].

BCL6 proto-oncogene is the transcription factor implicated in the pathogenesis of human B-cell chronic lymphocytic leukaemia. Deregulation of BCL6 expression or increased degradation are pathogenic events in many lymphomas. Mutation of the multi-subunit SCF FBXO11 E3 ligase is associated with the development of B-cell lymphomas. FBXO11 is the substrate recognition component of the ligase and functions as a tumour suppressor by targeting BCL6 transcription factor for proteasomal degradation [114, 115]. BCL6 binds to specific DNA sequences and down regulates transcription of a variety of genes involved in B-cell development, differentiation and activation including the transcription of STAT-dependent IL-4 responses of B-cells [116].

FANCL is a multisubunit E3 ubiquitin-protein ligase that monoubiquitylates FANCD2 and FANCI, a key step in the DNA damage repair pathway. FANCL contains a RING finger-like PHD domain with E3 activity [78, 117]. Mutations in 8 out of 13 components of the FANCL complex cause Fanconi anaemia (FA). Individuals with this condition experience severe multiple congenital and haematological abnormalities including predisposition to development of a variety of cancers [118]. Individuals may experience congenital skeletal anomalies especially of thumb and forearm. They also display endocrine abnormalities, short stature correlating to growth hormone production and hyperthyroidism among several other anomalies [119].

Mutations in von Hippel–Lindau (VHL), the substrate recognition component of Cullin RING E3 ubiquitin ligase (CRL2VHL), results in an autosomal-dominant familial VHL syndrome. This implies that a mutation in just one copy of the VHL gene in each cell is enough to increase risk of developing VHL disease. Substrates for CRL2VHL E3 ligase tumour suppressor include the hypoxia-inducible factor (HIF) family of transcription factors (HIF1-3α) which bind the VHL subunit when hydroxylated on two proline residues by prolyl hydroxylase. This results in ubiquitylation of HIF1-3α and subsequent proteasomal degradation which would otherwise accumulate and cause inappropriate upregulation of hypoxia-inducible genes such as Transforming growth factor alpha (TGF-α), Vascular endothelial growth factor A (VEGF-A), etc. leading to hyperangiogenesis in VHL mutant individuals [120]. VHL disease is characterised by benign and malignant tumours mostly clear cell kidney and lung carcinomas [121]. Hemangioblastomas that develop in the brain and spinal cord cause headaches, vomiting, weakness, and a loss of muscle coordination (ataxia). If left untreated, hemangioblastomas may result in blindness, permanent brain damage or death. Pheochromocytomas develop and affect the adrenal glands, which may produce excess hormones thereby causing high blood pressure.

Angelman syndrome is a complex genetic disorder caused by a disruption of the UBE3A gene, which encodes the wild-type ubiquitin ligase E6 associated protein (E6-AP/UBE3) [122]. The disease occurs in individuals with a loss of expression or mutations in E6-AP. Angelman syndrome is characterised by intellectual disability, severe speech impairment, a tendency to jerky movement (ataxia), recurrent seizures (epilepsy) and a small head size [123, 124]. Even though investigations have established that E6-AP plays an essential role in the proteasome-dependent degradation of several identified cellular substrates and therefore promote Angelman syndrome, to date, the molecular mechanisms behind the disease pathology is poorly understood [125]. Identified E6-AP substrates include Sox9, C/EBPα, α-synuclein, p27, promyelocytic leukaemia (PML) tumour suppressor, annexin A1, amplified in breast cancer 1 (AIB1), hHR23A, etc. [126]. This indicates that a functionally defective E6-AP mutant cannot initiate degradation of substrates thereby promoting development of Angelman syndrome.

3-M syndrome is a rare autosomal-recessive growth retardation disorder associated with mutations of the Cullin7 (Cul7) E3 ligase [127]. 3-M syndrome is characterised by severe pre- and postnatal growth retardation, large head circumference, facial dysmorphia and skeletal abnormalities including dwarfism even though affected individuals exhibit normal intelligence [128]. Cul7 is a member of the Cullin family of proteins, which function as scaffolds in the formation of numerous E3 ligases with RING proteins, adaptor proteins and substrate recognition receptors [129]. The specific role and substrates of Cul7 are mostly unknown. Recently however, evidence of CUL7's involvement in pivotal growth-regulatory signalling have begun to emerge. Cul7 interacts with both Skp1-Fbx29 heterodimer and the ROC1 RING-finger protein to form the Cul7 E3 ligase complex which ubiquitylates proteins for proteasomal and lysosomal degradation [130]. The insulin receptor substrate 1 (IRS-1), a critical mediator of the insulin/insulin-like growth factor 1 signalling has been identified as a proteolytic target of the Cul7 E3 ligase [131]. Additionally, mammalian Eag1 potassium (Eag1 K^+) channels, which are widely expressed in the brain, are novel targets of Cul7 E3 ligase [130]. Mutant Eag1 K^+ channels are associated with congenital neurodevelopmental anomalies. Cul7 E3 ligase has also been implicated in the proteasomal degradation of cyclin D1 [132]. Cyclin D1 proto-oncogene is a vital regulator of cell cycle progression from G1 to S phase in several different cell types. Accumulation of cyclin D is associated with development and progression of cancer and deregulation of its expression is linked to resistance to hormone therapy in breast cancer [133]. Consequently, in many cancers, impaired activity of Cul7 is essentially responsible for cellular elevated levels of cyclin D1.

Ubiquilin (UBQLN), a ubiquitin binding protein, is a ubiquitin-like protein that shares a high degree of sequence similarity with ubiquitin across several species. UBQLNs contain both *N*-terminal ubiquitin-like and *C*-terminal UBA (ubiquitin associated) domains and are capable of functionally recruiting E3s and linking to the ubiquitylation machinery to proteasomal degradation of targeted proteins [134, 135]. Mutations in UBQLNs may be associated with lesions in Alzheimer's and Parkinson's disease. UBQLNs have also been shown to modulate accumulation of presenilin proteins, which play an important role in the generation of ß amyloid precursor protein (APP). Presenilin is the sub-component of gamma secretase whose role is endoproteolysis of APP [136]. UBQLN-1 is also a molecular chaperone for APP. Therefore, its presence prevents aggregation of APP [137]. Consequently, low levels of UBQLN-1 increase malformation of APP thereby triggering Alzheimer's disease [138].

The list of diseases associated with E3 ligase dysfunction is not exhaustive. More insights are available in the literature of which each apparently addresses a specified disease in more detail. It is anticipated that new discoveries in this regard will emerge in the future.

3.2 Potential therapeutic benefits of E3 ligases

There remains a significant unmet need for novel therapeutic strategies for genetic disorders such as Alzheimer's, Amyotrophic Lateral Sclerosis, Arthritis, cancer, etc. Owing to the critical cellular signalling role of the ubiquitin system in protein degradation, activation, subcellular localization and beyond, various targets of the ubiquitylation pathway are being earmarked to have potential for development of drugs to treat malignant cancers among several other diseases. Among these targets include the E3 ubiquitin ligases [41]. So far, some E3 ligase inhibitors including Velcade (bortezomib), Ninlaro (ixazomib), and Kyprolis (carfilzomib) have proved effective and have been approved by the US FDA for the treatment of multiple myeloma suggesting that specific E3 inhibitors are promising in anti-cancer therapy and beyond. This breakthrough has inspired researchers to probe more aspects of the ubiquitylation system for therapeutics [139]. Ideally, a potential cancer treatment target should be playing an essential role in carcinogenesis, it should overexpress in cancer cells and its activity inhibition or expression should induce growth suppression and/or apoptosis in cancer cells [140]. In addition, an ideal druggable candidate should be an enzyme or a G Protein Coupled Receptor (GPCR) and therefore druggable. More importantly, it is either not expressed or is only expressed at low levels in normal cells and its inhibition has minimal effect on normal cell growth and function. Inhibition of such a target would achieve a maximal therapeutic index with minimal toxicity [140]. Being enzymes and mostly natural tumour suppressors, many E3 ligases function within these criteria and are therefore attractive targets for therapeutic intervention of cancers. Several studies have explored and validated the therapeutic potential of E3 ligases [58, 71, 72]. In this section, the rationale for selected E3 ligases with pharmacological potential have been reviewed.

Firstly, one therapeutically significant pathway is the PINK1-parkin cascade for mitochondrial quality control. PINK1 and parkin appear to offer multiple therapeutic targets for the treatment of Parkinson's disease and several other neurodegenerative disorders [141, 142]. Unidentified and putative members of this cascade may also be therapeutically relevant. Parkin displays low basal activity under normal physiological conditions [82]. Any means by which wild-type parkin activity may be increased for instance by development of small molecule activators could be significant in promoting mitophagy to downregulate progression of Parkinson's disease and other PINK1-parkin pathway-mediated neurodegenerative diseases. In *PINK1* mutant patients, PINK1 by-pass biomolecules that either mimic phospho-Ser65-ubiquitin or can disrupt the serpentine autoinhibited tertiary structure of parkin could augment its E3 ligase activity and hence, its neuroprotective functions [143]. Investigations revealed that primary neuron cells and post-mortem brain tissue from Parkinson's disease patients carrying pathogenic mutant PINK1 were largely devoid of phospho-Ser65-ubiquitin signal because phosphorylation of ubiquitin at Ser65 is dependent on PINK1 kinase activity [144]. Their data further indicated that phospho-Ser65-ubiquitin accumulates with stress, disease or age in individuals bearing functional PINK1 and therefore highlight the significance of phospho-Ser65-ubiquitin as a potential candidate for the development of biomarkers and therapeutics. *In vitro* mutagenic investigations have shown that mitochondrial depolarisation and PINK1-dependent recruitment of parkin to mitochondria is significantly enhanced by mutations at Trp403 in repressor element (REP) or Phe463 in RING2 of parkin [145]. This knowledge is pharmacologically relevant. The REP linker of parkin is anchored through Trp403 with RING1 and prevents E2 binding to its site on RING1. REP must dissociate from RING1 to enhance E2 binding. In addition, RING0 must dissociate from RING2 to expose the active site Cys431 [62, 146]. Therefore, any biomolecule capable of binding tightly to the pockets occupied by the amino acid side chains can disrupt parkin's autoinhibited structural state and induce UbcH7 binding and discharging thereby enhancing E3 ligase activity. Scientists have also shown that the DUBs, USP30 and USP15 function antagonistically to the PINK1-parkin dependent mitophagy making inhibitors of these enzymes prime candidates for designing drugs that will enhance PINK1-parkin dependent mitophagy [147, 148]. In contrast, USP8 promotes parkin-mediated mitophagy and thus agonists of this DUB will enhance mitophagy and neuroprotection [149]. Mutations of PINK1's kinase domain are also observed with Parkinson's disease patients where parkin translocation to mitochondria or mitochondrial aggregation does not occur [95].

As previously mentioned in Section 3.1, mdm2 is a direct downstream target of p53. P53, is a well-studied tumour suppressor which is often mutated in more than 50% of human cancers. This is because p53 induces growth arrest and apoptosis upon activation by different stimuli e.g., DNA damage [150]. Upon induction by p53, mdm2 in turn acts as an E3 ubiquitin ligase to ubiquitylate p53 for proteasomal degradation. This action reduces p53 levels and consequently inhibits p53-mediated cell apoptosis [151]. As such, inhibition of mdm2 E3 ligase activity is a potential approach to increasing p53 levels in order to induce cell apoptosis in human cancer cells harbouring wild-type p53. Mdm2 is usually expressed in low levels in normal cells but is overexpressed in several human cancers such as breast carcinomas, soft tissue sarcomas, oesophageal carcinomas, lung carcinomas, glioblastomas and malignant melanomas and will represent an excellent pharmacological candidate for further research [152].

Pirh2 has recently been found to be a major E3 ligase partnering mdm2 to target (tetrameric)p53 for proteasomal degradation. Further, the p53-Pirh2 complex promotes Twist1 degradation leading to inhibition of epithelial-mesenchymal transition in ovarian cancer [153]. Epithelial-mesenchymal transition is critical in cancer metastasis and chemoresistance implying that Pirh2 specific inhibition might be therapeutically relevant. Some IAPs (Inhibitor of Apoptosis Proteins) that are overexpressed in most common human cancers (correlating chemoresistance) also represent important therapeutic targets for drug development [154]. Examples of IAPs include XIAP, cIAP-1, cIAP-2, Ts-IAP, NAIP, Survivin, Livin/ML-IAP, and Apollon/Bruce. IAPs are characterised by the presence of BIR (baculoviral IAP repeat) domain(s) required for suppression of apoptosis. Additionally, some members of the family may contain a RING finger domain at the *C*-terminus essential for ubiquitylation and subsequent proteasomal degradation of the apoptosis inducer proteins, caspases and Smac [155]. Indeed, it has been demonstrated that overexpression of IAPs suppress apoptosis [156]. Therefore, IAPs contribute to their anti-apoptosis function by promoting proteasomal degradation of pro-apoptotic caspases and Smac proteins. Targeting specific IAPs' E3 ubiquitin ligase activity inhibition thus, has potential for anti-cancer drug development [140]. Additionally, research has shown overexpression of components of SCF of RING E3 ligases in several human cancers. For instance, Cul4A is reported to be overexpressed in breast cancers [157]. Skp2 is likewise overexpressed in other human cancers e.g., squamous, colorectal, gastric, and prostate carcinomas, small cell lung carcinoma, and breast cancer. Skp1-Cullin-F Box components are the substrate recognition sites of the RING E3 ligases and recognise a variety of substrates involved in critical cellular processes such as the cell cycle. Examples of SCF targets for ubiquitin-mediated proteasomal degradation are p27 and cyclin E which are down regulators of different sites of the cell cycle [44]. Furthermore, the role of Nedd4-1 in lung cancers has been studied (see 3.1 above). NEDD4-1 expression is up-regulated in lung adenocarcinoma compared with normal cells. It has also been demonstrated that NEDD4-1 silencing reduces non-small cell lung cancer cells *in vitro* as well as tumour growth *in vivo* [158]. Overexpression of Nedd4-1 is associated with lymph node metastasis and chemoresistance, as such NEDD4-1 is a potential drug target since specific inhibitors of NEDD4-1 will likely promote cancer cell apoptosis. NEDD4-1 E3 ligase activity inhibition therefore has potential for lung cancer treatment [108].

Cbl E3 ligase (Section 3.1) ubiquitylates EGFR thereby reducing EGFR signalling in glioblastoma. Specific inhibitors of Cbl could helpfully target upregulating EGFR signalling in anti-glioblastoma therapy. As the MAPK pathway is commonly mutated in many cancers resulting from increase in cellular proliferation, differentiation, migration, and invasion, specific inhibitors of E3 ligases of the MAPK pathway (e.g., TRIM9, SCFFBXO31, KBTBD7, LZTR1) may have potential for anti-cancer therapy [72].

Research suggests that ubiquitylation of Plasmodium falciparum proteins play essential roles in parasite development. Recent data indicate ubiquitylation of several essential proteins (e.g., merozoite pellicle proteins involved in erythrocyte invasion, exported proteins, and histones) of the human malaria parasite, *Plasmodium falciparum*, which suggest potential for the use of small-molecule inhibitors of the ubiquitin-mediated degradation machinery for the development of anti-malarial drugs [159]. The data further showed that some commercially available inhibitors of the ubiquitylation process e.g., the UBA1 inhibitor MLN7243, is a potent inhibitor and blocked schizont differentiation into merozoites by interrupting nuclear division and intracellular structural formation. Identification of the associated E3 ligases will predictably present a varied number of druggable targets with potential for malaria treatment [159]. More research will unveil the most relevant E3s to target in this regard.

Finally, besides small-molecule E3 ligase inhibitors, PROTACs (PROteolysis-TArgeting Chimeras) appear to have significant therapeutic potential. Reports indicate that small-molecule inhibitors have limitations [160]. For instance, small-molecule inhibitors are limited to molecules that have an active site (enzymes and receptors e.g., RTKs). The PROTAC technology has emerged to overcome these limitations and to facilitate the 75% of human proteome e.g., transcription factors, scaffolding proteins, and non-enzymatic proteins which are signal effector proteins but lack active sites and are thus undruggable. The emerging characteristics of PROTACs such as induction of substrate selectivity, rapid, profound, and sustained proteasomal degradation and consequential induction of robust inhibition of downstream signals coupled with overcoming resistance to small molecule inhibitors have been extensively reviewed [72, 160]. A PROTAC is a small heterobifunctional molecule consisting of an E3 binding domain and a substrate binding domain covalently joined together by a linker. This spatial arrangement enhances recruitment of the E3 enzyme in proximity with the specific substrate (e.g., an oncoprotein) for ubiquitylation and subsequent proteasomal degradation thereby inhibiting downstream signals and subsequently down regulating the cell cycle or inducing apoptosis. PROTAC technology therefore will utilise the E3-mediated ubiquitin proteasome mechanistic pathway to treat disease. It is a very promising alternative technique where E3 inhibitors are limited or will present less efficiency.

4. Conclusion and perspectives

The biological importance of E3 ubiquitin ligases cannot be overemphasised because protein ubiquitylation is crucial in the regulation of numerous cellular processes. E3 ubiquitin ligases have therefore recently emerged as significant future therapeutic opportunities for drug development for treatment of several human diseases associated with ubiquitin-mediated signalling. Regrettably, the mechanisms by which the ubiquitin system regulates cellular signalling and pathogenesis remain largely unknown. Many questions remain unanswered considering the number of E3 ligases (over 600) in the human genome and lack of the most relevant technologies to assess these principles, coupled with the extreme complexity of ubiquitin signalling processes. Blocking protein–protein interactions is problematic, yet it is apparently the most effective treatment option for utilising the ubiquitin system. This option relies on blocking the E3 ligase at specific substrate recognition sites. Hypothetically, targeting rapid screening of small specific molecular inhibitors of E3s which have potential to selectively stabilise specific downstream cellular proteins regulated by specific E3s while avoiding unwanted effects on other cellular proteins will achieve less toxicity. Therefore, a complete understanding of the mechanisms involved in protein substrate recognition by E3 ligases and functions, as well as easy identification of aberrant entities within the ubiquitin pathway will be instrumental in understanding the aetiology of associated diseases. With the current advances in proteomics technology, more E3 substrates are being identified and more insights to E3 regulatory roles in many diseases are being better understood. Though new technologies such as the siRNA, for validation of many E3s, the Fluorescence Resonance Energy Transfer (FRET) for High throughput (HTS) assays for screening inhibitors of ubiquitin transfer from E2s to E3s, and electrochemiluminescence (ECL)-Based HTS for screening inhibitors of the ubiquitylation machinery have facilitated screening of compounds against E3s and improved research in this aspect so far, these methods are themselves not devoid of challenges. Development of more efficient, cheaper, and simpler techniques will fast-track understanding of the ubiquitin system and the drug discovery process. Predictably, E3 ubiquitin ligases will present one of the most efficacious targets for anti-cancer drug discovery and for other diseases in the future.

Acknowledgements

We thank IntechOpen Ltd. UK, for an opportunity to contribute to a book project.

Conflict of interest

The authors declare no conflict of interest.

Author details

Julius Tieroyaare Dongdem[1*] and Cletus Adiyaga Wezena[2]

1 Department of Biochemistry and Molecular Medicine, School of Medicine, University for Development Studies, Tamale, Ghana

2 Department of Physiology and Biophysics, School of Medicine, University for Development Studies, Tamale, Ghana

*Address all correspondence to: julius.dongdem@uds.edu.gh

References

[1] Hershko A, Ciechanover, A. The ubiquitin system. Annu Rev Biochem. 1998;67:425-479.

[2] Pickart CM. Ubiquitin enters the new millennium. Mol Cell. 2001;8:499-504.

[3] Kirisako T, Kamei K, Murata S, Kato M, Oto, FH, Kanie M, Sano S, Tokunaga F, Tanaka K, Iwai K. A ubiquitin ligase complex assembles linear polyubiquitin chains. EMBO J. 2006;25:4877-4887.

[4] McDowell GS, Philpott A. Non-canonical ubiquitylation: mechanisms and consequences. Int J Biochem Cell Biol. 2013;45:1833-1842.

[5] Vittal V, Shi L, Wenzel DM, Scaglione KM, Duncan ED, Basrur V, Elenitoba-Johnson KS, Baker D, Paulson HL, Brzovic PS, Klevit RE. Intrinsic disorder drives N-terminal ubiquitination by Ube2w. Nat Chem Biol. 2015;11:83-89.

[6] Roark R, Itzhaki L, Philpott, A. Complex regulation controls Neurogenin3 proteolysis. Biol Open. 2012;12:1264-1272.

[7] Tokarev AA, Munguia J, Guatelli JC. Serine-threonine ubiquitination mediates downregulation of BST-2/tetherin and relief of restricted virion release by HIV-1 Vpu. J Virol. 2011;85:5-63.

[8] Schulman BA, Harper JW. Ubiquitin-like protein activation by E1 enzymes: the apex for downstream signaling pathways. Nat Rev Mol Cell Biol. 2009;10:319-331.

[9] Wijk SJ, Timmers, HT. The family of ubiquitin-conjugating enzymes (E2s): deciding between life and death of proteins. FASEB J. 2010;24(4):981-993.

[10] Dikic I, Robertson M. Ubiquitin ligases and beyond. BMC Biol. 2012;10:22 http://www.biomedcentral.com/1741-7007/10/22

[11] Zheng N, Shabek N. Ubiquitin Ligases: Structure, Function, and Regulation. Annu Rev Biochem. 2017;86:129-157.

[12] Clague MJ, Heride C, Urbe S. The demographics of the ubiquitin system. Trends Cell Biol. 2015;25:417-426.

[13] Li W, Bengtson MH, Ulbrich A, Matsuda A, Reddy VA, Orth A, Chanda SK, Batalo S, Joazeiro CA. Genome-wide and functional annotation of human E3 ubiquitin ligases identifies MULAN, a mitochondrial E3 that regulates the organelle's dynamics and signaling. PLoS ONE. 2008;3:e1487. doi: 10.1371/journal.pone.0001487.

[14] Groettrup M, Pelzer C, Schmidtke G, Hofmann K. Activating the ubiquitin family: UBA6 challenges the field. Trends Biochem Sci. 2008;33:230-237.

[15] Marblestone JG, Butt S, McKelvey DM, Sterner DE, Mattern MR, Nicholson B, Eddins MJ. Comprehensive ubiquitin E2 profiling of ten ubiquitin E3 ligases. Cell Biochem Biophys. 2013;67:161-167.

[16] Grossman SR, Deato ME, Brignone C, Chan HM, Kung AL, Tagami H, Nakatani Y, Livingston DM. Polyubiquitination of p53 by a ubiquitin ligase activity of p300. Sci. 2003;300:342-344.

[17] Shi D, Pop MS, Kulikov R, Love IM, Kung AL, Grossman, SR. CBP and p300 are cytoplasmic E4 polyubiquitin ligases for p53. Proc Natl Acad Sci USA. 2009;106:16275-16280.

[18] Johnson ES. Ubiquitin branches out. Nat Cell Biol. 2002;4:295-298.

[19] Ciechanover A, Heller H, Elias S, Haas AL, Hershko A. ATP-dependent conjugation of reticulocyte proteins with the polypeptide required for protein degradation. Proc Natl Acad Sci USA. 1980;77:1365-1368.

[20] Hershko A, Ciechanover A, Heller H, Haas AI, Rose, IA. Proposed role of ATP in protein breakdown: Conjugation of proteins with multiple chains of the polypeptide of ATP-dependent proteolysis. Prc Natl Acad Sci USA. 1980;77:1783-1786.

[21] Pickart CM, Fushman, D. Polyubiquitin chains: polymeric protein signals. Curr Opin Chem Biol. 2004;8:610-616.

[22] Hoeller D, Dikic I. How the proteasome is degraded. Proc Natl Acad Sci USA. 2016;113:13266-13268.

[23] Klaips CL, Jayaraj GG, Hart FU. Pathways of cellular proteostasis in aging and disease. J. Cell Biol. 2018;217:51-63.

[24] Liu WJ, Ye L, Huang WF, Guo LJ, Xu, .G, Wu HL, Yang C, Liu HF. p62 links the autophagy pathway and the ubiqutin–proteasome system upon ubiquitinated protein degradation. Cell Mol Biol Lett. 2016 21:29 doi 10.1186/s11658-016-0031-z.

[25] Hershko A, Ciechanover A. The ubiquitin system for protein degradation. Annu Rev Biochem. 1992;61:761-807.

[26] Fushman D, Walker O, Exploring the linkage dependence of polyubiquitin conformations using molecular modeling. J Mol Biol. 2010;395:803-814.

[27] Pickart CM, Eddins MJ. Ubiquitin: structures, functions, mechanisms. Biochim Biophys. 2004;1695:55-72.

[28] Amma I, Sommer T, Wolf DH. Protein quality control and elimination of protein waste: The role of the ubiquitin–proteasome system. Biochimica et Biophysica Acta. 2014;1843:182-196.

[29] Swatek KN, Komander D. Ubiquitin modifications. Cell Res. 2016;26:399-422.

[30] Bergink S, Jentsch S. Principles of ubiquitin and SUMO modifications in DNA repair. Nature. 2009;458:461-467.

[31] Haglund K, Di Fiore PP, Dikic I. Distinct monoubiquitin signals in receptor endocytosis. Trends Biochem. Sci. 2003;28:598-603.

[32] Hicke L. Protein regulation by monoubiquitin. Nat Rev Mol Cell Biol. 2001;2:195-201.

[33] Sarcevic B, Mawson A, Baker RT, Sutherland RL. Regulation of the ubiquitin-conjugating enzyme hHR6A by CDK-mediated phosphorylation. EMBO J. 2002;21:2009-2018.

[34] Haglund K, Dikic I. The role of ubiquitylation in receptor endocytosis and endosomal sorting. J Cell Sci. 2012;125(2):265-275.

[35] Thrower JS, Hoffman L, Rechsteiner M, Pickart CM. Recognition of the polyubiquitin proteolytic signal. EMBO J. 2000 19:94-102.

[36] Passmore L, Barford D. Getting into position: the catalytic mechanisms of protein ubiquitylation. Biochem. J. 2004;379 (3):513-525.

[37] Xia ZP, Sun L, Chen X, Pineda G, Jiang X, Adhikari A, Zeng W, Chen ZJ. Direct activation of protein kinases by unanchored polyubiquitin chains. Nature. 2009;461:114-119.1

[38] Ikeda F, Deribe YL, Skanland SS, Stieglitz B, Grabbe C, Franz-Wachtel M, van Wijk SJL, Goswami P, Nagy V, Terzic J, Tokunaga F, Androulidaki A, Nakagawa T, Pasparakis M, Iwai K, Sundberg JP, Schaefer L, Rittinger K, Macek B, Dikic I. SHARPIN forms a linear ubiquitin ligase complex regulating NF-kappaB activity and apoptosis. Nature. 2011;471:637-641.

[39] Tokunaga F, Sakata S, Saeki Y, Satomi Y, Kirisako T, Kamei K, Nakagawa T, Kato M, Murata S, Yamaoka S, Yamamoto M, Akira S, Takao T, Tanaka K, Iwai K. Involvement of linear polyubiquitylation of NEMO in NF-kappaB activation. Nat. Cell Biol. 2009;11:123-132.

[40] Tokunaga F, Nakagawa T, Nakahara M, Saeki Y, Taniguchi M, Sakata S, Tanaka K, Nakano H, Iwai K. SHARPIN is a component of the NF-kappaB-activating linear ubiquitin chain assembly complex. Nature. 2011;471:633-636.

[41] Berdsen CE, Wolberger C. New insights into ubiquitin E3 ligases mechanisms. Nat struct. Biol. 2014; 21 (4) doi:10.1038/nsmb.2780.

[42] Skaar JR, Pagano M. Control of cell growth by the SCF and APC/C ubiquitin ligases. Curr Opin Cell Biol. 2009;21:816-824.

[43] Ozkan E, Yu H, Deisenhofer J. Mechanistic insight into the allosteric activation of a ubiquitin-conjugating enzyme by RING-type ubiquitin ligases. Proc Natl Acad Sci U S A. 2005;102:18890-18895.

[44] Metzger MB, Hristova VA, Weissman AM. HECT and RING finger families of E3 ubiquitin ligases at a glance. J Cell Sci. 2012;125:531-537.

[45] Aravind L, Koonin EV. The U-box is a modified RING finger; a common domain in ubiquitination. Curr. Biol. 2014;10:R132-R134.

[46] Cyr DM, Hohfeld J, Patterson C. Protein quality control: U-box-containing E3 ubiquitin ligases join the fold. Trends Biochem Sci. 2002;27:368-375.

[47] Sharma B, Taganna J. Genome-wide analysis of the U-box E3 ubiquitin ligase enzyme gene family in tomato. Sci. Rep. 2020;10:9581. https://doi.org/10.1038/s41598-020-66553-1.

[48] Saha V, Chaplin, T, Gregorini A, Ayton P, Young BD, The leukemia-associated-protein (LAP) domain, a cysteine-rich mo tif, is present in a wide range of proteins, including MLL, AF10, and MLLT6 proteins. Proc. Natl. Acad. Sci. USA, 1995;92:9737-9741.

[49] Capili AD, Schultz DC, Rauscher IF, Borden KL. Solution structure of the PHD domain from the KAP-1 corepressor: structural determinants for PHD, RING and LIM zinc-binding domains. EMBO J. 2001;20:165-177.

[50] Pascual J, Martinez-Yamout M, Dyson HJ, Wright PE. Structure of the PHD zinc finger from human Williams-Beuren syndrome transcription factor. J. Mol. Biol. 2000;304:723-729.

[51] Borden KL, Freemont PS. The RING finger domain: a recent example of a sequence-structure family. Curr. Opin. Struct. Biol. 1996;6:395-401.

[52] Lu Z, Xu S, Joazeiro C, Cobb MH, Hunter T. The PHD domain of MEKK1 acts as an E3 ubiquitin ligase and mediates ubiquitination and degradation of ERK1/2. Mol Cell. 2002;9:945-956.

[53] Aasland R, Gibson TJ, Stewart AF. The PHD finger: implications for

chromatin-mediated transcriptional regulation. Trends Biochem. Sci. 1995;20:56-59.

[54] Bernassola F, Karin M, Ciechanover A, Melino G. The HECT family of E3 ubiquitin Ligases: Multiple players in cancer development. Cancer Cell. 2008;14:10-21.

[55] Verdecia MA, Joazeiro CA, Wells NJ, Ferrer JL, Bowman ME, Hunter T, Noel JP. Conformational flexibility underlies ubiquitin ligation mediated by the WWP1 HECT domain E3 ligase. Mol. Cell. 2003;11:249-259.

[56] Huang L, Kinnucan E, Wang G, Beaudenon S, Howley PM, Huibregtse JM, Pavletich NP. Structure of an E6AP-UbcH7 complex: insights into ubiquitination by the E2-E3 enzyme cascade. Sci. 1999;286:1321-1326.

[57] Li W, Bengtson MH, Ulbrich A, Matsuda A, Reddy VA, Orth A, Chanda SK, Batalov S, Joazeiro CA. Genome-wide and functional annotation of human E3 ubiquitin ligases identifies MULAN, a mitochondrial E3 that regulates the organelle's dynamics and signaling. PLoS ONE. 2008;3:e1487. doi: 10.1371/journal.pone.0001487.

[58] Meng Y, Qui L, Zhang S, Han J. The emerging roles of E3 ubiquitin ligases in ovarian cancer chemoresistance. Cancer Drug Resist. 2021 4:365-381.

[59] Rotin D, Kumar S. Physiological functions of the HECT family of ubiquitin ligases. Nat Rev Mol Cell Biol. 2009;10:398-409.

[60] Berndsen CE, Wolberger C. (2014). New insights into ubiquitin E3 ligase mechanism. Nat Struct Mol Biol. 2014;21:301-307.

[61] Lazarou M, Narendra DP, Jin SM, Tekle E, Banerjee S, Youle RJ. PINK1 drives parkin self-association and HECT-like E3 activity upstream of mitochondrial binding. J Cell Biol. 2013;200:163-172.

[62] Riley, BE, Lougheed JC, Callaway K, Velasquez M, Brecht E, Nguyen L, Shaler T, Walker D, Yang Y, Regnstrom K, Diep L, Zhang Z, Chiou S, Bova M, Artis DR, Yao N, Baker J, Yednock T Johnston JA. Structure and function of parkin E3 ubiquitin ligase reveals aspects of RING and HECT ligases. Nat Commun. 2013;4:1982 doi: 10.1038/ncomms2982.

[63] Stieglitz B, Morris-Davies AC, Koliopoulos MG, Christodoulou E, Rittinger K. LUBAC synthesizes linear ubiquitin chains via a thioester intermediate. EMBO Rep. 2012;13:840-846.

[64] Wenzel DM, Lissounov A, Brzovic PS, Klevit RE. UBCH7 reactivity profile reveals parkin and HHARI to be RING/HECT hybrids. Nature. 2011;474(7349):105-108.

[65] Wenzel D, Klevit R. Following Ariadne's thread: a new perspective on RBR ubiquitin ligases. BMC Biol. 2012;10:24.

[66] Morreale FE, Walden H. SnapShot: Types of ubiquitin ligases. Cell. 2016;165. http://dx.doi.org/10.1016/j.cell.2016.03.00

[67] Beasley SA, Hristova VA, Shaw GS. Structure of the parkin in-between-ring domain provides insights for E3-ligase dysfunction in autosomal recessive Parkinson's disease. Proc Natl Acad Sci USA. 2007;104(9):3095-3100.

[68] Hristova VA, Beasley SA, Rylett RJ, Shaw GS. Identification of a novel Zn2+-binding domain in the autosomal recessive juvenile Parkinson-related

E3 ligase parkin. J. Biol. Chem. 2009;284(22):14978-14986.

[69] Ho SR, Mahanic CS, Lee YJ, Lin WC. RNF144A, an E3 ubiquitin ligase for DNA-PKcs, promotes apoptosis during DNA damage. Proc Natl Acad Sci USA. 2014;111:E2646-E2645.

[70] Ciechanover A, Iwai K. The Ubiquitin System: From Basic Mechanisms to the Patient Bed. Life. 2004;56:193-201.

[71] Gao F, Fan Y, Zhou B, Guo W, Jiang X, Shi J, Ren C. The functions and properties of cullin-5, a potential therapeutic target for cancers. Am J Transl Res. 2020;12(2):618-632.

[72] Humphreys LM, Smith P, Chen Z, Fouad S, D'Angiolella V. The role of E3 ubiquitin ligases in the development and progression of glioblastoma. Cell Death Differ. 2021;28:522-537.

[73] Zhu L, Li Y, Zhou L, Yang G, Wang Y, Han J, Li L, Zhan S. Role of RING-Type E3 ubiquitin ligases in inflammatory signalling and inflammatory bowel disease. Mediators Inflamm, 2020; Article ID 5310180 https://doi.org/10.1155/2020/5310180

[74] Chen R, Chen Y, Huang T. Ubiquitin-mediated regulation of autophagy. J Biomed Sci. 2019;26:80. doi:10.1186/s12929-019-0569-y.

[75] Popovic D, Vucic D, Dikic I. Ubiquitination in disease pathogenesis and treatment. Nat Med. 2014;20:1242-1253.

[76] George AJ, Hoffiz YC, Charles AJ, Zhu Y, Mabb AM. A Comprehensive Atlas of E3 ubiquitin ligase mutations in neurological disorders. Front. Genet. 2018;9:29. doi: 10.3389/fgene.2018.00029.

[77] Liu L, Wong CC, Gong B, Yu J. Functional significance and therapeutic implication of ring-type E3 ligases in colorectal cancer. Oncogene. 2018;37:148-159.

[78] Lipkowitz S, Weissman AM. RINGs of good and evil: RING finger ubiquitin ligases at the crossroads of tumour suppression and oncogenesis. Nat Rev Cancer. 2013;11(9):629-643.

[79] Shimura H, Hattori N, Kubo S, Mizuno Y, Asakawa S, Minoshima S, Shimizu N, Iwai K, Chiba T, Tanaka K, Suzuki T. Familial Parkinson disease gene product, parkin, is a ubiquitin-protein ligase. Nat Genet. 2000;25(3):302-305.

[80] Zhang Y, Gao J, Chung KKK, Huang H, Dawson VL, Dawson TM. Parkin functions as an E2-dependent ubiquitin–protein ligase and promotes the degradation of the synaptic vesicle-associated protein, CDCrel-1. PNAS. 2000;97(24):13354-13359.

[81] Venderova, K, Park DS. Programmed cell death in Parkinson's disease. Cold Spring Harb Perspect Med. 2012;2:pii:a009365.

[82] Matsuda N, Sato S, Shiba K, Okatsu K, Saisho K, Gautier CA, Sou YS, Saiki S, Kawajiri S, Sato F, Kimura M, Komatsu M, Hattori N, Tanaka K. PINK1 stabilized by mitochondrial depolarization recruits parkin to damaged mitochondria and activates latent parkin for mitophagy. J. Cell Biol. 2010;189:211-221.

[83] Sarraf SA, Raman M, Guarani-Pereira V, Sowa ME, Huttlin EL, Gygi SP, Harper JW. Landscape of the PARKIN-dependent ubiquitylome in response to mitochondrial depolarization. Nature. 2013;496:372-376.

[84] Kitada T, Asakawa S, Hattori N, Matsumine H, Yamamura Y, Minoshima S, Yokochi M, Mizuno Y, Shimizu N. Mutations in the parkin gene cause autosomal recessive juvenile Parkinsonism. Nature. 1998;392:605-608.

[85] Trubana D, Houa X, Caulfielda TR, Fiesela FC, Springera W. PINK1, parkin, and mitochondrial quality control: what can we learn about Parkinson's Disease pathobiology? J Parkinson Dis. 2017;7:13-29.

[86] Durcan TM, Fon EA. The three 'P's of mitophagy: PARKIN, PINK1, and post-translational Modifications. Genes and Dev. 2016;29:989-999.

[87] Koyano F, Okatsu K, Kosako H, Tamura Y, Go E, Kimura M, Kimura Y, Tsuchiya H, Yoshihara H, Hirokawa T, Endo T, Fon EA, Trempe J-F, Saeki Y, Tanaka K, Matsuda N. Ubiquitin is phosphorylated by PINK1 to activate parkin. Nature. 2014;510:162-166.

[88] Martin I, Dawson VL, Dawson TM. Recent advances in the genetics of parkinson's disease. Annu Rev Genomics Hum Genet. 2011;12:301-325.

[89] Nuytemans K, Theuns J, Cruts M, and van Broeckhoven C. Genetic etiology of Parkinson disease associated with mutations in the SNCA, PARK2, PINK1, PARK7, and LRRK2 genes: a mutation update. Hum Mutat. 2010;1(7):763-780.

[90] Abbas N, Lucking CB, Ricard S, Durr A, Bonifati V, Michele G, Bouley S, Vaughan JR, Gasser T, Marconi R, Broussolle E, Brefel-Courbon C, Harhangi BS, Oostra BA, Fabrizio E, Böhme GA, Pradier L, Wood NW, Filla A, Meco G, Denefle P, Agid Y, Brice A. A wide variety of mutations in the parkin gene are responsible for autosomal recessive parkinsonism in Europe. Hum Mol Genet. 1999;8:567-574.

[91] Khan NL, Graham E, Critchley P, Schrag AE, Wood NW, Lees AJ, Bhatia KP, Quinn N. Parkin disease: a phenotypic study of a large series of cases. Brain. 2003;126:1279-1292.

[92] Klein C, Pramstaller PP, Kis B, Page CC, Kann M, Leung J. Woodward, H., Castellan, C.C., Scherer, M., Vieregge, P., Breakefield, X.O., Kramer, P.L. and Ozelius, LJ. Parkin deletions in a family with adult-onset, tremor-dominant parkinsonism: expanding the phenotype. Ann Neurol. 2000;48:65-71.

[93] Valente EM, Abou-Sleiman PM, Caputo V, Muqit MM, Harvey K, Gispert S, Ali Z, Del Turco D, Bentivoglio AR, Healy DG, Albanese A, Nussbaum R, González-Maldonado R, Deller T, Salvi S, Cortelli P, Gilks WP, Latchman DS, Harvey RJ, Dallapiccola B, Auburger G, Wood NW. Hereditary early-onset Parkinson's disease caused by mutations in PINK1. Sci. 2004;304:1158-1160.

[94] Cruts M, Theuns J, van Broeckhoven C. Locus-specific mutation databases for neurodegenerative brain diseases. Hum Mutat. 2012;33:1340-1344.

[95] Song S, Liu W, Wang Z, Cao MX, Jiang M. E3 ubiquitin ligase RNF170 inhibits innate immune responses by targeting and degrading TLR3 in murine cells. Cell. Mol. Immunol. 2020;17(8):865-874.

[96] Albensi, BC. What is Nuclear factor kappa B (NF-*k*B) Doing in and to the mitochondrion. Front. Cell Dev. Biol. 2019; doi:10.3389/fcell.2019.00154

[97] Yu Q, Zhang S, Chao K, Feng R, Wang H, Li M, Chen B, He Y, Zeng Z, Chen M. E3 ubiquitin ligase RNF183 is a novel regulator in inflammatory bowel mediators of inflammation disease, J. Crohn's Colitis, 2016;10(6):713-725.

[98] Jin J, Xiao Y, Hu H, Zou Q, Li Y, Gao Y, Ge W, Cheng X, Sun, S-C. Proinflammatory TLR signalling is regulated by a TRAF2-dependent proteolysis mechanism in macrophages. Nat. Commun. 2015;6(1):5930.

[99] Lee SY. Temozolomide resistance in glioblastoma multiforme. Genes Dis. 2016;3:198-210.

[100] Frattini V, Trifonov V, Chan JM, Castano A, Lia M, Abate F, Castano A, Lia M, Abate F, Keir ST, Ji AX, Zoppoli P, Niola F, Danussi C, Dolgalev I, Porrati P, Pellegatta S, Heguy A, Gupta G, Pisapia DJ, Canoll P, Bruce JN, McLendon RE, Yan H, Aldape K, Finocchiaro G, Mikkelsen T, Privé GG, Bigner DD, Lasorella A, Rabadan R, Iavarone A. The integrated landscape of driver genomic alterations in glioblastoma. Nat Genet. 2013;45:1141-1149.

[101] Steklov M, Pandolfi S, Baietti MF, Batiuk A, Carai P, Najm P, Steklov M, Pandolfi S, Baietti MF, Batiuk A, Carai P, Najm P, Zhang M, Jang H, Renzi F, Cai Y, Asbagh LA, Pastor T, De Troyer M, Simicek M, Radaelli E, Brems H, Legius E, Tavernier J, Gevaert K, Impens F, Messiaen L, Nussinov R, Heymans S, Eyckerman S, Sablina AA. Mutations in LZTR1 drive human disease by dysregulating RAS ubiquitination. Sci. 2018;362:1177-1182.

[102] Serova O, Montagna M, Torchard D, Narod SA, Tonin P, Sylla B, Lynch HT, Feunteun J, Lenoir GM. A high incidence of BRCA1 mutations in 20 breast-ovarian cancer families. Am J Hum Genet. 1996;58(1):42-51.

[103] Fakharzadeh SS, Trusko SP, George DL. Tumorigenic potential associated with enhanced expression of a gene that is amplified in a mouse tumor cell line. EMBO J. 1991;10:1565-1569.

[104] Rizzardi AE, Rosener NK, Koopmeiners JS, Vogel RI, Metzger GJ, Forster CL, Marston LO, Tiffany JR, McCarthy JB. Evaluation of protein biomarkers of prostate cancer aggressiveness. BMC Cancer 2014;14:244.

[105] Hsieh SC, Kuo SN, Zheng YH, Tsai MH, Lin YS, Lin JH. The E3 ubiquitin ligase SIAH2 is a prosurvival factor overexpressed in oral cancer. Anticancer Res 2013;33:4965-4973.

[106] Qiu, J. Geng, Y-C, Meng F. SIAH 2 expression predicts chemoresistance and poor clinical outcomes in patients with epithelial ovarian cancer. Int J Clin Exp Pathol. 2016;9:8432-8439.

[107] Yang Q, Zhao J, Cui M, Gi S, Wang W, Han X. NEDD4L expression is decreased in ovarian epithelial cancer tissues compared to ovarian non-cancer tissue. J Obstet Gynaecol Res. 2015;41 (12):1959-1964.

[108] He H, Huang C, Chenc Z, Huangd H, Wange X, Chen J. An outlined review for the role of Nedd4-1 and Nedd4-2 in lung disorders. Biomed. Pharmacother. 2020;125:109983. https://doi.org/10.1016/j.biopha.2020.109983.

[109] Cai J, Li R, Xu X, Zhang L, Lian R, Fang R, Huang Y, Feng X, Liu X, Li X, Zhu X, Zhang H, Wu J, Zeng M, Song E, He Y, Yin Y, Li J, Li, M. CK1α suppresses lung tumour growth by stabilizing PTEN and inducing autophagy, Nat. Cell Biol. 2018;20:465-478.

[110] Yanpallewar S, Wang T, Koh DC, Quarta E, Fulgenzi G, Tessarollo L. Nedd4-2 haploinsufficiency causes hyperactivity and increased sensitivity to inflammatory stimuli. Sci. 2016;6:32957.

[111] Gille T, Randrianarison-Pellan N, Goolaerts A, Dard N, Uzunhan Y, Ferrary E, Hummler E, Clerici C, and

Planès C. Hypoxia-induced inhibition of epithelial Na(+) channels in the lung. Role of Nedd4-2 and the ubiquitin-proteasome pathway. Am. J. Respir. Cell Mol. Biol. 2014;50:526-537.

[112] Bradley A, Zheng H, Ziebarth A, Sakarti W, Branham-O'Connor M, Blumer JB, Liu Y, Kistner-Griffin E, Rodriguez-Aguayo C, Lopez-Berestein G, Sood AK, Landen CN, Eblen ST. EDD enhances cell survival and cisplatin resistance and is a therapeutic target for epithelial ovarian cancer. Carcinogenesis, 2014;35:1100-1109.

[113] Eblen ST, Bradley A. MOAP-1, UBR5 and cisplatin resistance in ovarian cancer. Transl Cancer Res. 2017;6:18-21.

[114] Chiorazzi M, Rui L, Yang Y, Ceribelli M, Tishbi N, Maurer CW, Ranuncolo SM, Zhao H, Xu W, Chan WC, Jaffe ES, Gascoyne RD, Campo E, Rosenwald A, Ott G, Delabi, J, Rimsza LM, Shaham S, Staudt LM. Related F-box proteins control cell death in *Caenorhabditis elegans* and human lymphoma. Proc Natl Acad Sci USA. 2013;110(10):3943-3948.

[115] Duan S, Cermak L, Pagan JK, Rossi M, Martinengo C, di Celle PF, Chapuy B, Shipp M, Chiarle R, Pagano M. FBXO11 targets BCL6 for degradation and is inactivated in diffuse large B-cell lymphomas. Nature. 481(7379):90-93.

[116] Kaplan MH, Schindler U, Smiley ST and Grusby MJ. Stat6 Is Required for Mediating Responses to IL-4 and for the development of Th2 cells. Immunity. 1996;4:313-319.

[117] Moldovan GL, D'Andrea AD, How the fanconi anemia pathway guards the genome. Annu. Rev. Genet. 2009;43:223-249.

[118] Sareen A, Chaudhury I, Adams N, Sobeck A. Fanconi anemia proteins FANCD2 and FANCI exhibit different DNA damage responses during S-phase. Nucleic Acids Res. 2012;40(17):8425-8439.

[119] Elder DA, D'Alessio DA, Eyal O, Mueller R, Smith FO, Kansra AR, Rose SR. Abnormalities in glucose tolerance are common in children with fanconi anemia and associated with impaired insulin secretion. Pediatr Blood Cancer. 2008;51(2):256-260.

[120] Hoffman MA, Ohh M, Yang H, Klco JM, Ivan M, Kaelin WG. von Hippel–Lindau protein mutants linked to type 2C VHL disease preserve the ability to downregulate HIF. Hum. Mol. Genet. 2001;10:1019-1027.

[121] Maher ER, Webster AR, Richards FM, Green JS, Crossey PA, Payne SJ, Moore AT. Phenotypic expression in von Hippel-Lindau disease: correlations with germline VHL gene mutations. J Med Genet. 1996;33:328-332.

[122] Sadikovic, B., Fernandes, P., Zhang, V.W., Ward, P.A., Miloslavskaya, I., Rhead, W., Rosenbaum, R., Gin R, Roa B, Fang P. Mutation Update for UBE3A variants in Angelman syndrome. Hum Mutat. 2014;35(12):1407-1417.

[123] Dagli A, Buiting K, Williams CA. Molecular and clinical aspects of Angelman Syndrome. Mol Syndromol. 2012;2:100-112.

[124] Williams CA, Driscoll DJ, Dagli AI. Clinical and genetic aspects of Angelman syndrome. Genet Med. 2010;12:385-395.

[125] Tomaić V, Banks L. Angelman syndrome-associated ubiquitin ligase UBE3A/E6AP mutants interfere with the

proteolytic activity of the proteasome. Cell Death Dis. 2015;6:e1625. doi:10.1038/cddis.2014.572.

[126] Hattori T, Kishino T, Stephen S, Eberspaecher H, Maki S, Takigawa M, de Crombrugghe B, Yasuda H. E6-AP/UBE3A protein acts as a ubiquitin ligase toward SOX9 protein. J. Biol. Chem. 2013;288:35138-35148.

[127] Huber C, Dias-Santagata D, Glaser A, O'Sullivan J, Brauner R, Wu K, Xu X, Pearce K, Wang R, Uzielli, ML, Dagoneau N, Chemaitilly W, Superti-Furga A, Dos Santos H, Mégarbané A, Morin G, Gillessen-Kaesbach G, Hennekam R, van der Burgt I, Black GC, Clayton PE, Read A, Le Merrer M, Scambler PJ, Munnich A, Pan Q, Winter R, Cormier-Daire V. Identification of mutations in CUL7 in 3-M syndrome. Nat Genet. 2005;37:1119-1124.

[128] Huber C, Delezoide A, Guimiot F, Baumann C, Malan V, Le Merrer M, Da Silva DB, Bonneau D, Chatelain P, Chu C, Clark R, Cox H, Edery P, Edouard T, Fano V, Gibson K, Gillessen-Kaesbach G, Giovannucci-Uzielli M-L, Graul-Neumann L.M, van Hagen J-M, van Hest L, Horovitz D, Melki J, Partsch C-J, Plauchu H, Rajab A, Rossi M, Sillence D, Steichen-Gersdorf E, Stewart H, Unger S, Zenker M, Munnich A, Cormier-Daire V. A large-scale mutation search reveals genetic heterogeneity in 3M syndrome. Eur J Hum Genet. 2009;17(3):395-400.

[129] Chen Z, Sui J, Zhang F, Zhang C. Cullin family proteins and tumorigenesis: Genetic association and molecular mechanisms. J Cancer. 2015;6(3):233-242.

[130] Hsu P-H, Ma Y-T, Fang Y-C, Huang J-J, Gan Y-L, Chang P-T, Jow G-M, Tang C-Y, Jeng C-J. Cullin 7 mediates proteasomal and lysosomal degradations of rat Eag1 potassium channels. Sci Rep. 2017;7:40825. doi: 10.1038/srep40825.

[131] Xu X, Sarikas A, Dias-Santagata DC, Dolios G, Lafontant PJ, Tsai SC, Zhu W, Nakajima H, Nakajima HO, Field LJ, Wang R, Pan ZQ. The CUL7 E3 ubiquitin ligase targets insulin receptor substrate 1 for ubiquitin-dependent degradation. Mol Cell. 2008;30(4):403-414.

[132] Okabe H, Lee SH, Phuchareon J, Albertson DG, McCormick F, Tetsu O. A critical role for FBXW8 and MAPK in cyclin D1 degradation and cancer cell proliferation. PLoS ONE. 2006;1(1):e128. doi.org/10.1371/journal.pone.0000128.

[133] Connell-Crowley L, Harper JW, Goodrich DW. Cyclin D1/Cdk4 regulates retinoblastoma protein-mediated cell cycle arrest by site-specific phosphorylation. Mol Biol Cell. 1997;8(2):287-301.

[134] Hjerpe R, Bett JS, Keuss MJ, Solovyova A, McWilliams TG, Johnson C, Sahu I, Varghese J, Wood N, Wightman M, Osborne G, Bates GP, Glickman MH, Trost M, Knebel A, Marchesi F, Kur T. UBQLN2 mediates autophagy-independent protein aggregate clearance by the proteasome. Cell. 2016;166(4):935-949.

[135] Ko SH, Uehara T, Tsuruma K, Nomura Y. Ubiquilin interacts with ubiquitylated proteins and proteasome through its ubiquitin-associated and ubiquitin-like domains. FEBS Lett. 2004;566(1-3):110-114.

[136] Massey LK, Mah AL, Monteiro MJ. Ubiquilin regulates presenilin endoproteolysis and modulates gamma-secretase components, Pen-2 and nicastrin. Biochem J. 2005;391:513-525.

[137] Stieren ES, El Ayadi A, Xiao Y, Siller E, Landsverk ML, Oberhauser AF, Barral JM, Boehning D. Ubiquilin-1 is a molecular chaperone for the Amyloid Precursor Protein. J. Biol. Chem. 2011;286(41):35689-35698.

[138] Haapasalo A, Viswanathan J, Bertram L, Soininen H, Tanzi RE, Hiltunen M. Emerging role of Alzheimer's disease-associated ubiquilin-1 in protein aggregation. Biochem. Soc. Trans. 2010;38:150-155.

[139] Richardson PG, Barlogie B, Berenson J, Singhal S, Jagannath S, Irwin D, Rajkumar SV, Srkalovic G, Alsina M, Alexanian R, Siegel D, Orlowski RZ, Kuter D, Limentani SA, Lee S, Hideshima T, Esseltine D-L, Kauffman M, Adams J, Schenkein DP, Anderson KC. A phase 2 study of bortezomib in relapsed, refractory myeloma. N Engl J Med. 2003;348:2609-2617.

[140] Sun, Y. Targeting E3 Ubiquitin Ligases for cancer therapy. Cancer Biol. Ther. 2003;2 (6):623-629.

[141] Arano T, Imai Y. Mitophagy Regulated by the PINK1-Parkin Pathway, Cell death-autophagy. Apoptosis and necrosis. Ntuli TM. InTechOpen. 2015;doi:10.5772/61284. https//www.intechopen.com/chapters/49196.

[142] Barodia SK, Creed RB, Goldberg MS. Parkin and PINK1 functions in oxidative stress and neurodegeneration. Brain Res Bull. 2017;133:51-59.

[143] Seirafi M, Kozlov G, Gehring K. Parkin structure and function. FEBS J. 2015;282(11):2076-2088.

[144] Fiesel FC, Ando M, Hudec R, Hill AR, Castanedes-Casey MC, Caulfield TR, Moussaud-Lamodière EL, Stankowski JN, Bauer PO, Lorenzo-Betancor O, Ferrer I, Arbelo JM, Siuda J, Chen L, Dawson VL, Dawson TM, Wszolek ZK, Ross OA, Dickson DW, Springer W. (Patho-) physiological relevance of PINK1-dependent ubiquitin phosphorylation. EMBO Rep. 2015;16(9):1114-1130.

[145] Trempe JF, Sauve V, Grenier K, Seirafi M, Tang MY, Menade M, Al-Abdul-Wahid S, Krett J, Wong K., Kozlov G, Nagar B, Fon, EA, Gehring K. Structure of parkin reveals mechanisms for ubiquitin ligase activation. Sci. 2013;340:1451-1455.

[146] Wauer T, Komander D. Structure of the human parkin ligase domain in an autoinhibited state. EMBO J. 2013;32:2099-2112.

[147] Bingol B, Tea JS, Phu L, Reichelt M, Bakalarski CE, Song Q, Foreman O, Kirkpatrick DS, Sheng M. The mitochondrial deubiquitinase USP30 opposes parkin-mediated mitophagy. Nature. 2014;510:370-375.

[148] Cornelissen T, Haddad D, Wauters F, Van Humbeec C, Mandemakers W, Koentjoro B, Sue C, Gevaert K, De Strooper B, Verstreken P, Vandenberghe W. The deubiquitinase USP15 antagonizes parkin-mediated mitochondrial ubiquitination and mitophagy. Hum Mol Genet. 2014;23:5227-5242.

[149] Durcan TM, Fon EA. The three 'P's of mitophagy: PARKIN, PINK1, and post-translational Modifications. Genes and Dev. 2016;29:989-999.

[150] Greenblatt MS, Bennett WP, Hollstein M, Harris CC. Mutations in the p53 tumor suppressor gene: Clues to cancer etiology and molecular pathogenesis. Cancer Res. 1994;54:4855-4878.

[151] Honda R, Yasuda H. Activity of MDM2, a ubiquitin ligase, toward p53 or itself is dependent on the RING finger domain of the ligase. Oncogene. 2000;19(11):1473-1476.

[152] Wang ZH. MDM2 oncogene as a novel target for human cancer therapy. Curr Pharm Des. 2000;6:393-416.

[153] Yang-Hartwich Y, Tedja R, Roberts CM, Goodner-Bingham J, Cardenas C, Gurea M, Sumi NJ, Alvero AB, Glackin CA, Mor G. p53-Pirh2 Complex Promotes Twist1 Degradation and Inhibits EMT. Mol Cancer Res. 2019;17:153-164.

[154] Deveraux QL, Reed, JC. IAP family proteins—suppressors of apoptosis. Genes Dev. 1999;13:239-252.

[155] Hu S, Yang X. Cellular Inhibitor of Apoptosis 1 and 2 Are Ubiquitin Ligases for the Apoptosis Inducer Smac/DIABLO. J Biol Chem. 2003;78:10055-10060.

[156] Yang Y, Fang S, Jensen JP, Weissman AM, Ashwell JD. Ubiquitin protein ligase activity of IAPs and their degradation in proteasomes in response to apoptotic stimuli. Sci. 2000;88:874-877.

[157] Signoretti S, Di Marcotullio L, Richardson A, Ramaswamy S, Isaac B, Rue M, Ramaswamy S, Isaac B, Rue M, Monti F, Loda M, Pagano, M. Oncogenic role of the ubiquitin ligase subunit Skp2 in human breast cancer. J Clin Invest. 2002;110:633-641.

[158] Song YH, Zhang CQ, Chen FF, Lin XY. Upregulation of neural precursor cell expressed developmentally downregulated 4-1 is associated with poor prognosis and chemoresistance in lung adenocarcinoma, Chin. Med. J. (Engl.). 2018;131:16-24.

[159] Green JL, Wu Y, Enchev V, Lasonder E, Prommaban A, Kunzelmann S, Prommaban A, Kunzelmann S, Christodoulou C, Grainger M, Truongvan N, Bothe S, Sharma V, Song W, Pinzuti I, Uthaipibull C, Srichairatanakool S, Birault V, Langsley, G, Schindelin H, Stieglitz B, Snijders AP, Holderet AA. Ubiquitin activation is essential for schizont maturation in Plasmodium falciparum blood-stage development. PLoS Pathog. 2020;16(6):e1008640. https://doi.org/10.1371/journal.ppat.1008640.

[160] An A, Fu L. An emerging and promising approach for the development of targeted therapy drugs. EBioMedicine, 2018;36:553-556.

Chapter 8

Ubiquitination Enzymes

Toshiyuki Habu and Jiyeong Kim

Abstract

Posttranslational protein modifications by mono- or polyubiquitination are involved in diverse cellular signaling pathways and tightly regulated to ensure proper function of cellular processes. Three types of enzymes, namely ubiquitin-activating enzymes (E1), ubiquitin-conjugating enzymes (E2), and ubiquitin-protein ligases (E3), contribute to ubiquitination. Combinations of E2 and E3 enzymes determine ~ the fate of their substrates via ubiquitination. The seven lysine residues of ubiquitin, Lys6, Lys11, Lys27, Lys29, Lys33, Lys48, and Lys63, can serve as attachment sites for other ubiquitin molecules. Lys48 (K48)-linked polyubiquitination facilitates recognition of the conjugated protein by proteasome molecules and subsequent proteolytic degradation of the target protein. By contrast, Lys63 (K63)-linked polyubiquitination appears to be involved in polyubiquitin signaling in critical cellular processes, such as DNA repair, regulation of the I-kappaB kinase/NF-kappaB cascade, or T cell receptor signaling, but not protein degradation. In this review, we describe the properties of ubiquitin modification enzymes and the structural interplay among these proteins.

Keywords: E1 ubiquitin, ubiqutin-activating enzyme, E2 ubiquitin-conjugating enzyme, E3 ubiquitin ligase

1. Introduction

Very large-scale studies of protein ubiquitination have been conducted over the past two decades. Ubiquitin modification is mediated by three types of enzyme activity, mediated out by E1 ubiquitin-activating enzymes (UBA; also referred to as UAE or E1 enzymes; EC 6.2.1.45), E2 ubiquitin-conjugating enzymes (UBC; also termed E2 ubiquitin-carrier proteins or E2 enzymes; EC 2.3.2.23), and ubiquitin-protein ligases (E3 enzymes). To better understand the molecular mechanisms underlying ubiquitin modification, this review focuses on the structural interactions between ubiquitin modification enzymes and their functions.

2. Types of ubiquitin and ubiquitin-like proteins (UBLs)

Ubiquitin is a small, highly conserved 76 amino acids polypeptide found throughout eukaryotic cells, that modifies cellular proteins. Two mammalian ubiquitin genes, *UBB* and *UBC*, encode polyubiquitin and another two genes, *RPS27A* and *UBA52* encode ubiquitins fused with ribosomal proteins [1]. Ubiquitin is produced as precursor peptides that are proteolytically processed by deubiquitinating enzymes into active forms with C-terminal glycine residues. The C-terminal glycine (Gly76) and seven lysine residues (Lys6, Lys11, Lys27, Lys29, Lys33, Lys48, and Lys63) of ubiquitin are essential for ubiquitin modification. There is 96% sequence identity between human and yeast ubiquitin, and the two glycine and seven lysine residues are conserved throughout the eukaryotic kingdom.

Ubiquitin-like proteins (UBLs) do not share sequence homology with ubiquitin but also function as protein modifiers. A number of UBLs have been reported, including SUMO1/SMT3, SUMO2–4 [2–4], Neural-precursor-cell-expressed developmentally downregulated protein-8 (NEDD8)/RUB1 [3, 5, 6], ISG15 [6, 7], ATG8/APG8 [8], ATG12/APG12 [9], URM1 [3], and homologous to ubiquitin 1 (HUB1) [10]. During protein modification by ubiquitin and UBLs, specific activating, conjugating, and ligase enzymes, catalyze attachment of the modifier to target proteins. Similar to ubiquitin, UBLs are also produced as precursors, and deubiquitinating enzymes expose their C-terminal glycine residues to activate them, although HUB1 lacks a C-terminal diglycine sequence.

The SUMO-1 protein has only 18% sequence identity with ubiquitin, but contains the ββαββαβ fold structure characteristic of the ubiquitin protein family [2]. The hydrophobic cores of SUMO-1 and ubiquitin are similar; however, the overall charge surface topology of SUMO-1 differs significantly from that of ubiquitin or other UBLs [2, 11]. The selective modifications mediated by the four SUMO homologs, SUMO-1, SUMO-2, SUMO-3, and SUMO-4 [12–16], remain to be determined. In addition, SUMO has an N-terminal extension (approximately 20 amino acids) not present in ubiquitin, which is required for SUMO function [17]. A consensus motif and lysine residues involved in SUMOylation are present in SUMO-2, SUMO-3, and SUMO-4 and well-conserved; however, these SUMO proteins do not have Lys residue counterparts of ubiquitin Lys 48 and 63.

NEDD8 shows 58% sequence identity and 80% sequence similarity to ubiquitin polypeptide. By contrast, the ATG12 sequence is unrelated to that of ubiquitin, and forms a covalent protein complex with ATG5, which is required for autophagy; this reaction requires the C-terminal glycine residue of ATG12 and a lysine residue in ATG5 [18].

The ISG15/ubiquitin cross-reacting protein (*UCRP*) gene comprises two exons and encodes a 17 kDa polypeptide [19, 20]. The immature polypeptide is cleaved at its carboxy terminus, generating a mature 15 kDa product that terminates with an LRLRGG motif that is also found in ubiquitin. The tertiary structure of ISG15 also resembles ubiquitin, despite having only approximately 30% sequence homology [21, 22]. ISG15 is induced by type I interferon and serves many functions, acting as an extracellular cytokine and an intracellular protein modifier [23, 24].

HLA-F adjacent transcript 10 (FAT10; also known as ubiquitin D) also bears two ubiquitin-like domains. HUB1 has only 22% sequence identity with ubiquitin and possesses an invariant C-terminal double-tyrosine motif, unlike the double glycine residues present in ubiquitin and other UBLs [25].

3. Ubiquitin modification enzymes

3.1 E1 ubiquitin-activating enzymes (UBA)

E1 UBA enzymes adenylate the C-terminal glycine residue (Gly 76) of ubiquitin polypeptides, coupling them with ATP. The C-terminal ubiquitin polypeptide glycine residue is linked to AMP via an acyl-phosphate bond, and the adenylated ubiquitin polypeptide is linked with the sulfhydryl side chain of a cysteine residue (Cys 632 in human UBA1) in the E1 enzyme catalytic center. A thioester intermediate (S-ubiquitinyl-(E1 UAE)-L-Cys) is synthesized in this two-step reaction, along with AMP and diphosphate (Reaction (1)).

$$\begin{aligned}&\text{ATP} + \text{Ubiquitin} + [\text{E1 UBA}]\text{-L-cysteine}\\&< = > \text{AMP} + \text{diphosphate} + \text{S-ubiquitinyl-}[\text{E1 UBA}]\text{-L-cysteine.}\end{aligned} \tag{1}$$

In human, ten UBA orthologues have been identified that can activate ubiquitin or UBLs. Ubiquitin-like modifier-activating enzyme 1 (UBA1) is mainly responsible for ubiquitin-activation and can also activate the NEDD8 UBL peptide [26–28]. UBA2 (or UBLE1B) is also known as SUMO-activating enzyme subunit 2 (SAE2), and activates the SUMO UBL peptide as heterodimer with SAE1 [29]. UBA3 (or UBE1C) encodes the NEDD8-activating enzyme E1 catalytic subunit and forms a heterodimer with NAE1 (or APPBP1, an amyloid-beta precursor protein binding protein), which activates NEDD8. UBA5 activates UFM1 (ubiquitin-fold modifier 1) [30], while UBA6 (alternatively UBE1L2) is an E1 enzyme involved in UBL activation [31, 32]. Autophagy related 7 (ATG7) is an E1 enzyme for UBLs including ATG12 and ATG8. NEDD8-activating enzyme E1 regulatory subunit (NAE1 or APPBP1) is an E1 enzyme for NEDD8, along with UBA3 [33]. Ubiquitin-like modifier-activating enzyme 1 Y (UBA1Y) is encoded by the Y chromosome and expressed specifically during spermatogenesis [34–37]. The UBL, FAT10, is activated by UBA1 and UBA6. UBA7 is induced by interferon-α and β and involved in ISG15 induction.

UBA structures consist of an adenylation domain that interacts with ATP and UBLs, a catalytic domain comprising a Cys residue that binds to UBLs, and a C-terminal ubiquitin fold domain (UFD) required for binding to E2 enzymes.

In a study of mammalian UBA1 with a temperature-sensitive (ts) mutation, cells expressing the ts-UBA1 mutant exhibited cell cycle arrest at the G_2/M phase transition, as well as dramatically decreased ubiquitin conjugation [38, 39]. UBA1-knockdown in human cells also leads to reduced cell proliferation [40]. Furthermore, cells expressing the ts-UBA1 mutant show reduced receptor tyrosine kinase endocytosis and degradation [41]. In addition, mice lacking UBA3 are characterized by a mitotic defect in G_1/G_0 transition, that causes accumulation of SCF ligase targets, including Cyclin E and β-catenin.

3.2 E2 ubiquitin-conjugating enzymes (UBC)

E2 UBC enzymes transthiolate activated ubiquitin from S-ubiquitinyl-[E1 UBA] to themselves. A thioester linkage is formed between an E2 UBC and ubiquitin via the C-terminal glycine of ubiquitin, and the sulfhydryl side chain of a Cys residue in the E2 UBC catalytic center [42] (Reaction (2)).

$$\begin{aligned}&\text{S-ubiquitinyl-}[\text{E1 UBA}]\text{-L-cysteine} + [\text{E2 UBC}]\text{-L-cysteine}\\&< = > [\text{E1 UBA}]\text{-L-cysteine} + \text{S-ubiquitinyl-}[\text{E2 UBC}]\text{-L-cysteine}\end{aligned} \tag{2}$$

The transthiolation reaction involving S-ubiquitinyl-[E1 UBA]-L-cysteine and E2 UBC is strongly stimulated by occupancy of the nucleotide-binding site by either adenylated ubiquitin or ATP alone [43]. Ubiquitin transfer to the target protein is assisted by E3 ubiquitin ligases. Homologous to E6-AP C-terminus (HECT) domain family E3 ligases transfer ubiquitin to the target via a Cys residue in the E3 ligase. By contrast, Really Interesting New Gene (RING) family E3 ligases transfer ubiquitin directly to target proteins. The properties of specific E2 UBC enzymes determine the ubiquitin moiety and substrate specificity of E3 ligases. Indeed, the specificity of interactions with E2 reflect E3 substrate specificity. Amino acids surrounding the Cys residue are evolutionarily conserved among E2 UBCs, and referred to as the ubiquitin-conjugating (UBC) domain or the core catalytic domain [44]. The UBC domain folds an N-terminal helix (H1), a four-stranded β-meander structure (S1–S4), a short 3_{10}-helix (H2), and three C-terminal helices (H3–H5) [45, 46]. Amino acid sequence variations in the UBC domain contribute to specific interactions with E1 UBAs, E3 ligases, and target proteins [47–51].

E2 UBCs are divided into four classes based on structural differences [52]: class I E2 enzymes consist of only a UBC domain; class II E2 enzymes contain additional C-terminal extension residues; class III E2 enzymes have N-terminal extension residues; and class IV E2s have both N- and C-terminal extensions. Class II UBC2 and UBC3 proteins have acidic C-terminal extensions, which mediate a preference for binding to basic substrates, including histones [53–55]. The acidic extension is also required to contact basic canyon residues of the Cul1 subunit of the SCF RING subcomplex (ROC1-CUL1) [56–58]. UBC6 processes C-terminal extensions, to promote ER localization [59, 60]. Class II UBCs include: E2-25K (yeast UBC1) [61], UBC4 [62], UBCH6, UBCH7 [63], UBE2E1 [64], UBE2E2 [65], and UBE2E3 [66, 67].

UBL-specific E2 UBC enzymes process proteins for ubiquitin-like modification. UBC9 is an E2 UBC enzyme specific for the UBL, SUMO, and binds directly to SUMO substrates through a specific short consensus amino acid motif, Y-K-X-[D/E], where Y is any bulky hydrophobic amino acid, including isoleucine, leucine, or valine; K is the lysine residue which is modified by SUMO; X is any residue; D is aspartic acid; and E is glutamic acid [68]. UBC9 contains N- and C-terminal extension residues within the UBC domain, and non-conserved residues in the H1 helix and the insertion β-sheet (S1/2) are required for both interaction with UBA enzymes and formation of the SUMO-thioester bond [69, 70]. ATG3/AUT1 is a dedicated E2 UBC for ATG8 [71]. UBC12 is an E2 UBC specific for NEDD8, which interacts with the NEDD8 E1 UBA via its UBC domain [72, 73], and includes a unique N-terminal region that docks to the E1 enzyme, UBA3, but not to other UBAs. The E2 for ISG15, UBCH8, takes part in reactions involving both UBEL1, the E1 for ISG15, and UBA1, the E1 for ubiquitin [7].

3.3 E3 ubiquitin ligases

E3 ubiquitin ligases are also referred to as ubiquitin-protein ligases, E3 ligases, or E3 enzymes. Ubiquitin is covalently bonded to the ε-amino group of a lysine residue within the substrate protein via an isopeptide bond. The last step in this binding is mediated by E3 ubiquitin ligases, which determine the substrate specificity by ~ to target proteins. E3 ubiquitin ligases transfer ubiquitin linked with a UBC to the ϵ-amino group of a lysine residue of the target protein. An isopeptide bond is formed between the C-terminal glycine residue of ubiquitin and an ϵ-amino group of a lysine residue in the target protein.

E3 ubiquitin ligases are divided into three major classes: HECT type (Section 3.3.1), RING-type (Section 3.3.3), and U-box E3 ligases (Section 3.3.4). HECT type E3 ubiquitin ligases form a thioester intermediate with the active-site cysteine of the E3, following the formation of an isopeptide bond between the C-terminal glycine residue of ubiquitin and the ϵ-amino group of a lysine residue in the target protein. RBR-type E3 ubiquitin ligases (Section 3.3.2) mediate similar reactions to HECT type E3 ligases. RING-type and U-box E3 ligases mediate different reactions from HECT and RBR-type E3 ligases, in which ubiquitin is transferred from ubiquitinyl-UBC directly to the target protein without formation of a thioester intermediate. Multi-subunit RING-type E3 ligases (Section 3.3.3.1) form complexes with a scaffold protein and a contain recognition modules that bind to substrates.

3.3.1 HECT-type E3 ubiquitin transferases (EC 2.3.2.26)

HECT-type E3 ligases transfer ubiquitin from an E2 ubiquitin-conjugating enzyme (EC 2.3.2.23) to a cysteine residue in the HECT domain in the C-terminal region of an E3 ligase (Reaction (3)). The activated ubiquitin from S-ubiquitinyl-[E3 ligase]-L-cysteine is transferred from the intermediate to the target protein (Reaction (4)). The C-terminal glycine residue of the received ubiquitin is linked with the ε-amino chain of a lysine residue of the acceptor protein, forming an isopeptide bond. Importantly, the HECT domain forms a thioester intermediate with ubiquitin, unlike other E3 ligases.

$$\begin{aligned}&\text{S-ubiquitinyl-[E2 UBC]-cysteine} + \left[\text{HECT-type E3}\right]\text{-cysteine}\\&< = > \text{[E2 UBC]-cysteine} + \text{S-ubiquitinyl-}\left[\text{HECT-type E3}\right]\text{-cysteine}\end{aligned} \tag{3}$$

$$\begin{aligned}&\text{S-ubiquitinyl-}\left[\text{HECT-type E3}\right]\text{-cysteine} + \text{[acceptor protein]-lysine}\\&< = > \left[\text{HECT-type E3}\right]\text{-cysteine} + \text{N6-ubiquitinyl-[acceptor protein]-lysine}\end{aligned} \tag{4}$$

$$\begin{aligned}&\text{S-ubiquitinyl-[E2 UBC]-L-cysteine} + \text{[acceptor protein]-L-lysine}\\&< = > \text{[E2 UBC]-L-cysteine} + \text{N(6)-ubiquitinyl-[acceptor protein]-L-lysine}\end{aligned} \tag{5}$$

HECT type E3 ligases catalyze a thioester bond between a C-terminal glycine residue of ubiquitin and themselves and then transfer the ubiquitin to a substrate protein. HECT type E3 ligase family proteins possess a well-conserved, approximately 350 residue, catalytic HECT domain close to their C-terminal region [74, 75]. The HECT domain has a bi-lobal structure comprising an approximately 250 residue N-lobe, required for the binding to UBC-ubiquitin complex, and a C-lobe of around 100 residues, required for ubiquitin transfer [74, 76, 77]. Various linker sequences between the two HECT domain lobes mediate the properties of HECT type E3 ligases in accepting ubiquitin from E2 enzymes and transferring it to a target substrate.

The HECT type E3 ligase, E6-AP, can ubiquitinate p53 in the presence of human papillomavirus E6 protein [75, 78, 79], and another HECT E3 enzyme NEDD4 ubiquitinates SMAD proteins, thereby regulating transcription factors mediating TGFβ signaling [80], the P63 tumor antigen [81], and MDM2 [82]. HECW1 [83], HECW2 [84], WWP1 [85], HERC1 [86], HERC2 [87], and ITCH [88] also belong to the HECT type E3 ubiquitin ligase family.

3.3.2 RBR-type E3 ubiquitin transferase (EC 2.3.2.31)

RBR-type E3 ubiquitin transferases possess two RING finger domains, each of which is separated by an internal IBR (In Between RING) motif. These E3 ligases bind to the Cullin-RING ubiquitin Ligase (CRL) complex (see Cullin-type E3 NDD8 transferase), within which a neddylated cullin scaffold protein and a substrate recognition module are required for ubiquitin transfer. The first RING domain binds S-ubiquitinyl- [E2 UBC]-cysteine and transfers the ubiquitin to an internal Cys residue in the second RING domain (Reaction (6)), followed by transfer of the ubiquitin from the Cys residue in the second RING domain to a Lys in the acceptor protein (Reaction (7)). RBR-type ligases stimulate a cycling ubiquitination reaction via the S-ubiquitinyl-[E2 UBC]-cysteine in the first RING domain [88]. RBR-type ligase activity depends on the neddylation of the cullin protein in the CRL complex. RBR-type E3 ubiquitin ligases include Parkin, Parc, RNF19, RNF144, RNF216 RFA1 HOIP, and HHARI [89, 90].

S-ubiquitinyl-[E2 UBC]-cysteine + [RBR-type E3]-cysteine
< = > [E2 UBC]-cysteine + S-ubiquitinyl-[RBR-type E3]-cysteine (6)

S-ubiquitinyl-[RBR-type E3]-cysteine + [acceptor protein]-lysine
< = > [RBR-type E3]-cysteine + N6-ubiquitinyl-[acceptor protein]-lysine (7)

S-ubiquitinyl-[E2 UBC]-cysteine + [acceptor protein]-lysine
< = > [E2 UBC]-cysteine + N (6)-ubiquitinyl-[acceptor protein]-lysine (8)

3.3.3 RING-type E3 ubiquitin transferases (EC 2.3.2.27)

RING E3 ubiquitin ligases (also referred to as RING E3 ligases or ubiquitin transferase RING E3 enzymes) transfer ubiquitin peptides directly from a ubiquitinyl-E2 UBC enzyme to an acceptor protein. The ε-amino group of a lysine residue of the target protein forms an isopeptide bond with the C-terminal glycine residue of ubiquitin (Reaction (9)). Unlike HECT E3 ligases, the RING-E3 domain does not create a catalytic thioester intermediate with ubiquitin through a Cys residue.

S-ubiquitinyl-[E2 UBC]-L-lysine + [acceptor protein]-L-lysine
< = > [E2 UBC]-L-cysteine + N (6)-ubiquitinyl-[acceptor protein]-L-lysine (9)

Human proteome analysis has identified approximately 580 genes encoding putative RING-type ubiquitin E3 ligase family proteins, which is more than the number of protein kinase genes (518) [91]. Among RING-type E3 ligase genes, 309 and 270 encode single and multi-subunit RING-type E3 ligase molecules, respectively. While RING-type E3 ubiquitin ligase family proteins do not form thioester intermediates with ubiquitin, they function as a scaffold for ubiquitin-charged UBC and the substrate. RING-type E3 ligases contain both a RING domain and a substrate-binding site, and almost half the RING proteins belong to multisubunit RING-type E3 ligases, which require an additional subunit for substrate recognition (see multisubunit RING-type ubiquitin ligases).

The RING domain was initially thought to function as a DNA binding domain because of the discovery of RING domain-containing proteins with DNA binding activity [92, 93]. RING-type E3 ligases were subsequently identified as interacting partners of the human E2 ubiquitin-conjugating enzyme UBCH5 [94], which has self-ubiquitination activity that depends on its RING domain sequence. The canonical RING domain structure consists of a Zn^{2+}-coordination complex and o series of Cys and His residues and mediates E2-dependent ubiquitylation. The coordination complex with two zinc ions forms a cross-brace structure. RING finger domains have consensus sequences that are classified into two different types, C_3HC_4-type (RING-HC) and $C_3H_2C_3$-type (RING-H2), according to the cysteine/histidine arrangement (where C = Cys and H = His) [95]. The C_3HC_5-type RING domain has different properties from the C_3HC_4 RING-HC finger [96], Casitas B-lineage Lymphoma (c-Cbl), which is a RING-HC type ligase. Ubiquitination activity modulates receptor tyrosine kinase signaling [97] and structural analysis of the c-Cbl-UbcH7-substrate tertiary complex showed that the interaction surface of the UbcH7 E2 enzyme is commonly used by both c-Cbl and HECT-type E3 ligases, where c-Cbl binds UbcH7 using both its RING domain and linker helix structure [50]. The amino acid residues involved in the interaction are structurally similar between E2 enzymes and E3 ligases.

BRCA1 forms a heterodimer with the RING-type ligase BARD. The dimerization of two RING-type E3 ligases results in upregulation of ubiquitination activity. By biochemical approaches, UbcH5c and UbcH7 enzymes were identified as candidate E2 enzymes for the BRCA1-BARD complex. Christensen et al. developed an excellent method for identifying E2-E3 pairing [98, 99], using a BRCA1-BARD fusion protein; BRCA1 can synthesize specific polyubiquitin chain linkages, depending on the presence of a paired E2 enzyme [98]. This approach has increased the identification of E2-E3 pairs; for example, RNF213, a RING-HC type E3 ligase and its paired UBC13E2 enzyme were identified using this method [100]. UBCH5b mutants, which can bind to E3 ligase, exhibit defective stimulation by E3 ligases [101]. Ubiquitin-charged E2 is conformationally activated by binding to the RING domain [101–103]. Furthermore, the interaction between E1 and E2 enzymes can direct substrate specificity, ubiquitin transfer, and polyubiquitin chain linkages.

Some RING-type ubiquitin ligase family members form hetero- or homo-multimers through the RING domain or its surrounding region. RING-RING complexes, including MDM2-HMDX, BRCA1-BARD1, and RING1-BMI1, form heterodimers. In heterodimers, one partner (HDMX, BRAD1, and BMI1) lacks ubiquitin ligase activity, while the other partner (MDM2, BRCA1, and RING1) has E2 UBC binding activity. Heterodimer formation leads to stabilization of E2-E3 binding, and in dimerizing E3 ligases, the five C-terminal residues of the RING domain are essential for both dimer formation and E3 activity [104–106].

TRAF2, cellular inhibitor of apoptosis (cIAP; officially known as BIRC2), SIAH, BIRC7, and RNF213 form homodimers [100, 107–111]. Dimeric BIRC7 recruits UBCH5B-ubiquitin and optimizes the donor ubiquitin configuration for transfer [112]. Homo- and hetero-dimerization of RING-type ubiquitin ligases may stabilize their interactions with ubiquitin-charged UBC E2 enzymes and optimize ubiquitin transfer activity.

3.3.3.1 Multisubunit RING protein complexes

Enzymes of the RING-type E3 ubiquitin ligase family do not bind directly to a substrate, but rather form a complex with a cullin scaffold protein and substrate recognition modules, referred to as CRL complexes. The SCF complex (SKP, Cullin, F-box containing complex) and anaphase-promoting complex/cyclosome (APC/C) (anaphase-promoting complex/cyclosome) are two major multisubunit RING containing complexes.

Ubiquitination by SCF and APC/C are implicated in the degradation of cell cycle proteins [113–116]. APC/C regulates mitosis and entry into the G1 phase of the cell cycle, and SCF controls S phase progression.

SCF E3 complexes comprise at least four different subunits, including the F-box protein, SKP adaptor protein, Cullin scaffold protein, and Rbx RING-type E3 ligase [116–119]. The F-box motif is a protein–protein interaction motif comprising approximately 50 amino acid residues. There is low sequence identity among F-box proteins, which recognize and bind substrate and bridge connections between adaptor proteins (including SKP1) and substrates. Phosphorylation of F-box proteins regulates their interactions with substrates. The SKP adaptor proteins, SKP2 (S-phase kinase-associated protein 2), β-TrCP (beta-transducin repeat-containing protein), FBW7, and FBXO4 are F-box proteins involved in cell-cycle regulation. Cullins are scaffold proteins for ubiquitin ligases; CUL1 is a subunit of the SCF complex, and the Cullin-homology domain at its C-terminus interacts with RING E3 ligases while the N-terminal region can interact with the adaptor protein, SKP2. Cullin family members function as adaptors for multisubunit RING-type E3 ligase complexes. The adaptor proteins SKP1 and CUL1 and the RING-type E3 ligase RBX1 form the CRL catalytic core complex.

APC/C is a multisubunit RING-type E3 ligase containing approximately 13 subunits. The Cullin subunit protein, APC2, and the RING H2 type E3 ligase, APC11, form the catalytic core domain [120–122]. TPR (tetratricopeptide residue) motif-containing subunits, including CDC16, CDC27, CDC23, and APC5, are thought to function as scaffold assembling proteins. Two co-activators, CDC20 (cell division cycle homologue 20) and FZR/CDH1 (Fzy-related/cell division cycle 20 related 1), bind to the CDC27 subunit of APC/C through their WD40 repeat and determine APC/C substrate specificity dependent on cell cycle to stages [115, 118, 123–128]. The APC10 subunit contributes to optimal co-activator-dependent substrate recognition and substrate affinity [129–131]. APC/C-mediated ubiquitination depends on destruction box, KEN box, and CRY box sequences in the substrate [132–140]. Assembly of these co-activators into the APC/C complex in G_1 or M phase during cell cycle is regulated by phosphorylation [141–144].

3.3.4 U-Box E3 ubiquitin ligases

The U-box domain displays a similar three-dimensional structure to the RING domain [145]. The U-box domain shows similarity to UFD2, which has a multiubiquitin chain elongation activity (known as E4 activity) [146]. Unlike the RING domain, the U-box domain does not form a coordination complex consisting of a central zinc ion through Cys residues; rather, the U-box domain structure is maintained by hydrogen bonding. The U-box domain has ubiquitin ligase activity, and the U-box protein, carboxyl terminus of HSC70-interacting protein (CHIP), also has E4 activity and includes tetratricopeptide repeat and U-box domains. The C-terminal U-box domain interacts with the molecular chaperones HSC70, HSP70, and HSP90 [147], in the presence of unfolded or misfolded proteins, where CHIP regulates protein quality control [148, 149].

U-box proteins have various structures; for example, ARC1, CMPG1, PUB13, and PUB20 contain armadillo which represents approximately 40 amino acids tandem repeats sequence. PUB23 has a serine/threonine kinase domain while PUB59 and PUB60 have WD40 repeats. These domains may coordinate the function of ubiquitin ligase activity by the U-box domain.

4. Conclusions

Research conducted over several decades has uncovered the cellular and biochemical functions involved in ubiquitin modification. Protein–protein networks and studies of complex structures have contributed to unraveling the biochemical mechanisms underlying ubiquitin modification. Identification of physiological E2 UBC-E3 ligase pairings has facilitated understanding of modification-types and associations. Deep understanding of the structures and biochemical processes involved in ubiquitin modification has contributed to determination of E3 ligase-substrate pairing and network construction. RING-type ubiquitin ligases comprise the largest gene family and are associated with various cellular processes and several diseases. Fundamental questions remain to be answered regarding the biological functions served by ubiquitin modification. Extensive further study of enzymes involved in ubiquitination and related processes has potential to contribute to the understanding of the pathogenesis of several diseases.

Acknowledgements

This work was supported by grants from the Ministry of Education, Culture, Sports, Science, and Technology of Japan to T. H.

Conflict of interest

The authors declare no conflict of interest.

Notes/thanks/other declarations

The authors declare no competing interests.

Author details

Toshiyuki Habu* and Jiyeong Kim
Department of Food Sciences and Nutrition, School of Food Sciences and Nutrition, Mukogawa Women's University, Nishinomiya, Hyogo, Japan

*Address all correspondence to: habu@mukogawa-u.ac.jp

References

[1] Redman KL, Rechsteiner M. 1989 Identification of the long ubiquitin extension as ribosomal protein S27a. Nature 338, 438–440. (doi:10.1038/338438a0)

[2] Bayer P, Arndt A, Metzger S, Mahajan R, Melchior F, Jaenicke R, Becker J. 1998 Structure determination of the small ubiquitin-related modifier SUMO-1. J Mol Biol 280, 275–286. (doi:10.1006/jmbi.1998.1839)

[3] Jentsch S, Pyrowolakis G. 2000 Ubiquitin and its kin: how close are the family ties? Trends Cell Biol 10, 335–342. (doi:10.1016/s0962-8924(00)01785-2)

[4] Schwartz DC, Hochstrasser M. 2003 A superfamily of protein tags: ubiquitin, SUMO and related modifiers. Trends Biochem Sci 28, 321–328. (doi:10.1016/s0968-0004(03)00113-0)

[5] Larsen CN, Wang H. 2002 The Ubiquitin Superfamily: Members, Features, and Phylogenies. J Proteome Res 1, 411–419. (doi:10.1021/pr025522n)

[6] Kerscher O, Felberbaum R, Hochstrasser M. 2006 Modification of Proteins by Ubiquitin and Ubiquitin-Like Proteins. Annu Rev Cell Dev Bi 22, 159–180. (doi:10.1146/annurev.cellbio.22.010605.093503)

[7] Arimoto K-I, Konishi H, Shimotohno K. 2008 UbcH8 regulates ubiquitin and ISG15 conjugation to RIG-I. Mol Immunol 45, 1078–1084. (doi:10.1016/j.molimm.2007.07.021)

[8] Tsukada M, Ohsumi Y. 1993 Isolation and characterization of autophagy-defective mutants of Saccharomyces cerevisiae. Febs Lett 333, 169–174. (doi:10.1016/0014-5793(93)80398-e)

[9] Schulman BA, Harper JW. 2009 Ubiquitin-like protein activation by E1 enzymes: the apex for downstream signalling pathways. Nat Rev Mol Cell Bio 10, 319–331. (doi:10.1038/nrm2673)

[10] Dittmar GAG, Wilkinson CRM, Jedrzejewski PT, Finley D. 2002 Role of a Ubiquitin-Like Modification in Polarized Morphogenesis. Science 295, 2442–2446. (doi:10.1126/science.1069989)

[11] Gill G. 2004 SUMO and ubiquitin in the nucleus: different functions, similar mechanisms? Gene Dev 18, 2046–2059. (doi:10.1101/gad.1214604)

[12] Azuma Y, Arnaoutov A, Dasso M. 2003 SUMO-2/3 regulates topoisomerase II in mitosis. J Cell Biology 163, 477–487. (doi:10.1083/jcb.200304088)

[13] Saitoh H, Hinchey J. 2000 Functional Heterogeneity of Small Ubiquitin-related Protein Modifiers SUMO-1 versus SUMO-2/3. J Biol Chem 275, 6252–6258. (doi:10.1074/jbc.275.9.6252)

[14] Melchior F. 2000 SUMO-NONCLASSICAL UBIQUITIN. Annu Rev Cell Dev Bi 16, 591–626. (doi:10.1146/annurev.cellbio.16.1.591)

[15] Vertegaal ACO, Ogg SC, Jaffray E, Rodriguez MS, Hay RT, Andersen JS, Mann M, Lamond AI. 2004 A Proteomic Study of SUMO-2 Target Proteins. J Biol Chem 279, 33791–33798. (doi:10.1074/jbc.m404201200)

[16] Zhao Y, Kwon SW, Anselmo A, Kaur K, White MA. 2004 Broad Spectrum Identification of Cellular Small Ubiquitin-related Modifier (SUMO) Substrate Proteins. J Biol Chem 279, 20999–21002. (doi:10.1074/jbc.m401541200)

[17] Bylebyl GR, Belichenko I, Johnson ES. 2003 The SUMO Isopeptidase Ulp2 Prevents Accumulation of SUMO Chains in Yeast. J Biol Chem 278, 44113–44120. (doi: 10.1074/jbc.m308357200)

[18] Klionsky DJ, Ohsumi Y. 1999 VACUOLAR IMPORT OF PROTEINS AND ORGANELLES FROM THE CYTOPLASM. Annu Rev Cell Dev Bi 15, 1–32. (doi:10.1146/annurev.cellbio.15.1.1)

[19] Blomstrom DC, Fahey D, Kutny R, Korant BD, Knight E. 1986 Molecular characterization of the interferon-induced 15-kDa protein. Molecular cloning and nucleotide and amino acid sequence. *J Biological Chem* 261, 8811–8816.

[20] Dzimianski JV, Scholte FEM, Bergeron É, Pegan SD. 2019 ISG15: It's Complicated. J Mol Biol 431, 4203–4216. (doi:10.1016/j.jmb.2019.03.013)

[21] Recht M, Borden EC, Knight E. 1991 A human 15-kDa IFN-induced protein induces the secretion of IFN-gamma. J Immunol Baltim Md *1950* 147, 2617–2623.

[22] Swaim CD, Scott AF, Canadeo LA, Huibregtse JM. 2017 Extracellular ISG15 Signals Cytokine Secretion through the LFA-1 Integrin Receptor. *Mol Cell* 68, 581-590.e5. (doi:10.1016/j.molcel.2017.10.003)

[23] D'Cunha J, Knight E, Haas AL, Truitt RL, Borden EC. 1996 Immunoregulatory properties of ISG15, an interferon-induced cytokine. Proc National Acad Sci 93, 211–215. (doi: 10.1073/pnas.93.1.211)

[24] Knight E, Cordova B. 1991 IFN-induced 15-kDa protein is released from human lymphocytes and monocytes. J Immunol Baltim Md *1950* 146, 2280–2284.

[25] Loeb KR, Haas AL. 1994 Conjugates of ubiquitin cross-reactive protein distribute in a cytoskeletal pattern. Mol Cell Biol 14, 8408–8419. (doi:10.1128/mcb.14.12.8408)

[26] Moudry P, Lukas C, Macurek L, Hanzlikova H, Hodny Z, Lukas J, Bartek J. 2014 Ubiquitin-activating enzyme UBA1 is required for cellular response to DNA damage. Cell Cycle 11, 1573–1582. (doi:10.4161/cc.19978)

[27] Correale S *et al.* 2014 Structural Model of the hUbA1-UbcH10 Quaternary Complex: In Silico and Experimental Analysis of the Protein-Protein Interactions between E1, E2 and Ubiquitin. Plos One 9, e112082. (doi: 10.1371/journal.pone.0112082)

[28] Leidecker O, Matic I, Mahata B, Pion E, Xirodimas DP. 2012 The ubiquitin E1 enzyme Ube1 mediates NEDD8 activation under diverse stress conditions. Cell Cycle 11, 1142–1150. (doi:10.4161/cc.11.6.19559)

[29] Okuma T, Honda R, Ichikawa G, Tsumagari N, Yasuda H. 1999 In VitroSUMO-1 Modification Requires Two Enzymatic Steps, E1 and E2. Biochem Bioph Res Co 254, 693–698. (doi:10.1006/bbrc.1998.9995)

[30] Komatsu M *et al.* 2004 A novel protein-conjugating system for Ufm1, a ubiquitin-fold modifier. Embo J 23, 1977–1986. (doi:10.1038/sj.emboj.7600205)

[31] Pelzer C, Kassner I, Matentzoglu K, Singh RK, Wollscheid H-P, Scheffner M, Schmidtke G, Groettrup M. 2007 UBE1L2, a Novel E1 Enzyme Specific for Ubiquitin. J Biol Chem 282, 23010–23014. (doi:10.1074/jbc.c700111200)

[32] Jin J, Li X, Gygi SP, Harper JW. 2007 Dual E1 activation systems for ubiquitin

differentially regulate E2 enzyme charging. Nature 447, 1135–1138. (doi:10.1038/nature05902)

[33] Bohnsack RN, Haas AL. 2003 Conservation in the Mechanism of Nedd8 Activation by the Human AppBp1-Uba3 Heterodimer. J Biol Chem 278, 26823–26830. (doi:10.1074/jbc.m303177200)

[34] Ayusawa D, Kaneda S, Itoh Y, Yasuda H, Murakami Y, Sugasawa K, Hanaoka F, Seno T. 1992 Complementation by a Cloned Human Ubiquitin-Activating Enzyme E1 of the S-Phase-Arrested Mouse FM3A Cell Mutant with Thermolabile E1. Cell Struct Funct 17, 113–122. (doi:10.1247/csf.17.113)

[35] Cook JC, Chock PB. 1992 Isoforms of mammalian ubiquitin-activating enzyme. J Biological Chem 267, 24315–24321.

[36] Handley PM, Mueckler M, Siegel NR, Ciechanover A, Schwartz AL. 1991 Molecular cloning, sequence, and tissue distribution of the human ubiquitin-activating enzyme E1. Proc National Acad Sci 88, 258–262. (doi:10.1073/pnas.88.1.258)

[37] Lévy N, Navarro A, Bishop CE, Mitchell MJ. 2000 The ubiquitin-activating enzyme E1 homologous genes on the mouse Y Chromosome (Ube1y) represent one functional gene and six partial pseudogenes. Mamm Genome 11, 164–168. (doi:10.1007/s003350010031)

[38] Finley D, Ciechanover A, Varshavsky A. 1984 Thermolability of ubiquitin-activating enzyme from the mammalian cell cycle mutant ts85. Cell 37, 43–55. (doi:10.1016/0092-8674(84)90299-x)

[39] Ciechanover A, Finley D, Varshavsky A. 1985 Mammalian cell cycle mutant defective in intracellular protein degradation and ubiquitin-protein conjugation. Prog Clin Biol Res 180, 17–31.

[40] Schlabach MR *et al.* 2008 Cancer Proliferation Gene Discovery Through Functional Genomics. Science 319, 620–624. (doi:10.1126/science.1149200)

[41] Strous GJ, Kerkhof P, Govers R, Ciechanover A, Schwartz AL. 1996 The ubiquitin conjugation system is required for ligand-induced endocytosis and degradation of the growth hormone receptor. Embo J 15, 3806–3812. (doi:10.1002/j.1460-2075.1996.tb00754.x)

[42] David Y, Ziv T, Admon A, Navon A. 2010 The E2 Ubiquitin-conjugating Enzymes Direct Polyubiquitination to Preferred Lysines. J Biol Chem 285, 8595–8604. (doi:10.1074/jbc.m109.089003)

[43] Pickart CM, Kasperek EM, Beal R, Kim A. 1994 Substrate properties of site-specific mutant ubiquitin protein (G76A) reveal unexpected mechanistic features of ubiquitin-activating enzyme (E1). J Biological Chem 269, 7115–7123.

[44] Burroughs AM, Jaffee M, Iyer LM, Aravind L. 2008 Anatomy of the E2 ligase fold: Implications for enzymology and evolution of ubiquitin/Ub-like protein conjugation. J Struct Biol 162, 205–218. (doi:10.1016/j.jsb.2007.12.006)

[45] Cook WJ, Jeffrey LC, Carson M, Chen Z, Pickart CM. 1992 Structure of a diubiquitin conjugate and a model for interaction with ubiquitin conjugating enzyme (E2). J Biol Chem 267, 16467–16471. (doi:10.1016/s0021-9258(18)42026-1)

[46] Cook WJ, Jeffrey LC, Xu Y, Chau V. 1993 Tertiary structures of class I ubiquitin-conjugating enzymes are

highly conserved: Crystal structure of yeast Ubc4. Biochemistry-us 32, 13809–13817. (doi:10.1021/bi00213a009)

[47] Hamilton KS, Ellison MJ, Barber KR, Williams RS, Huzil JT, McKenna S, Ptak C, Glover M, Shaw GS. 2001 Structure of a Conjugating Enzyme-Ubiquitin Thiolester Intermediate Reveals a Novel Role for the Ubiquitin Tail. Structure 9, 897–904. (doi:10.1016/s0969-2126(01)00657-8)

[48] Dominguez C, Bonvin AMJJ, Winkler GS, Schaik FMA van, Timmers HThM, Boelens R. 2004 Structural Model of the UbcH5B/CNOT4 Complex Revealed by Combining NMR, Mutagenesis, and Docking Approaches. Structure 12, 633–644. (doi:10.1016/j.str.2004.03.004)

[49] Huang L, Kinnucan E, Wang G, Beaudenon S, Howley PM, Huibregtse JM, Pavletich NP. 1999 Structure of an E6AP-UbcH7 Complex: Insights into Ubiquitination by the E2-E3 Enzyme Cascade. Science 286, 1321–1326. (doi:10.1126/science.286.5443.1321)

[50] Zheng N, Wang P, Jeffrey PD, Pavletich NP. 2000 Structure of a c-Cbl–UbcH7 Complex RING Domain Function in Ubiquitin-Protein Ligases. Cell 102, 533–539. (doi:10.1016/s0092-8674(00)00057-x)

[51] Bernier-Villamor V, Sampson DA, Matunis MJ, Lima CD. 2002 Structural Basis for E2-Mediated SUMO Conjugation Revealed by a Complex between Ubiquitin-Conjugating Enzyme Ubc9 and RanGAP1. Cell 108, 345–356. (doi:10.1016/s0092-8674(02)00630-x)

[52] Jentsch S. 1992 The Ubiquitin-Conjugation System. Annu Rev Genet 26, 179–207. (doi:10.1146/annurev.ge.26.120192.001143)

[53] Sharon G, Raboy B, Parag HA, Dimitrovsky D, Kulka RG. 1991 RAD6 gene product of Saccharomyces cerevisiae requires a putative ubiquitin protein ligase (E3) for the ubiquitination of certain proteins. J Biological Chem 266, 15890–15894.

[54] RABOY B, KULKA RG. 1994 Role of the C-terminus of Saccharomyces cerevisiae ubiquitin-conjugating enzyme (Rad6) in substrate and ubiquitin-protein-ligase (E3-R) interactions. Eur J Biochem 221, 247–251. (doi:10.1111/j.1432-1033.1994.tb18735.x)

[55] Kolman CJ, Toth J, Gonda DK. 1992 Identification of a portable determinant of cell cycle function within the carboxyl-terminal domain of the yeast CDC34 (UBC3) ubiquitin conjugating (E2) enzyme. Embo J 11, 3081–3090. (doi:10.1002/j.1460-2075.1992.tb05380.x)

[56] Kleiger G, Saha A, Lewis S, Kuhlman B, Deshaies RJ. 2009 Rapid E2-E3 Assembly and Disassembly Enable Processive Ubiquitylation of Cullin-RING Ubiquitin Ligase Substrates. Cell 139, 957–968. (doi:10.1016/j.cell.2009.10.030)

[57] Kleiger G, Hao B, Mohl DA, Deshaies RJ. 2009 The Acidic Tail of the Cdc34 Ubiquitin-conjugating Enzyme Functions in Both Binding to and Catalysis with Ubiquitin Ligase SCFCdc4. J Biol Chem 284, 36012–36023. (doi:10.1074/jbc.m109.058529)

[58] Choi Y-S, Wu K, Jeong K, Lee D, Jeon YH, Choi B-S, Pan Z-Q, Ryu K-S, Cheong C. 2010 The Human Cdc34 Carboxyl Terminus Contains a Non-covalent Ubiquitin Binding Activity That Contributes to SCF-dependent Ubiquitination. J Biol Chem 285, 17754–17762. (doi:10.1074/jbc.m109.090621)

[59] Sommer T, Jentsch S. 1993 A protein translocation defect linked to ubiquitin

conjugation at the endoplasmic reticulum. Nature 365, 176–179. (doi:10.1038/365176a0)

[60] Yuan L, Lv Z, Adams MJ, Olsen SK. 2021 Crystal structures of an E1–E2–ubiquitin thioester mimetic reveal molecular mechanisms of transthioesterification. Nat Commun 12, 2370. (doi:10.1038/s41467-021-22598-y)

[61] Merkley N, Shaw GS. 2004 Solution Structure of the Flexible Class II Ubiquitin-conjugating Enzyme Ubc1 Provides Insights for Polyubiquitin Chain Assembly. J Biol Chem 279, 47139–47147. (doi:10.1074/jbc.m409576200)

[62] Matuschewski K, Hauser H-P, Treier M, Jentsch S. 1996 Identification of a Novel Family of Ubiquitin-conjugating Enzymes with Distinct Amino-terminal Extensions. J Biol Chem 271, 2789–2794. (doi:10.1074/jbc.271.5.2789)

[63] Nuber U, Schwarz S, Kaiser P, Schneider R, Scheffner M. 1996 Cloning of Human Ubiquitin-conjugating Enzymes UbcH6 and UbcH7 (E2-F1) and Characterization of Their Interaction with E6-AP and RSP5. J Biol Chem 271, 2795–2800. (doi:10.1074/jbc.271.5.2795)

[64] Banka PA, Behera AP, Sarkar S, Datta AB. 2015 RING E3-Catalyzed E2 Self-Ubiquitination Attenuates the Activity of Ube2E Ubiquitin-Conjugating Enzymes. J Mol Biol 427, 2290–2304. (doi:10.1016/j.jmb.2015.04.011)

[65] Kimura M, Hattori T, Matsuda Y, Yoshioka T, Sumi N, Umeda Y, Nakashima S, Okano Y. 1997 cDNA cloning, characterization, and chromosome mapping of UBE2E2 encoding a human ubiquitin-conjugating E2 enzyme. Cytogenet Genome Res 78, 107–111. (doi:10.1159/000134639)

[66] Ito K, Kato S, Matsuda Y, Kimura M, Okano Y. 1999 cDNA cloning, characterization, and chromosome mapping of UBE2E3 (alias UbcH9), encoding an N-terminally extended human ubiquitin-conjugating enzyme. Cytogenet Genome Res 84, 99–104. (doi:10.1159/000015229)

[67] Plafker SM, Plafker KS, Weissman AM, Macara IG. 2004 Ubiquitin charging of human class III ubiquitin-conjugating enzymes triggers their nuclear import. J Cell Biology 167, 649–659. (doi:10.1083/jcb.200406001)

[68] Johnson ES. 2004 PROTEIN MODIFICATION BY SUMO. Annu Rev Biochem 73, 355–382. (doi:10.1146/annurev.biochem.73.011303.074118)

[69] Silver ET, Gwozd TJ, Ptak C, Goebl M, Ellison MJ. 1992 A chimeric ubiquitin conjugating enzyme that combines the cell cycle properties of CDC34 (UBC3) and the DNA repair properties of RAD6 (UBC2): implications for the structure, function and evolution of the E2s. Embo J 11, 3091–3098.

[70] Sullivan ML, Vierstra RD. 1991 Cloning of a 16-kDa ubiquitin carrier protein from wheat and Arabidopsis thaliana. Identification of functional domains by in vitro mutagenesis. *J Biological Chem* 266, 23878–23885.

[71] Ichimura Y *et al.* 2000 A ubiquitin-like system mediates protein lipidation. Nature 408, 488–492. (doi:10.1038/35044114)

[72] Huang DT, Miller DW, Mathew R, Cassell R, Holton JM, Roussel MF, Schulman BA. 2004 A unique E1-E2 interaction required for optimal conjugation of the ubiquitin-like protein NEDD8. Nat Struct Mol Biol 11, 927–935. (doi:10.1038/nsmb826)

[73] VanDemark AP, Hill CP. 2004 Grabbing E2 by the tail. Nat Struct Mol Biol 11, 908–909. (doi:10.1038/nsmb1004-908)

[74] Wang Y, Argiles-Castillo D, Kane EI, Zhou A, Spratt DE. 2020 HECT E3 ubiquitin ligases – emerging insights into their biological roles and disease relevance. *J Cell Sci* 133, jcs228072. (doi:10.1242/jcs.228072)

[75] Huibregtse JM, Scheffner M, Beaudenon S, Howley PM. 1995 A family of proteins structurally and functionally related to the E6-AP ubiquitin-protein ligase. Proc National Acad Sci 92, 2563–2567. (doi:10.1073/pnas.92.7.2563)

[76] Berndsen CE, Wolberger C. 2014 New insights into ubiquitin E3 ligase mechanism. Nat Struct Mol Biol 21, 301–307. (doi:10.1038/nsmb.2780)

[77] Zheng N, Shabek N. 2015 Ubiquitin Ligases: Structure, Function, and Regulation. Annu Rev Biochem 86, 1–29. (doi:10.1146/annurev-biochem-060815-014922)

[78] Scheffner M, Werness BA, Huibregtse JM, Levine AJ, Howley PM. 1990 The E6 oncoprotein encoded by human papillomavirus types 16 and 18 promotes the degradation of p53. Cell 63, 1129–1136. (doi:10.1016/0092-8674(90)90409-8)

[79] Werness, Levine A, Howley P. 1990 Association of human papillomavirus types 16 and 18 E6 proteins with p53. *Science (New York, N.Y.)* 248, 76–9. (doi:10.1126/science.2157286)

[80] Ingham RJ, Gish G, Pawson T. 2004 The Nedd4 family of E3 ubiquitin ligases: functional diversity within a common modular architecture. Oncogene 23, 1972–1984. (doi:10.1038/sj.onc.1207436)

[81] Bakkers J, Camacho-Carvajal M, Nowak M, Kramer C, Danger B, Hammerschmidt M. 2005 Destabilization of ΔNp63α by Nedd4-Mediated Ubiquitination Ubc9-Mediated Sumoylation, and Its Implications on Dorsoventral Patterning of the Zebrafish Embryo. Cell Cycle 4, 790–800. (doi:10.4161/cc.4.6.1694)

[82] Xu C, Fan CD, Wang X. 2015 Regulation of Mdm2 protein stability and the p53 response by NEDD4-1 E3 ligase. Oncogene 34, 281–289. (doi:10.1038/onc.2013.557)

[83] Li Y, Ozaki T, Kikuchi H, Yamamoto H, Ohira M, Nakagawara A. 2008 A novel HECT-type E3 ubiquitin protein ligase NEDL1 enhances the p53-mediated apoptotic cell death in its catalytic activity-independent manner. Oncogene 27, 3700–3709. (doi:10.1038/sj.onc.1211032)

[84] Miyazaki K, Ozaki T, Kato C, Hanamoto T, Fujita T, Irino S, Watanabe K, Nakagawa T, Nakagawara A. 2003 A novel HECT-type E3 ubiquitin ligase, NEDL2, stabilizes p73 and enhances its transcriptional activity. Biochem Bioph Res Co 308, 106–113. (doi:10.1016/s0006-291x(03)01347-0)

[85] Peschiaroli A, Scialpi F, Bernassola F, Sherbini ESE, Melino G. 2010 The E3 ubiquitin ligase WWP1 regulates ΔNp63-dependent transcription through Lys63 linkages. Biochem Bioph Res Co 402, 425–430. (doi:10.1016/j.bbrc.2010.10.050)

[86] Holloway A, Simmonds M, Azad A, Fox JL, Storey A. 2015 Resistance to UV-induced apoptosis by β-HPV5 E6 involves targeting of activated BAK for proteolysis by recruitment of the HERC1 ubiquitin ligase. Int J Cancer 136, 2831–2843. (doi:10.1002/ijc.29350)

[87] Wu W, Sato K, Koike A, Nishikawa H, Koizumi H, Venkitaraman AR, Ohta T. 2010 HERC2 Is an E3 Ligase That Targets BRCA1 for Degradation. Cancer Res 70, 6384–6392. (doi:10.1158/0008-5472.can-10-1304)

[88] Gao K *et al.* 2015 The E3 ubiquitin ligase Itch and Yap1 have antagonistic roles in the regulation of ASPP2 protein stability. Febs Lett 589, 94–101. (doi: 10.1016/j.febslet.2014.11.030)

[89] Marín I. 2010 Diversification and Specialization of Plant RBR Ubiquitin Ligases. Plos One 5, e11579. (doi:10.1371/journal.pone.0011579)

[90] Fernandez MA *et al.* 2019 RBR-type E3 ligases and the Ub-conjugating enzyme UBC26 regulate ABA receptor levels and signaling. Plant Physiol 182, pp.00898.2019. (doi:10.1104/pp.19.00898)

[91] Li W, Bengtson MH, Ulbrich A, Matsuda A, Reddy VA, Orth A, Chanda SK, Batalov S, Joazeiro CAP. 2008 Genome-Wide and Functional Annotation of Human E3 Ubiquitin Ligases Identifies MULAN, a Mitochondrial E3 that Regulates the Organelle's Dynamics and Signaling. Plos One 3, e1487. (doi:10.1371/journal.pone.0001487)

[92] Freemont PS, Hanson IM, Trowsdale J. 1991 A novel gysteine-rich sequence motif. Cell 64, 483–484. (doi: 10.1016/0092-8674(91)90229-r)

[93] Lovering R *et al.* 1993 Identification and preliminary characterization of a protein motif related to the zinc finger. Proc National Acad Sci 90, 2112–2116. (doi:10.1073/pnas.90.6.2112)

[94] Lorick KL, Jensen JP, Fang S, Ong AM, Hatakeyama S, Weissman AM. 1999 RING fingers mediate ubiquitin-conjugating enzyme (E2)-dependent ubiquitination. Proc National Acad Sci 96, 11364–11369. (doi:10.1073/pnas.96.20.11364)

[95] Borden KL, Freemont PS. 1996 The RING finger domain: a recent example of a sequence—structure family. Curr Opin Struc Biol 6, 395–401. (doi:10.1016/s0959-440x(96)80060-1)

[96] Katoh M. 2001 Molecular Cloning and Characterization of RNF26 on Human Chromosome 11q23 Region, Encoding a Novel RING Finger Protein with Leucine Zipper. Biochem Bioph Res Co 282, 1038–1044. (doi:10.1006/bbrc.2001.4671)

[97] Levkowitz G, Waterman H, Zamir E, Kam Z, Oved S, Langdon WY, Beguinot L, Geiger B, Yarden Y. 1998 c-Cbl/Sli-1 regulates endocytic sorting and ubiquitination of the epidermal growth factor receptor. Gene Dev 12, 3663–3674. (doi:10.1101/gad.12.23.3663)

[98] Christensen DE, Brzovic PS, Klevit RE. 2007 E2–BRCA1 RING interactions dictate synthesis of mono- or specific polyubiquitin chain linkages. *Nature Structural and Molecular Biology* 14, nsmb1295. (doi:10.1038/nsmb1295)

[99] Christensen DE, Klevit RE. 2009 Dynamic interactions of proteins in complex networks: identifying the complete set of interacting E2s for functional investigation of E3-dependent protein ubiquitination. FEBS Journal 276, 5381–5389. (doi:10.1111/j.1742-4658.2009.07249.x)

[100] Habu T, Harada KH. 2021 UBC13 is an RNF213-associated E2 ubiquitin-conjugating enzyme, and Lysine 63-linked ubiquitination by the RNF213-UBC13 axis is responsible for angiogenic activity. Faseb Bioadvances (doi: 10.1096/fba.2019-00092)

[101] Özkan E, Yu H, Deisenhofer J. 2005 Mechanistic insight into the allosteric activation of a ubiquitin-conjugating enzyme by RING-type ubiquitin ligases. P Natl Acad Sci Usa 102, 18890–18895. (doi:10.1073/pnas.0509418102)

[102] Seol JH *et al.* 1999 Cdc53/cullin and the essential Hrt1 RING–H2 subunit of SCF define a ubiquitin ligase module that activates the E2 enzyme Cdc34. Gene Dev 13, 1614–1626. (doi:10.1101/gad.13.12.1614)

[103] Metzger MB, Pruneda JN, Klevit RE, Weissman AM. 2014 RING-type E3 ligases: Master manipulators of E2 ubiquitin-conjugating enzymes and ubiquitination. Biochimica et Biophysica Acta (BBA) - Molecular Cell Research 1843, 47–60. (doi:10.1016/j.bbamcr.2013.05.026)

[104] Linke K, Mace PD, Smith CA, Vaux DL, Silke J, Day CL. 2008 Structure of the MDM2/MDMX RING domain heterodimer reveals dimerization is required for their ubiquitylation in trans. Cell death and differentiation 15, 841–848. (doi:10.1038/sj.cdd.4402309)

[105] Wu-Baer F, Lagrazon K, Yuan W, Baer R. 2003 The BRCA1/BARD1 Heterodimer Assembles Polyubiquitin Chains through an Unconventional Linkage Involving Lysine Residue K6 of Ubiquitin. Journal of Biological Chemistry 278, 34743–34746. (doi:10.1074/jbc.C300249200)

[106] Buchwald G, Stoop P van der, Weichenrieder O, Perrakis A, Lohuizen M van, Sixma TK. 2006 Structure and E3-ligase activity of the Ring-Ring complex of polycomb proteins Bmi1 and Ring1b. The EMBO journal 25, 2465–2474. (doi:10.1038/sj.emboj.7601144)

[107] Yin Q *et al.* 2009 E2 interaction and dimerization in the crystal structure of TRAF6. Nature structural & molecular biology 16, 658–666. (doi:10.1038/nsmb.1605)

[108] Liew CW, Sun H, Hunter T, Day CL. 2010 RING domain dimerization is essential for RNF4 function. Biochem J 431, 23–29. (doi:10.1042/bj20100957)

[109] Mace PD, Linke K, Feltham R, Schumacher F-R, Smith CA, Vaux DL, Silke J, Day CL. 2008 Structures of the cIAP2 RING Domain Reveal Conformational Changes Associated with Ubiquitin-conjugating Enzyme (E2) Recruitment*. J Biol Chem 283, 31633–31640. (doi:10.1074/jbc.m804753200)

[110] Park YC, Burkitt V, Villa AR, Tong L, Wu H. 1999 Structural basis for self-association and receptor recognition of human TRAF2. Nature 398, 533–538. (doi:10.1038/19110)

[111] Polekhina G, House CM, Traficante N, Mackay JP, Relaix F, Sassoon DA, Parker MW, Bowtell DDL. 2002 Siah ubiquitin ligase is structurally related to TRAF and modulates TNF-α signaling. Nat Struct Biol 9, 68–75. (doi:10.1038/nsb743)

[112] Dou H, Buetow L, Sibbet GJ, Cameron K, Huang DT. 2012 BIRC7–E2 ubiquitin conjugate structure reveals the mechanism of ubiquitin transfer by a RING dimer. Nat Struct Mol Biol 19, 876–883. (doi:10.1038/nsmb.2379)

[113] Koepp DM, Harper JW, Elledge SJ. 1999 How the Cyclin Became a Cyclin Regulated Proteolysis in the Cell Cycle. Cell 97, 431–434. (doi:10.1016/s0092-8674(00)80753-9)

[114] Murray A. 1995 Cyclin Ubiquitination: The destructive end of mitosis. Cell 81, 149–152. (doi:10.1016/0092-8674(95)90322-4)

[115] King RW, Deshaies RJ, Peters J-M, Kirschner MW. 1996 How Proteolysis Drives the Cell Cycle. Science 274, 1652–1659. (doi:10.1126/science.274.5293.1652)

[116] Deshaies RJ. 1999 SCF AND CULLIN/RING H2-BASED UBIQUITIN LIGASES. Annu Rev Cell Dev Bi 15, 435–467. (doi:10.1146/annurev.cellbio.15.1.435)

[117] Kipreos ET, Pagano M. 2000 The F-box protein family. *Genome Biol* 1, reviews3002.1. (doi:10.1186/gb-2000-1-5-reviews3002)

[118] Nakayama KI, Nakayama K. 2006 Ubiquitin ligases: cell-cycle control and cancer. Nat Rev Cancer 6, 369–381. (doi:10.1038/nrc1881)

[119] Cope GA, Deshaies RJ. 2003 COP9 Signalosome A Multifunctional Regulator of SCF and Other Cullin-Based Ubiquitin Ligases. Cell 114, 663–671. (doi:10.1016/s0092-8674(03)00722-0)

[120] Gmachl M, Gieffers C, Podtelejnikov AV, Mann M, Peters J-M. 2000 The RING-H2 finger protein APC11 and the E2 enzyme UBC4 are sufficient to ubiquitinate substrates of the anaphase-promoting complex. Proc National Acad Sci 97, 8973–8978. (doi:10.1073/pnas.97.16.8973)

[121] Zachariae W, Shevchenko A, Andrews PD, Ciosk R, Galova M, Stark MJR, Mann M, Nasmyth K. 1998 Mass Spectrometric Analysis of the Anaphase-Promoting Complex from Yeast: Identification of a Subunit Related to Cullins. Science 279, 1216–1219. (doi:10.1126/science.279.5354.1216)

[122] Leverson JD, Joazeiro CAP, Page AM, Huang H, Hieter P, Hunter T. 2000 The APC11 RING-H2 Finger Mediates E2-Dependent Ubiquitination. Mol Biol Cell 11, 2315–2325. (doi:10.1091/mbc.11.7.2315)

[123] Kramer ER, Scheuringer N, Podtelejnikov AV, Mann M, Peters J-M. 2000 Mitotic Regulation of the APC Activator Proteins CDC20 and CDH1. Mol Biol Cell 11, 1555–1569. (doi:10.1091/mbc.11.5.1555)

[124] Visintin R, Prinz S, Amon A. 1997 *CDC20* and *CDH1*: A Family of Substrate-Specific Activators of APC-Dependent Proteolysis. Science 278, 460–463. (doi:10.1126/science.278.5337.460)

[125] Vodermaier HC, Gieffers C, Maurer-Stroh S, Eisenhaber F, Peters J-M. 2003 TPR Subunits of the Anaphase-Promoting Complex Mediate Binding to the Activator Protein CDH1. Curr Biol 13, 1459–1468. (doi:10.1016/s0960-9822(03)00581-5)

[126] Kramer ER, Gieffers C, Hölzl G, Hengstschläger M, Peters J-M. 1998 Activation of the human anaphase-promoting complex by proteins of the CDC20/Fizzy family. Curr Biol 8, 1207-12S4. (doi:10.1016/s0960-9822(07)00510-6)

[127] Raff JW, Jeffers K, Huang J. 2002 The roles of Fzy/Cdc20 and Fzr/Cdh1 in regulating the destruction of cyclin B in space and time. J Cell Biology 157, 1139–1149. (doi:10.1083/jcb.200203035)

[128] Prinz S, Hwang ES, Visintin R, Amon A. 1998 The regulation of Cdc20 proteolysis reveals a role for the APC components Cdc23 and Cdc27 during S phase and early mitosis. Curr Biol 8, 750–760. (doi:10.1016/s0960-9822(98)70298-2)

[129] Passmore LA, McCormack EA, Au SWN, Paul A, Willison KR, Harper JW, Barford D. 2003 Doc1

mediates the activity of the anaphase-promoting complex by contributing to substrate recognition. Embo J 22, 786–796. (doi:10.1093/emboj/cdg084)

[130] Au SWN, Leng X, Harper JW, Barford D. 2002 Implications for the Ubiquitination Reaction of the Anaphase-promoting Complex from the Crystal Structure of the Doc1/Apc10 Subunit. J Mol Biol 316, 955–968. (doi: 10.1006/jmbi.2002.5399)

[131] Carroll CW, Morgan DO. 2002 The Doc1 subunit is a processivity factor for the anaphase-promoting complex. Nat Cell Biol 4, 880–887. (doi:10.1038/ncb871)

[132] Pfleger CM, Lee E, Kirschner MW. 2001 Substrate recognition by the Cdc20 and Cdh1 components of the anaphase-promoting complex. Gene Dev 15, 2396–2407. (doi:10.1101/gad.918201)

[133] Burton JL, Solomon MJ. 2001 D box and KEN box motifs in budding yeast Hsl1p are required for APC-mediated degradation and direct binding to Cdc20p and Cdh1p. Gene Dev 15, 2381–2395. (doi:10.1101/gad.917901)

[134] Kraft C, Vodermaier HC, Maurer-Stroh S, Eisenhaber F, Peters J-M. 2005 The WD40 Propeller Domain of Cdh1 Functions as a Destruction Box Receptor for APC/C Substrates. Mol Cell 18, 543–553. (doi:10.1016/j.molcel.2005.04.023)

[135] Tian W, Li B, Warrington R, Tomchick DR, Yu H, Luo X. 2012 Structural analysis of human Cdc20 supports multisite degron recognition by APC/C. Proc National Acad Sci 109, 18419–18424. (doi:10.1073/pnas.1213438109)

[136] Zur A, Brandeis M. 2002 Timing of APC/C substrate degradation is determined by fzy/fzr specificity of destruction boxes. Embo J 21, 4500–4510. (doi:10.1093/emboj/cdf452)

[137] Glotzer M, Murray AW, Kirschner MW. 1991 Cyclin is degraded by the ubiquitin pathway. Nature 349, 132–138. (doi:10.1038/349132a0)

[138] Ayad NG, Rankin S, Ooi D, Rape M, Kirschner MW. 2005 Identification of Ubiquitin Ligase Substrates by In Vitro Expression Cloning. Methods Enzymol 399, 404–414. (doi:10.1016/s0076-6879(05)99028-9)

[139] Pfleger CM, Kirschner MW. 2000 The KEN box: an APC recognition signal distinct from the D box targeted by Cdh1. Gene Dev 14, 655–665.

[140] Reis A, Levasseur M, Chang H, Elliott DJ, Jones KT. 2006 The CRY box: a second APCcdh1-dependent degron in mammalian cdc20. Embo Rep 7, 1040–1045. (doi:10.1038/sj.embor.7400772)

[141] Zachariae W, Schwab M, Nasmyth K, Seufert W. 1998 Control of Cyclin Ubiquitination by CDK-Regulated Binding of Hct1 to the Anaphase Promoting Complex. Science 282, 1721–1724. (doi:10.1126/science.282.5394.1721)

[142] Harper JW, Burton JL, Solomon MJ. 2002 The anaphase-promoting complex: it's not just for mitosis any more. Gene Dev 16, 2179–2206. (doi:10.1101/gad.1013102)

[143] Yu H. 2007 Cdc20: A WD40 Activator for a Cell Cycle Degradation Machine. Mol Cell 27, 3–16. (doi:10.1016/j.molcel.2007.06.009)

[144] Baker DJ, Dawlaty MM, Galardy P, Deursen JM van. 2007 Mitotic regulation of the anaphase-promoting complex. Cell Mol Life Sci 64, 589–600. (doi:10.1007/s00018-007-6443-1)

[145] Aravind L, Koonin EV. 2000 The U box is a modified RING finger — a common domain in ubiquitination. Curr Biol 10, R132–R134. (doi:10.1016/s0960-9822(00)00398-5)

[146] Hatakeyama S, Yada M, Matsumoto M, Ishida N, Nakayama K-I. 2001 U Box Proteins as a New Family of Ubiquitin-Protein Ligases. J Biol Chem 276, 33111–33120. (doi:10.1074/jbc.m102755200)

[147] Ballinger CA, Connell P, Wu Y, Hu Z, Thompson LJ, Yin L-Y, Patterson C. 1999 Identification of CHIP, a Novel Tetratricopeptide Repeat-Containing Protein That Interacts with Heat Shock Proteins and Negatively Regulates Chaperone Functions. Mol Cell Biol 19, 4535–4545. (doi:10.1128/mcb.19.6.4535)

[148] Jiang J, Ballinger CA, Wu Y, Dai Q, Cyr DM, Höhfeld J, Patterson C. 2001 CHIP Is a U-box-dependent E3 Ubiquitin Ligase IDENTIFICATION OF Hsc70 AS A TARGET FOR UBIQUITYLATION. J Biol Chem 276, 42938–42944. (doi:10.1074/jbc.m101968200)

[149] Hatakeyama S, Nakayama KI. 2003 U-box proteins as a new family of ubiquitin ligases. Biochem Bioph Res Co 302, 635–645. (doi:10.1016/s0006-291x(03)00245-6)

Chapter 9

E3 Ligase for CENP-A (Part 1)

Yohei Niikura and Katsumi Kitagawa

Abstract

CENP-A is a centromere-specific histone H3 variant that is required to ensure kinetochore assembly for proper chromosome segregation and its function is highly conserved among different species including budding yeast, *Saccharomyces cerevisiae*. The budding yeast *Saccharomyces cerevisiae* has genetically defined point centromeres, unlike other eukaryotes. Although, most eukaryotic centromeres are maintained epigenetically, currently only budding yeast *S. cerevisiae* centromeres are known to be genetically specified by DNA sequence, The small size and sequence specificity of the budding yeast centromere has made yeast a powerful organism for its study in many aspects. Many post-translational modifications (PTMs) of CENP-A and their functions have been recently reported, and studies with budding yeast are providing insights into the role of CENP-A/Cse4 PTMs in kinetochore structure and function. Multiple functions are controlled especially by ubiquitylation and sumoylation by E3 ligases that control CENP-A protein has initially emerged in the budding yeast as an important regulatory mechanism. Here we focus on what is known about the budding yeast E3 ligases for CENP-A/Cse4 ubiquitylation and sumoylation and their biological functions and significance.

Keywords: CENP-A, Cse4, Cnp1, E3 ligase, centromere, kinetochore, ubiquitylation, sumoylation, epigenetics, Psh1, Siz1 and Siz2, Slx5 and Slx8, CUL3/RDX, SCF, APC, CUL4A/RBX1/COPS8, DAXX (fruit fly DLP), Scm3, CAF-1 complex, CAL1, HJURP, Mis18 (human Mis18α and Mis18β) and Mis16 (human RbAp46 and RbAp48)

1. Introduction

The mechanistic process to establish centromeric chromatin of budding yeast and its structures have been actively studied [1–3]. In contrast to most eukaryotic centromeres that span megabases of DNA, in the budding yeast, *Saccharomyces cerevisiae* point centromeres are comprised of ~125 bp of DNA and are conserved among all 16 chromosomes [3]. There are three conserved centromere-determining elements (CDE) consisting of CDEI-III [1–3]. Although, most eukaryotic centromeres are maintained epigenetically, currently only budding yeast *S. cerevisiae* centromeres are known to be genetically specified by DNA sequence. The CDEIII consensus (TGTTT(T/A) TGNTTTCCGAAANNNAAAAA) binds to the CBF3 complex via a conserved CCG motif that is essential for centromere function, and the small size and sequence specificity of the budding yeast centromere has made yeast a powerful organism for its study in many aspects [1].

In *S. cerevisiae*, all pre-existing CENP-A^{Cse4} is replaced by newly synthesized CENP-A^{Cse4} during the S phase [4]. Centromeric assembly of CenH3 requires the adaptor protein, suppressor of chromosome mis-segregation (Scm3) in budding and fission yeasts [5–9], as well as the Holliday junction recognition protein (HJURP) in humans [10, 11]. Scm3/HJURP directly interacts with CenH3 and is essential for the assembly and maintenance of a functional kinetochore [5–12]. Scm3 recognizes CENP-A^{Cse4} through the centromere-targeting domain (CATD) in the histone fold and mediates its incorporation into chromatin in vivo and in vitro [1].

Early studies showed that Scm3 is required for G2/M progression and Cse4 localization at centromeres. Scm3 contains 2 essential protein domains: a Leu-rich nuclear export signals and a heptad repeat domain that is widely conserved in fungi [5–11]. Localization of Cse4 to centromeres and the assembly activity is dependent on an evolutionarily conserved central core motif in Scm3 [13]. Camahort et al. showed that Scm3 is required throughout the whole cell cycle as well as the loading period for Cse4 [5, 14]. Consistent with these findings, Xiao et al. showed that Scm3 has an N-terminal nonspecific DNA binding domain for AT-rich DNA and a central histone chaperone domain (Cse4/H4 binding domain, CBD) that promotes specific loading of Cse4/H4 [15]. Moreover, Xiao et al. demonstrated that Scm3-GFP is enriched at centromeres in all cell cycle phases in live cells, and their results of ChIP analysis showed that Scm3 occupies centromere DNA throughout the cell cycle, even when Cse4 and H4 are temporarily dislodged in the S phase, suggesting Scm3 is a critical factor for recruitment of Cse4/H4 as well as maintenance of an H2A/H2B-deficient centromeric nucleosome [15]. Luconi et al. showed that Scm3 signals are present at centromeres when metaphase begins, and enriched in anaphase [14, 16] as observed for Scm3 in fission yeast *S. pombe* [7, 14]. However, HJURP is recruited to centromeres during early G1 [10, 11] (see also next chapter, section (4.1)).

Currently, the structure of budding yeast centromeric CENP-A^{Cse4}-containing nucleosomes remains controversial among different research groups as in other species [17]. Dechassa et al. performed structural analysis and showed that the substitution of H3 with Cse4 results in octameric nucleosomes that organize DNA in a left-handed superhelix [18]. Cse4-nucleosomes exhibit an open conformation with weakly bound terminal DNA segments and do not preferentially form nucleosomes on its cognate centromeric DNA. The Cse4-specific octameric nucleosomes do not contain Scm3 as a stably bound component. Cho et al. reported the structure of a complex formed by an N-terminal fragment of Scm3 with the histone-fold domains of Cse4, and H4, which were all purified from the budding yeast *Kluyveromyces lactis* [19]. They described the structure of a (Cse4: H4)(2) heterotetramer; comparison with the structure of the Scm3:Cse4:H4 complex shows that tetramer formation and DNA-binding require displacement of Scm3 from the nucleosome core. Previously published structures of the Scm3 histone complex demonstrated that Scm3 binds only one copy of Cse4–H4 [20]. Dechassa et al. further showed that Scm3 deposits Cse4–H4 through a dimer intermediate onto DNA to form a (Cse4–H4)2–DNA complex (tetrasome) [20] (**Figure 1**, right). Recently, the budding yeast $Yta7^{ATAD2}$ (the homolog of AAA^+ ATPase and bromodomain factor ATAD2/ANCCA, which is overexpressed in many types of cancers) was shown to collaborate with Scm3 to deposit Cse4 at the centromere [25].

Recently, the importance of centromeric long non-coding RNA (cenRNA) for centromere integrity has been suggested in various species [35–37] including budding yeast [38–40]. Ling et al. reported that all the budding yeast centromere express long noncoding RNAs (cenRNAs), especially in S phase and induction of cenRNAs

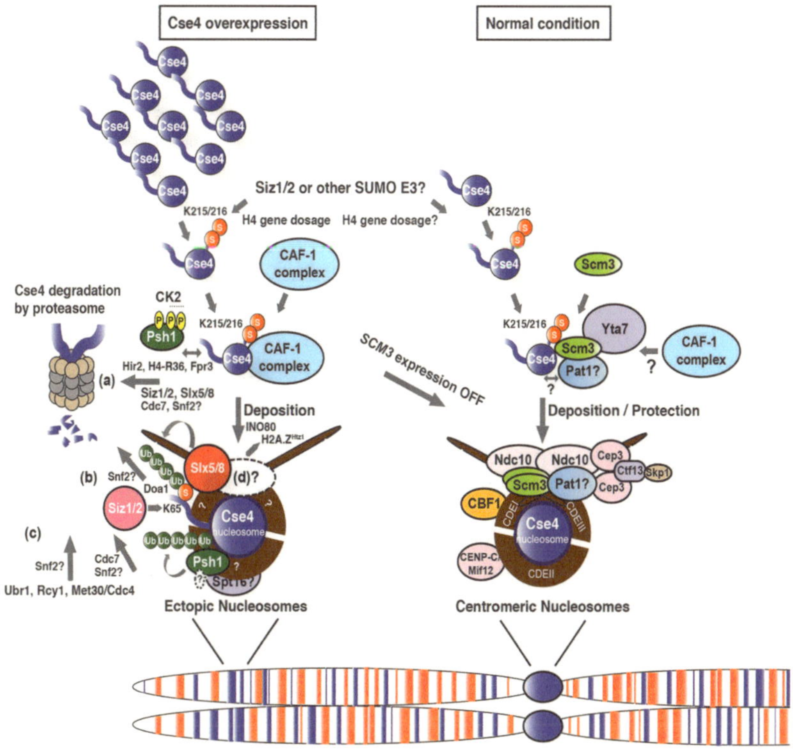

Figure 1.
Mechanistic scheme for Saccharomyces Cerevisiae CENP-A^{Cse4} pathways. (Right) In normal conditions, sumoylation of Cse4 K215/216 facilitates deposition into chromatin [21]. Centromeric Cse4 is protected by Scm3 from Psh1-mediated degradation [22, 23]. Interaction of Pat1 with Scm3 is required for its maintenance at kinetochores and Pat1 affects the structure of CEN chromatin and protects Cse4 from Psh1-mediated ubiquitylation [24]. Yta7 cooperates with Scm3 to deposit Cse4 at the centromere [25]. Note that histone H4 including one in Cse4–H4 that binds one copy of Scm3 [20] is omitted for simplicity. (Right-center) The functional role of Cse4 K215/216 sumoylation is distinct from that of Cse4 K65 sumoylation [21], although it is not yet clear if Siz1/Siz2 also target Cse4 C-terminal K215/216 as Cse4 N-terminal K65 (see also left) or a different unknown SUMO E3 target Cse4 K215/216. (Center) The interaction of sumoylated Cse4 K215/216 with CAF-1 promotes centromeric localization of overexpressed Cse4 only under conditions when Scm3 is depleted (SCM3 expression OFF) [26]. However, CAF-1 function in normal conditions is not clear, although CAF-1 promotes ubiquitylation of free Cse4, opposite to the effect of Scm3 (see also left and right). (Left-center) The histone H4 gene dosage promotes Cse4 sumoylation and mislocalization to noncentromeric regions [27], but its effect on Cse4 K215/216 sumoylation [21] to facilitate the deposition of overexpressed (or endogenous) Cse4 into CEN is not clear. (Left) When Cse4 is overexpressed, Psh1 promotes the degradation of free and ectopically incorporated Cse4. Ohkuni et al. suggested two independent pathways prevent the stable incorporation of Cse4 into non-centromeric chromatin [28]. (i) The first pathway depends on the interaction of Psh1 with Cse4. (ii) The second pathway requires Cse4 K65 sumoylation by Siz1/Siz2 and subsequent Cse4 K65-ubiquitination by Slx5. Both pathways contribute to (a) regulate soluble pools of Cse4 to prevent its mislocalization and/or (b) facilitate proteolysis of non-centromeric chromatin-bound Cse4. The Cse4 K65 sumoylation occurs downstream of Cse4 K215/ 216 sumoylation, i.e., after Cse4 is incorporated into chromatin [21]. Psh1 is phosphorylated by the Cka2 subunit of casein kinase 2 (CK2) to promote its E3 activity for Cse4, and Cse4 misincorporation is prevented by the intact Psh1-Cse4 association [29], but the status of Psh1's post-translational modifications, including the phosphorylation in ectopic Cse4-nucleosomes, is not yet elucidated. Hir2 [30], histone H4-R36 [31], and structural change in Cse4 by Fpr3 [32] might be important for the interaction between Cse4 and Psh1. (c) Ubr1, Rcy1, and Met30/Cdc4 E3 ligases could be in a Psh1-independent proteolysis pathway [33, 34], but the mechanistic details are unknown. (d) Factors/components that stabilize ectopically incorporated Cse4 are not yet clear. Preceding post-translational modifications before ubiquitylation or sumoylation of Cse4 and other proposed factors relevant to Psh1 function (e.g., Snf2, Doa1, Spt16, Pat1, Hir2, Cse4 MIMAS motif, Cdc7, Ubp8 [SAGA-DUB; not shown in this cartoon], etc.) and multiple E3s (Psh1, Ubr1, Rcy1, and Met30/Cdc4) are summarized in ***Table 1****. While there are three conserved centromere-determining elements (CDE) consisting of CDEI-III [1–3], DNA sequence elements required for non-centromeric Cse4 nucleosome or its presence itself is unclear. Histone H4, including one in Cse4–H4 that binds one copy of Scm3 [20], is omitted for simplicity. This figure is partly adapted from Ohkuni et al. [21, 28].*

coincides with Cse4 loading time and is dependent on DNA replication [38]. The cenRNA is tightly regulated and repressed by the kinetochore protein Cbf1 and histone H2A variant H2A.Z^{Htz1}, and de-repressed during the S phase of the cell cycle, suggesting that an appropriate level of cenRNAs is essential for point centromere activity [38]. Interestingly, when they knocked down all cenRNAs from the endogenous chromosomes, but not the cenRNA from the circular minichromosome, they still observed an increase in minichromosome loss, suggesting that cenRNA functions in trans to regulate centromere activity. Chen et al. independently demonstrated that budding yeast cenRNA is negatively regulated by Cbf1 and binding of the Pif1 DNA helicase to the centromeres, which happens in mid–late S phase, occurred at about the same time as Cbf1 loss from the centromere [40]. These data suggest that Pif1 may facilitate this loss by its known ability to displace proteins from DNA. Ling et al. further showed that budding yeast cenRNAs are cryptic unstable transcripts (CUTs) that can be degraded by the nuclear RNA decay pathway suggesting that cenRNA can serve important cellular functions when it exists at the right time with the right level [39]. Together, these results in budding yeast indicate that the regulation of cenRNA is an essential factor for centromere structure and function.

1.1 Identification of Psh1 E3 ligase and its function

CENP-A (CenH3) proteolysis has also been reported in senescent human cells [41] or upon infection with herpes simplex virus 1 [42]. However, little had been known about the actual mechanisms that regulate CENP-A (CenH3) proteolysis. Collins et al. initially reported that the levels of the budding yeast CenH3, Cse4, are regulated by ubiquitin-proteasome-mediated proteolysis in 2004 [43]. They isolated a dominant lethal mutant, *CSE4-351*, and showed that the Cse4-351 mutant protein is stable and localized to euchromatin, suggesting that proteolysis prevents Cse4 euchromatic localization. They also constructed wild-type Cse4 fused to a degron signal, and showed that the soluble Cse4 protein was rapidly degraded, but the centromere-bound Cse4 was stable. These data indicate that centromere localization protects Cse4 from degradation. In 2010, two groups reported that budding yeast (*Saccharomyces cerevisiae*) CENP-A homolog, Cse4, is ubiquitylated by an E3 ubiquitin ligase called Psh1 (named for Pob3/Spt16/ histone associated [44]) [22, 45] (**Figure 1** and **Table 1**). Both groups identified independently that Psh1 leads to the degradation of Cse4 controlling the cellular level of Cse4 via ubiquitylation and proteolysis.

Hewawasam et al. performed TAP purification of Psh1 and identified Cse4 as well as several other kinetochore proteins by multidimensional protein identification technology analysis [22]. They described that Psh1 consists of three main domains: (i) a RING finger, (ii) a zinc finger, and (iii) a highly acidic domain [22, 23]. They performed co-immunoprecipitation using whole-cell extracts and showed that the RING finger of Psh1 is important to interact with Cse4. They also performed a pulse-chase assay and demonstrated that both RING and zinc fingers are critical for efficient control of Cse4 levels. They demonstrated the specificity of the ubiquitylation activity of Psh1 toward Cse4 in vitro and identified the sites of ubiquitylation. Mutation of these lysine sites prevents ubiquitylation of Cse4 by Psh1 in vitro and stabilizes Cse4 in vivo. Elimination of the Cse4-specific chaperone Scm3 destabilizes Cse4, and the addition of Scm3 to the Psh1-Cse4 ubiquitylation reaction prevents Cse4 ubiquitylation. Meanwhile, the deletion of Psh1 stabilizes Cse4. These data suggest that Scm3 and Psh1 might compete

lo	CENP-A homolog	E3 ligase (ubiquitylation or sumoylation)	Function	Preceding PTMs before ubiquitylation or sumoylation	Other proposed factor relevant to E3 function
Saccharomyces cerevisiae	Cse4	Psh1 (ubiquitylation)	Proteasomal degradation to remove non-centromeric CENP-A	P134 isomerization by Fpr3	Scm3, Snf2, Doa1, Fpr3, Spt16. Phosphorylation of Psh1 by Cka2, Pat1, Histone H4-R36, Gene dosage of histone H4 (HHF1 and HHF2), CAF-1, Hir2, Cdc7, Ubp8 (deubiquitylation, SAGA-DUB), Cse4 MIMAS motif
		Slx5/8 (vertebrate RNF4) (ubiquitylation)	Proteasomal degradation to remove non-centromeric CENP-A (Slx5-mediated Cse4 proteolysis could be independent of Psh1)	K65 sumoylation by Siz1/2	K65 sumoylation by Siz1/2
		Siz1/2 (sumoylation)	Proteasomal degradation to remove non-centromeric CENP-A	N.D.	N.D. (The effect of SUMO-proteases Ulp2/SENP6, on CenH3 was not confirmed.)
		Ubr1, Rcy1 (ubiquitylation)	Proteasomal degradation of Cse4	N.D.	N.D.
		Met30/Cdc4 (ubiquitylation)	Proteasomal degradation of Cse4 (Met30/Cdc4-mediated Cse4 proteolysis could be independent of Psh1)	N.D.	N.D.

Table 1.
E3 ligase for budding yeast (Saccharomyces Cerevisiae) CENP-A^{Cse4}.

for binding to Cse4. Cse4 that is not associated with Scm3 may be targeted by Psh1 for proteolysis, but Cse4 in a complex with Scm3 may be protected [23] (**Figure 1**, right) (see also next chapter, section 4.2). Cse4 overexpression is toxic without Psh1, and Cse4 is found at ectopic locations. Therefore, they suggested that the E3 activity of Psh1 prevents the mislocalization of Cse4 (**Figure 1**, left).

Ranjitkar et al. also identified Psh1 by mass spectrometry analysis after purification of 3xFLAG-Cse4^{16R} that is not ubiquitylated in vivo [45]. They demonstrated that Cse4 overexpression causes growth defects on *psh1*-deleted (*psh1*Δ) cells and results in euchromatic localization of overexpressed Cse4. In immunoprecipitation analysis, they detected that full-length Cse4 and the histone fold domains (HFD)-Cse4 associate with Psh1, but the N-terminal domains (NTD)-Cse4 does not interact with Psh1. However, greater levels of full-length Cse4 associated with Psh1 compared to HFD-Cse4 were observed. These data suggest that the Cse4 N-terminus might contribute to the interaction of Cse4 with Psh1 in vivo. Because the CATD is critical for Psh1 binding to Cse4, they analyzed the stability of the chimeric proteins. Myc-Cse4-CATD levels in wild-type and *psh1* mutant cells after repressing transcription and translation were assessed. The degradation of $H3^{CATD}$ was dependent on Psh1 in contrast to the Cse4 chimera lacking the CATD, suggesting that the Cse4 CATD is a key regulator of its stability and facilitates Psh1 to distinguish Cse4 from histone H3. Therefore, they proposed a new role of the CATD in maintaining the exclusive localization of Cse4 by preventing its mislocalization to euchromatin via Psh1-mediated degradation.

However, the new findings of E3 ligase, Psh1, by these two groups left these open questions and stimulated other researchers to study the Psh1-mediated ubiquitylation and degradation of Cse4 as well as CENP-A homologs of other species.

i. Why does deletion of *PSH1* not show a phenotype unless Cse4 is overexpressed [22, 45]? These data may suggest additional Cse4 regulatory mechanisms. In agreement with this concept, Cse4 is not completely stabilized when Psh1 is deleted and a lysine-free mutant of Cse4 is still degradable [22, 23, 43, 45]. Thus, it seems plausible that there are other destabilization mechanisms not yet discovered [22, 23, 43, 45].

ii. No Psh1 ortholog in other eukaryotes is yet identified. Because the RING and zinc fingers are highly conserved motifs in many proteins from yeast to human, it is difficult to verify such an ortholog. It is also unclear whether the ubiquitin–proteasome pathway that controls CENP-A proteolysis is conserved among different species.

iii. Can Psh1 be the unique E3 ligase in yeast? Is it possible to identify other E3 ligases that ubiquitylate Cse4 in the same or different function? Is the function of the Cse4 ubiquitylation restricted only to proteolysis?

iv. Are any other post-translational modifications of Cse4 involved in upstream or downstream functions of Cse4 ubiquitylation?

v. What is the genome-wide misincorporation pattern of Cse4? How does the pattern change in the presence and absence of Psh1? Does Cse4 misincorporation affect promotor function and transcriptional regulation?

vi. What is the molecular mechanism for the selective recognition and ubiquitylation of Cse4 by Psh1? Are other components required for such activities, or are other PTMs of Cse4 involved?

vii. What are the deubiquitylase and deubiquitylation mechanisms of Cse4?

In the following sections, answers to some of these questions are further described.

1.2 Additional Cse4 regulatory mechanisms and factors that are required for proper ubiquitylation of Cse4 in vivo

1.2.1 SWI/SNF complex

Gkikopoulos et al. had identified DNA sequences to which the *S. cerevisiae* ATP-dependent SWI/SNF chromatin remodeling complex is bound genome-wide to gain insight into that complex [46], and they observed that the complex is enriched at the centromeres of each chromosome. In their study, partial redistribution of the Cse4 to sites on chromosome arms was observed by deletion of the gene encoding the Snf2 subunit of the complex (**Figure 1** and **Table 1**). Cultures of *snf2Δ* yeast were found to progress through mitosis slowly, and this slow progress depends on the mitotic checkpoint protein Mad2; defects in chromosome segregation were observed in the absence of Mad2. Chromatin organization at centromeres is less distinct in the absence of Snf2, and especially hypersensitive sites flanking the Cse4-containing nucleosomes are less prominent. In addition, SWI/SNF complex was especially effective in the dissociation of Cse4 containing chromatin in their nucleosome reconstitution and remodeling assay in vitro. Taken together, these data suggest a role for Snf2 in the maintenance of point centromeres involving the removal of Cse4 from ectopic sites, rather than via directing incorporation of Cse4 at centromeres.

1.2.2 A novel role of the N-terminus of Cse4

The aforementioned groups had shown interactions of Psh1 with the C-terminus CATD of Cse4 and ubiquitylation of Cse4 at its C-terminus in vitro [22, 45]. Further, Au et al. demonstrated a role for ubiquitination of the N-terminus of Cse4 in regulating Cse4 proteolysis [47]. They initiated their studies with a mutant *cse4*^{16KR} (16KR) and fusion mutants in which lysines (K) are mutated to arginines (R). Their results indicated that lack of ubiquitylation of the C-terminus due to KR mutations does not increase protein stability, while the mutations in the N-terminus do so significantly, suggesting that the N-terminus of Cse4 facilitates its proteolysis. They also investigated the role of Cse4 domains in directing Cse4 proteolysis using a genome-wide screen (a synthetic genetic array, SGA), and identified *DOA1* and *PSH1*. Their results using cse4KR mutants suggest that Psh1 is not the sole regulator of Cse4 proteolysis and that Doa1 facilitates Cse4 N-terminus-dependent proteolysis. We also note that N-terminal functions of CENP-A were described for some species [12, 28, 45, 47–55] (see also Sections 1.1, 1.4.1, 1.9 and next chapter, sections 2.1, 2.3, 2.4, 3.1, 4.1, 4.6, and 5.1).

1.2.3 Fpr3 peptidyl-prolyl cis-trans isomerase

Ohkuni et al. reported that the proline isomerase Fpr3 regulates Cse4 proteolysis [32] (**Figure 1** and **Table 1**). *FPR3* encodes a peptidylprolyl cis-trans isomerase (PPIase) which has a function in the meiotic recombination checkpoint pathway [56, 57]. In their study, the *fpr3Δ* or *fpr4Δ* strain displayed a significant chromosome missegregation phenotype. Cse4 protein levels were increased in *fprΔ* cells, and deletion of *FPR3* stabilized Cse4 protein levels in vivo. PPIase dead mutants (W363L and F402Y) stabilized Cse4 protein levels in vivo, suggesting that Fpr3 isomerization activity is necessary for Cse4 proteolysis. Interaction between Cse4 and Psh1 was diminished in *fpr3Δ* cells, and P134V mutation (a mutation of a putative target of Fpr3 isomerization) in Cse4 diminished the Psh1 interaction, suggesting that Fpr3 regulates the Cse4-Psh1 interaction. In summary, they proposed that structural change in Cse4 by Fpr3 might be important for the interaction between Cse4 and the E3 ubiquitin ligase Psh1 (**Figure 1**, left). Prolyl isomerization of fission yeast CENP-A^{Cnp1} was discussed in this paper [32] (see also next chapter, section 2).

1.2.4 FACT complex

Deyter et al. identified a role for the conserved chromatin-modifying complex FACT (facilitates chromatin transcription/transactions) in preventing Cse4 mislocalization to euchromatin by mediating its proteolysis [58]. They initially found that Psh1 cannot efficiently ubiquitylate Cse4 nucleosomes in vitro, suggesting that additional factors must facilitate Cse4 removal from chromatin in vivo. The Spt16 subunit (**Figure 1b** and **Table 1**) of the FACT complex binds to Psh1, and this interaction between Psh1 and Spt16 is critical for both Cse4 ubiquitylation and its exclusion from euchromatin. Therefore, a Psh1 mutant that cannot associate with FACT has reduced interaction with Cse4 in vivo. Collectively, they proposed a previously unknown mechanism to maintain centromere identity and genomic stability through the FACT-mediated degradation of ectopically localized Cse4.

1.2.5 Phosphorylation by casein kinase 2 (CK2)

Hewawasam et al. reported that Psh1 is phosphorylated by the Cka2 subunit of casein kinase 2 (CK2) to promote its E3 activity for Cse4 [29] (**Figure 1a** and **Table 1**). They first showed that the deletion of *CKA2* significantly stabilized Cse4. Consistently, Cse4 has stabilized in a Psh1 phospho-depleted mutant strain in which all identified phosphorylation sites (total 10 sites) were changed to alanines. However, they showed that phosphorylation of Psh1 did not control Psh1-Cse4 or Psh1-Ubc3(E2) interactions. Mislocalization of Cse4 was mild, although Cse4 was highly stabilized in a *cka2Δ* strain. These data suggest that Cse4 misincorporation was prevented by the intact Psh1-Cse4 association. Supporting that idea, Psh1 was also stabilized in a *cka2Δ* strain. However, some questions remain if the phosphorylation of Psh1 by Cka2 is required for its E3 activity to degrade "already" mis-incorporated Cse4 on the non-centromeric chromatin and the status of Psh1's PTMs in such non-centromeric chromatin (**Figure 1b**). Collectively, these results suggest that phosphorylation is important for Psh1-assisted control of Cse4 levels and that the Psh1-Cse4 association itself functions to prevent Cse4 misincorporation.

1.2.6 Pat1, a protector of Cse4 against Psh1

Mishra et al. showed that a kinetochore protein, Pat1 (**Figure 1**, right and **Table 1**), protects *CEN*-associated Cse4 from ubiquitylation to maintain the proper structure and function of the kinetochore [24]. In their study, *PAT1*-deletion (*pat1Δ*) strains exhibit increased ubiquitylation of Cse4 and faster turnover of Cse4 at kinetochores. Psh1 interacts with Pat1 and contributes to the increased ubiquitylation of Cse4 in *pat1Δ* strains. Their results showed that transient induction of *PSH1* in a wild-type strain resulted in phenotypes similar to a *pat1Δ* strain (e.g., a reduction in *CEN*-associated Cse4, increased Cse4 ubiquitylation, defects in the spatial distribution of Cse4 at kinetochores, and altered structure of *CEN* chromatin). Pat1 interacts with Scm3 (a Cse4-specific chaperone) and *pat1Δ* strains showed reduced levels of centromeric Scm3, suggesting that the interaction of Pat1 with Scm3 is required for its maintenance at kinetochores. In summary, these results suggest a new mechanism by which Pat1 affects the structure of *CEN* chromatin and protects Cse4 from Psh1-mediated ubiquitylation for faithful chromosome segregation.

1.3 Cse4 misincorporation affects promotor function and transcriptional regulation, and histone H4 facilitates the proteolysis of the Cse4

1.3.1 Regulation of Cse4 protein levels prevents misincorporation at promotor nucleosomes and transcriptional defects

One interesting question is if Cse4 misincorporation affects promotor function and transcriptional regulation. Hildebrand et al. addressed the genome-wide misincorporation pattern of Cse4 in the presence and absence of Psh1, performing chromatin immunoprecipitation analysis followed by high throughput sequencing [59]. They found that ectopic Cse4 mislocalized to intergenic regions of the genome. Mislocalized Cse4 is enriched at promoters that contain histone H2A. Z^{Htz1} nucleosomes flanking nucleosome-depleted regions (NDRs), however, Cse4 mislocalization does not depend on H2A.Z^{Htz1}. Instead, the chromatin remodeling inositol-requiring 80 (INO80) complex (INO80-C), which removes H2A.Z^{Htz1} from nucleosomes, contributes to the ectopic deposition of Cse4 [59] (**Figure 1**, left). However, the functional relationship of INO80-C with other factors (e.g., CAF-1 complex) for Cse4 ectopic deposition remains to be elucidated (**Figure 1**, left). Together, this transcriptional profiling revealed that mislocalized Cse4 significantly disturbs transcription in the absence of Psh1, suggesting that regulating centromeric nucleosome localization is important for ensuring accurate promoter function and transcriptional regulation.

1.3.2 Histone H4 facilitates the proteolysis of the Cse4

Because Cse4 proline residues though the Fpr3 regulation influence its degradation as reported by Ohkuni et al. [32] (see also Section 1.2.3), Deyter et al. hypothesized that additional features of the Cse4 nucleosome might be important for Cse4 proteolysis [31]. They initially asked whether histone H4 residues are important for Cse4 degradation, since Cse4 binds with high affinity to histone H4 before and after deposition on DNA, and they determined that Cse4 protein levels are stabilized in H4-R36A mutant cells and Cse4 is enriched in the euchromatin. Consistent with those data, they also demonstrated that H4-R36 is important for the interaction between Cse4 and Psh1 (**Figure 1** and **Table 1**). They also analyzed Psh1 localization in WT vs. H4- R36A cells at the 5′, 3′, and coding regions of two highly transcribed genes, *ADH1* and *PMA1*, because Psh1 interaction with FACT is important for Cse4 ubiquitylation and degradation, as previously reported [58] (see also Section 1.2.4). Their ChIP-qPCR revealed that Psh1 also shows a strong enrichment at the 3' UTRs of these genes in H4-R36A cells compared to wild-type cells, while the levels at the promoter and gene regions were similar to wild-type cells. These data suggest that altered Psh1 localization could contribute to the Cse4 stability phenotype in H4-R36A mutant cells.

This group previously had discovered that overexpressed Cse4 is mislocalized to nucleosomes in both tandem and divergent intergenic regions in the absence of Psh1, as shown earlier [59]. Therefore, they tested whether this is also true in the H4-R36A mutant cells by performing ChIP-qPCR. Cse4 mislocalization was negatively correlated with Psh1 enrichment in H4-R36A cells. Taken together, these data revealed H4-R36 is a key residue for efficient Cse4 degradation, likely by facilitating the interaction between Psh1 and Cse4.

1.3.3 Reduced gene dosage of histone H4 prevents Cse4 mislocalization and chromosomal instability

Eisenstatt et al. further utilized a genome-wide screen (SGA) to identify factors that facilitate the mislocalization of overexpressed Cse4 by characterizing suppressors of the *psh1Δ GAL-CSE4* SDL [27]. They found that deletions of histone H4 alleles (HHF1 or HHF2; **Table 1**), which were among the most major suppressors, also suppress *slx5Δ*, *cdc4-1*, *doa1Δ*, *hir2Δ*, and *cdc7-4 GAL-CSE4* SDL (**Table 1**). Defects in sumoylation and reduced mislocalization of overexpressed Cse4 are observed with a reduced dosage of H4, and these events lead to suppression of CIN when Cse4 is overexpressed (see about Cse4 sumoylation also in the following Section 1.4). *hhf1-20*, *cse4-102*, and *cse4-111* mutants, which have defective Cse4-H4 interactions, also show reduced sumoylation of Cse4 and do not cause *psh1Δ GAL-CSE4* SDL. Overall, these results identified a novel function of the histone H4 gene dosage in promoting Cse4 sumoylation and mislocalization to noncentromeric regions, which leads to CIN when Cse4 is overexpressed.

One question is how this H4 dosage balance affects the function of H4-R36 (see also Section 1.5.1). Deyter et al. reported that H4-R36 is a key residue for efficient Cse4 degradation, likely by facilitating the interaction between Psh1 and Cse4 [31]. This group also found that a basic residue at H4-R36, but not PTM (e.g., methylation) of the amino acid, is required to prevent sensitivity to Cse4 overexpression [31]. Then how is it possible that reduced dosage of H4 leads to sumoylation and reduced mislocalization of overexpressed Cse4? Eisenstatt et al. showed that deletion of either histone H4 allele resulted in reduced levels of sumoylated Cse4 and concluded that physiologic levels of histone H4 are required for Cse4 sumoylation [27] (see also Section 1.4). However, the level of sumoylation loss caused by the deletion of either histone H4 allele (in *hhf1Δ* or *hhf2Δ*) was similar to one caused by *cse4*^{16KR} mutant that should be total loss of sumoylation of Cse4. Ohkuni et al. proposed a model in which Cse4 K215/216 sumoylation facilitates the deposition of overexpressed Cse4 into CEN and non-CEN regions, respectively [21] (**Figure 1**). If a total loss of sumoylation is achieved in the H4 loss (in *hhf1Δ* or *hhf2Δ*) as a *cse4*^{16KR} mutant, and if the sumoylation of Cse4 K215/216 is required for centromeric deposition of Cse4 into chromatin as Ohkuni et al. suggest [21], how do the *hhf1Δ* or *hhf2Δ* strains keep centromeric Cse4 and survive? How H4 gene dosage on Cse4 K215/216 sumoylation facilitates deposition of overexpressed (or endogenous) Cse4 into CEN is not clear (**Figure 1**, left-center).

The mechanism by which other histones' PTMs and dosages are involved in the incorporation of CENP-A/CenH3 is highly interesting, but at the same time, it suggests many questions. A further question raised is whether H4 dosage affects heterotypic CENP-A-H3.3 nucleosomes (see also Section 1.6) or H3 dosage among species including humans? Results in both budding and fission yeast suggest that the balance among histones H3,H4 and CENP-A is important for centromeric chromatin assembly [60, 61]. In fission yeast, increasing cellular histone H3 levels relative to Cnp1 promotes accumulation of H3 and loss of Cnp1 from the central domain and leads to defects in kinetochore function, however, there does not appear to be an efficient mechanism for the active exclusion of histone H3 from the centromeric nucleosomes [60, 62]. If H4 dosage affects heterotypic CENP-A-H3.3 nucleosomes or H3 dosage, is there an indirect pathway through which H4 dosage affects CENP-A incorporation into chromatin through H3? The inter-regulation among different histones, including CENP-A/CenH3 for high(macro) and low(micro) order chromatin structures, must be intricate. However, this could make it difficult to elucidate the mechanisms of incorporation, maintenance, and inheritance of CENP-A/CenH3.

In fission yeast and human studies, Mis18 (human Mis18α and Mis18β homolog) and Mis16 (human RbAp46 and RbAp48 homolog) are required for loading of newly synthesized Cnp1/CENP-A into centromeric chromatin [63, 64] (see also next chapter, sections 2.1, 3.1, and 4.1). Mis16 and Mis18 are also required for the maintenance of the hypoacetylation of histone H4 specifically within the central domain of the centromere [64], and Mis16 homologs are components of several histone chaperone complexes [65]. Moreover, acetylation of histone H4 lysine 5 and 12 (H4K5ac and H4K12ac) within the pre-nucleosomal CENP-A-H4-HJURP complex mediated by the RbAp46/48-Hat1 complex is required for CENP-A deposition into centromeres in chicken and humans [66], consistent with the Hat1 role shown in *Drosophila melanogaster* [67] (see also next chapter, sections 3.1 and 4.1). In mouse studies, Mis18α interacts with DNMT3A/3B, and this interaction is required to maintain DNA methylation [68]. *Mis18α* deficiency leads to not only the reduction of DNA methylation, but altered histone H3 modifications, and uncontrolled non-coding transcripts in the centromere region (see also next chapter, section [4.1]). However, Mis16 and Mis18 proteins are absent from budding yeast *S. cerevisiae* with point centromeres [69]. In addition, how these proteins and H4 hypoacetylation facilitate the fission yeast Cnp1/CENP-A incorporation into chromatin is still not clear [62] (see also next chapter, section [2.1]).

1.4 Cse4 sumoylation

1.4.1 Slx5/RNF4 and Ulp2/SENP6

It is known that sumoylation is involved in multiple intercellular pathways, and a subset of polysumoylation-mediated polyubiquitylation processes lead to proteasome-mediated degradation [70, 71]. Such machineries of SUMO-dependent ubiquitylation and degradation of CENP-A are interesting and important issues. Recent research has revealed new insights about the sumoylation of Cse4.

Ohkuni et al. reported the first evidence that Cse4 is sumoylated by E3 ligases Siz1 and Siz2 in vivo and in vitro [28] (**Figure 1** and **Table 1**). Siz1 is the founding member of the Siz/PIAS (protein inhibitor of activated STAT) RING family of SUMO E3 ligases, and both Siz1 and Siz2 are normally bound to chromatin via their SAP domains [72]. The Siz/PIAS RING family is involved in the sumoylation of the septin protein group and several chromatin proteins including core histones and the replication clamp PCNA (proliferating cell nuclear antigen) [70, 72]. They showed that ubiquitylation of Cse4 by the small ubiquitin-related modifier (SUMO)-targeted ubiquitin ligase (STUbL), Slx5, is important for proteolysis of Cse4 and prevents mislocalization of Cse4 to euchromatin under normal physiologic conditions (**Figure 1b** and **Table 1**). Sumoylated Cse4 proteins are accumulated and protein stability of Cse4 is increased in *slx5*Δ strains, suggesting that sumoylation precedes ubiquitin-mediated proteolysis of Cse4 (**Figure 1b**). *slx5*Δ *psh1*Δ strains exhibit higher levels of Cse4 stability and mislocalization than either *slx5*Δ or *psh1*Δ strains, suggesting that Slx5-mediated Cse4 proteolysis is independent of Psh1 (**Figure 1b**). In addition to the Psh1-dependent ubiquitylation pathway, their results suggested a second pathway that requires sumoylation of Cse4 by Siz1/Siz2 and ubiquitination of sumoylated Cse4 by Slx5 to prevent its mislocalization and maintain genome stability.

Further, Ohkuni et al. identified lysine 65 (K65) in Cse4 as a site that regulates sumoylation and ubiquitin-mediated proteolysis of Cse4 through Slx5 [52] (**Figure 1b**). The abundance of sumoylated and ubiquitinated Cse4 in vivo is reduced in budding yeast strains expressing cse4 K65R. They also showed that the interaction of cse4 K65R with Slx5 is significantly reduced, and stability and mislocalization of cse4 K65R are increased under normal physiologic conditions. The stability of cse4 K65R in *psh1Δ* strains is increased, but not in *slx5Δ* strains. Therefore, they concluded that Slx5 targets sumoylated Cse4 K65 for ubiquitination-mediated proteolysis, but Psh1 does not (**Figure 1b**). Overall, they clarified the function and biological significance of Cse4 K65 in sumoylation, ubiquitin-mediated proteolysis, and localization of Cse4 for genome stability.

In humans, depletion of the human Slx5 homolog ring finger protein 4 (RNF4) contributes to sumoylation-dependent degradation of the CCAN protein CENP-I, while SENP6 (a member of a large family of Sentrin-specific protease enzymes that belongs to the yeast Ulp2 group) stabilizes CENP-I by antagonizing RNF4 [73]. However, depletion of SENP6 in HeLa cells leads to the loss of the CENP-H/I/K complex from the centromeres, but not an apparent reduction in centromeric CENP-A/B/C levels recognized by CREST sera [73]. Recent analyses by some groups also indicated that CENP-A was not a direct substrate of SENP6 [74, 75]. Differences among species of roles of sumoylation in the regulation of CENP-A stability are described later (see also next chapter, section [4]).

1.4.2 Deposition of Cse4 into chromatin through its C-terminal sumoylation

C-terminal sumoylation of Cse4 also contributes to the deposition of Cse4 into chromatin. Ohkuni et al. identified sumoylation sites lysine (K) 215/216 in the C-terminus of Cse4 and showed that sumoylation of Cse4 K215/216 facilitates its genome-wide deposition into chromatin when overexpressed [21] (**Figure 1**). Their results showed reduced levels of sumoylation of mutant Cse4 K215/216R/A [K changed to arginine (R) or alanine (A)] and reduced interaction of mutant Cse4 K215/216R/A with Scm3 and CAF-1 (**Figure 1** and **Table 1**) (see also Section 1.5.3) when compared to wild-type Cse4. Consistently, levels of Cse4 K215/216R/A in the chromatin fraction and localization to centromeric and noncentromeric regions were reduced. In addition, GAL-cse4 K215/216R does not exhibit synthetic dosage lethality (SDL) in these strains—unlike *GAL-CSE4*, which exhibits SDL in *psh1Δ*, *slx5Δ*, and *hir2Δ* strains. Thus, they clearly demonstrated that the deposition of Cse4 into chromatin is facilitated by its C-terminal sumoylation. Based on their data, they also updated a model in which Cse4 K215/216 sumoylation promotes its interaction with the histone chaperones Scm3 and CAF-1, facilitating the deposition of overexpressed Cse4 into *CEN* and non-*CEN* regions, respectively (**Figure 1**). Their results suggest the importance of the SUMO-interaction motif in Slx5's targets and histone chaperone proteins (Scm3 and CAF-1), and it will be interesting to test if this sumoylation machinery is conserved in humans and if human CENP-A sumoylation regulates its interaction with HUJRP and/or DAXX.

Further questions remain about the SUMO E3 ligase of Cse4 C-terminal K215/216 sumoylation. Sumoylation of Cse4 is barely detectable in a *siz1Δ siz2Δ* strain [21]. Do Siz1/Siz2 also target the Cse4 C-terminal K215/216, as they do the Cse4 N-terminal K65? Or do different E3 entities target Cse4 K215/216 (**Figure 1**, right-center)? If Siz1/Siz2 are required for Cse4 C-terminal K215/216 sumoylation for proper Cse4 deposition at centromeres, how do Siz1/Siz2 distinguish between Cse4 for centromeric deposition and Cse4 for degradation? Functional comparisons among different species (esp. budding yeast and human) of sumoylation in the regulation of CENP-A stability are also described later (see also next chapter, section 4).

1.5 More intricate E3 network of Cse4 and other chaperones that function in proteolysis and mislocalization of Cse4

1.5.1 Cse4 R37 methylation and Ubr2/Mub1 E3 ligase

Samel et al. reported that the absence of the E3 ubiquitin ligase Ubr2, as well as its adaptor protein Mub1, suppresses the synthetic growth defects (or lethality) caused by the absence of Cse4-R37 methylation in *cbf1Δ* or deletion mutants of Ctf19/CCAN complex [76]. Previously Ubr2 had been shown to control the levels of the MIND complex protein Dsn1 via ubiquitination and proteasome-mediated degradation [77]. Consistent with these results, Samel et al. found that overexpression of *DSN1* also led to suppression of growth defects (or lethality) caused by the absence of Cse4-R37 methylation. Collectively, they proposed that the absence of Cse4 R37 methylation reduces the recruitment of kinetochore proteins to centromeric chromatin, and that this can be compensated for by stabilizing the outer kinetochore protein Dsn1.

However, the relationship between Ubr2 and Psh1, and E3 activity of Ubr2 on Cse4 is still not clear, although the absence of both E3s suppressed the synthetic growth defects (or lethality) shown in their study. The authors stated that most likely increased levels of kinetochore proteins other than Dsn1 in *ubr2Δ* cells can also compensate for the absence of R37 methylation, since *ubr2Δ* controls the levels of other proteins, possibly also kinetochore proteins. On the other hand, others had previously suggested that the role of Ubr2 at kinetochores seems to be partially redundant with the E3 ubiquitin ligase Psh1 [78]. Samel et al. also hypothesized that the ubiquitin ligase Psh1 restricts the loading of inner kinetochore proteins, whereas Ubr2 regulates the loading of outer kinetochore proteins, indicating that the mechanism of the suppression by *ubr2Δ* and *psh1Δ* differs.

1.5.2 Multiple E3 ligases promote the degradation of Cse4

In addition to the aforementioned report of Slx5 by Ohkuni et al. [28], Cheng et al. demonstrated that 4 ubiquitin ligases (i.e., Ubr1, Slx5, Psh1, and Rcy1) (**Figure 1c** and **Table 1**) contribute in parallel to the Cse4 proteolysis and turnover in budding yeast cells [33]. Cse4 overexpression generates cellular toxicity and cell cycle delay in budding yeast cells lacking *PSH1*, but not in cells lacking *UBR1*. These data suggest different roles of these two degradation pathways, and that various ubiquitin ligases collaborate to check and control Cse4 protein levels.

On the other hand, Cheng et al. also noted the lack of clarity about how this different E3s collaborate [33]. Their finding also generated these questions:

i. How do these E3s specifically recognize Cse4?

ii. How do they work with other cellular cues and pathways (e.g., casein kinase 2, Siz1- and Siz2-mediated sumoylation, SWI/SNF remodeling enzymes, the FACT complex, and the proline isomerase Fpr3)?

iii. What is the functional role and mechanism of each degradation pathway?

Further study is required to address these intricate E3 networks of Cse4 as well as other kinetochore proteins. In addition, the ubiquitin ligase(s) involved in human CENP-A degradation still remains unclear, although the CUL4A complex was identified as an E3 ligase that is required for CENP-A deposition at the centromere [79] (see also next chapter, section [4.2]). As Cheng et al. noted, while Psh1 does not seem to have a mammalian counterpart, Ubr1 (human UBR1), Slx5 (human RNF4) and Rcy1 (human EXOC5) are known to have human homologs. It is highly interesting to test CENP-A turnover in mammalian cells deficient for these homologs and also to determine if the human homologs of these E3s are altered in CENP-A-related cancer cells. Analogous questions are also raised in Section 1.6.2.

1.5.3 CAF-1 chaperone

Hewawasam et al. reported that chromatin assembly factor-1 (CAF-1) (**Figure 1** and **Table 1**) controls Cse4 deposition in budding yeast (see also Section 1.4.2). CAF-1 is an evolutionarily conserved histone H3/H4 chaperone; its subunits were shown to interact with CenH3 in flies and human cells. Previously, it had been reported that subunits of CAF-1 are required for building functional kinetochores [80], for recruitment of CenH3/Cnp1 and Scm3 to centromeres in fission yeast, *S. pombe* [7, 64], and for regulating Cse4/H3 exchange kinetics [81]. Hewawasam et al. showed that yCAF-1 interacts with Cse4 and can assemble Cse4 nucleosomes in vitro, using both *CEN* and non-*CEN* plasmids [26] (**Figure 1**). In their study, when Cse4 is overexpressed, loss of yCAF-1 markedly reduces Cse4 deposition into chromatin genome-wide. They suggest that incorporation of Cse4 genome-wide may induce multifactorial effects on growth and gene expression. Loss of yCAF-1 rescues growth defects and some changes in gene expression caused by Cse4 genome-wide misincorporation that occur in the absence of Psh1-mediated proteolysis. The incorporation of Cse4 into promoter nucleosomes at transcriptionally active genes is dependent on yCAF-1. In summary, these findings suggest CAF-1 can act as a Cse4 chaperone, controlling the amount and the incorporation of Cse4 in chromatin.

Some questions remain about the relationships among Cse4, Psh1, Scm3, and CAF-1. The first question is about the role of CAF- 1 in Psh1-mediated proteolysis of Cse4: How does CAF-1 function in the process of Cse4 ubiquitylation by Psh1? Hewawasam et al. observed more ubiquitylation of Cse4 in the presence of CAF-1 compared with the absence of CAF-1 in vitro, suggesting CAF-1 can promote ubiquitylation of free Cse4, opposite to the effect of Scm3 that protects Cse4 from ubiquitylation by Psh1 in vitro [22]. They also tested CAF-1 interaction with Psh1, but their co-immunoprecipitation experiment in whole-cell extracts did not show any interactions. Thus, they speculated that soluble Cse4 bound to CAF-1 may expose ubiquitylation sites on Cse4, promoting ubiquitylation by Psh1.

The second question is whether CAF-1 could assemble Cse4 at centromere as Scm3. If so, do the roles of the two proteins simply overlap, or does each protein have a unique role in the process of Cse4 assembly at centromere? To test this question, Hewawasam et al. used the Scm3$^{on/off}$ strain, which can be toggled by galactose, along with copper-inducible Cse4 overexpression, so that Cse4 protein levels can be controlled by the concentration of copper [26]. Their results suggest that when Scm3 is absent and Cse4 levels are high, CAF1 may be a primary chaperone targeting Cse4 to the centromere (**Figure 1**, center). Meanwhile, in the fission yeast *S. pombe*, the CAF-1 subunit can recruit the Scm3 to centromeres [7]. Thus, Hewawasam et al. speculated that under normal conditions, Scm3 and CAF-1 both play important functions in the deposition of Cse4 at centromere; however, further study is required to reinforce this hypothesis (**Figure 1**, right).

The third question is how CAF-1 can be responsible for the mislocalization of Cse4/CENP-A in cancer development. The human CAF-1 subunit p60 was one of the overexpressed chaperones in CENP-A-overexpressing breast cancer cells [82], and ectopic CENP-A nucleosomes from colorectal cancer cells keep a subpopulation of structurally distinct hybrid (chimeric) nucleosomes containing both CENP-A and H3.3 [82, 83]. Misregulation of Scm3/HJURP causes chromosome instability in both yeast and humans [84], and many previous reports suggested the functional relevance of Scm3/HJURP with the development of a wide spectrum of cancers (e.g., colon, lung, liver, breast, pancreatic, brain cancer) [85–94]. As aforementioned, CAF-1 may cooperate with Psh1 and Scm3 to regulate proteolysis of Cse4, in the way that CAF-1 association with free Cse4 may promote ubiquitylation and proteolysis. If so, how do these two chaperons (CAF-1 and Scm3/HJURP) cooperate together in genomic stability and anti-cancer development? Further in-depth study is required to elucidate the collaboration among Psh1, Scm3, and CAF-1 in genomic stability and anti-cancer development.

1.6 A genome-wide screen (a synthetic genetic array, SGA) revealed other proteins that are required for proteolysis and proper localization of Cse4

1.6.1 HIR histone chaperone complex

Deletion for genes encoding the replication-independent histone chaperone HIR complex (HIR1, HIR2, HIR3, HPC2) and a Cse4-specific E3 ubiquitin ligase, PSH1, showed the highest SDL using a genome-wide synthetic genetic array (SGA) to identify gene deletions that exhibit SDL when Cse4 is overexpressed [30]. Thus, Ciftci-Yilmaz et al. performed functional analysis for Hir2 (**Figure 1** and **Table 1**) in proteolysis of Cse4 that prevents mislocalization of Cse4 to noncentromeric regions for genome stability. They demonstrated the interaction of Hir2 with Cse4 in vivo, and defects in Cse4 proteolysis and stabilization of chromatin-bound Cse4 appear in *hir2Δ* strains. The *hir2Δ* strains also exhibit mislocalization of Cse4 to noncentromeric regions with a preferential enrichment at promoter regions. They also found that Hir2 facilitates the interaction of Cse4 with Psh1. In the *hir2Δ* strain, defects in Psh1-mediated proteolysis increase Cse4 stability and lead to mislocalization of Cse4 compared to wild-type cells. Collectively, they identified a novel role for the HIR complex to prevent mislocalization of Cse4 by facilitating proteolysis of Cse4, thereby promoting genomic stability.

Analogous questions can be raised regarding CAF-1, especially about the functional relationships among different Cse4 chaperone proteins (e.g., Scm3, CAF-1, and Hir2) and their roles in cancer development. In the Psh1-mediated proteolysis of free Cse4 using whole-cell lysates, CAF-1 and Hir2 promote proteolysis and Scm3 inhibits it [22, 26, 30]. CAF-1 and Hir2 could be involved in the proteolysis of noncentromeric Cse4, but Scm3 in the anti-proteolysis of the centromeric Cse4. However, CAF-1 may promote centromeric localization of overexpressed Cse4 only under conditions when Scm3 is depleted (*SCM3* expression OFF; see also Section 1.5.3) [26]. Furthermore, the centromeric Cse4 level is decreased in *hir2Δ* strains, suggesting that Hir2 might have anti-proteolytic activity in centromeric Cse4. If the functions of these chaperones change with the expression level and localization pattern of CENP-A that they target, how do these chaperones sense the dosage change of CENP-A (between overexpression and normal levels of CENP-A) and distinguish centromeric and ectopic CENP-A?

Studying the real-time 3D structure of free CENP-A/CenH3 after post-translational modification and before incorporation into chromatin could be a key future direction. In budding yeast, Ohkuni et al. proposed that structural change in Cse4 caused by the proline isomerase Fpr3 might be important for the interaction between Cse4 and the E3 ubiquitin ligase Psh1 [32] (see also Section 1.2.3).

1.6.2 Skp1, Cullin, F-box (SCF)-Met30 and SCF-Cdc4-mediated proteolysis of Cse4

Au et al. identified two Skp1, Cullin, F-box (SCF) ubiquitin ligases with the evolutionarily conserved F-box proteins Met30 and Cdc4 (**Figure 1c** and **Table 1**) as essential genes required for Cse4 homeostasis through a genome-wide SGA screen [34]. They showed that Met30 and Cdc4 interact through the Met30-WD40 domain, and these two proteins cooperatively regulate proteolysis of endogenous Cse4 and prevent its mislocalization for faithful chromosome segregation (**Figure 1**). The interaction of Met30 with Cdc4 is independent of the Met30-D domain, which is essential for their homodimerization and ubiquitination of other substrates. Ubiquitin affinity pull-down assays showed that both Cdc4 and Met30 specifically target Cse4 for its ubiquitination. They suggest that Met30 is necessary for the interaction between Cdc4 and Cse4, and its defective interaction leads to stabilization and mislocalization of Cse4, which in turn promotes to CIN. They also provided the first direct link between Cse4 mislocalization and defects in kinetochore structure measured by the sensitivity against the restriction enzyme *DraI,* and collectively showed that proteolysis of Cse4 by SCF-Met30/Cdc4 prevents mislocalization and CIN.

Further studies are also required to address analogous questions as in Section 1.5.2: How does the Met30/Cdc4-pathway work with other cellular cues and multiple E3 pathways, including Psh1-dependent and independent proteolysis? Are the human homologs of these E3s (e.g., human FBXO24, TRAF7, etc.) altered in CENP-A-related cancer cells?

1.6.3 Dbf4-dependent kinase (DDK)-mediated proteolysis of Cse4

Eisenstatt et al. identified five alleles of CDC7 and DBF4 that encode the Dbf4-dependent kinase complex, which regulates DNA replication initiation in their SGA [95]. They found that cdc7–7 strains show defects in ubiquitin-mediated proteolysis of Cse4 and mislocalization of Cse4 [95]. Mutation of MCM5 (mcm5-bob1) bypasses the requirement of Cdc7 for replication initiation and rescues replication defects in a cdc7-7 strain. They demonstrated that mcm5-bob1 does not rescue the SDL and defects in proteolysis of overexpressed Cse4 (*GALCSE4*) in a cdc7-7 strain. These data suggest a DNA replication-independent role for Cdc7 in Cse4 proteolysis. Their results of the SDL phenotype, defects in ubiquitin-mediated proteolysis, and the mislocalization pattern of Cse4 in a cdc7-7 *psh1Δ* strain were similar to that in the cdc7-7 and *psh1Δ* strains. These data suggest that Cdc7 regulates Cse4 in a pathway that overlaps with Psh1. They propose a role for the Dbf4-dependent kinase complex as a regulator of Psh1-mediated proteolysis of Cse4 to prevent mislocalization of Cse4, independent of DNA replication initiation.

1.6.4 Reduced gene dosage of histone H4

Recently, Eisenstatt et al. further utilized SGA to identify factors that facilitate the mislocalization of overexpressed Cse4 by characterizing suppressors of the *psh1Δ GALCSE4 SDL* [27]. See Section 1.3.3 for more details of this study.

1.7 Cse4 deubiquitylase

Compared to the ubiquitylation mechanism of Cse4, there are relatively few studies on the deubiquitylation mechanism of Cse4. We should also consider how the deubiquitylation affects the localization and the function of Cse4 at both centromere and ectopic sites along chromosomes.

Canzonetta et al. investigated the role of Ubp8-driven deubiquitylation of the Cse4 in budding yeast [96]. Ubp8 is a component of the SAGA (Spt-Ada-Gcn5-acetyltransferase) complex, a multicomponent regulator of acetylation. The SAGA complex is also involved in deubiquitylation through its deubiquitylation (DUB) module, and one example of its activity is upon histone H2B [97, 98]. Canzonetta et al. demonstrated that the deubiquitylation process was inhibited and a short ubiquitin oligomer on Cse4 was accumulated by the loss of Ubp8. Such defective deubiquitylation caused by Ubp8 loss leads to chromosome instability and Cse4 protein degradation, and induces ectopic localization of the Cse4 outside the centromere.

1.8 Psh1 is also involved in proper plasmid segregation, and two distinct cellular effects by Psh1 and Cse4

Interestingly, there was a report suggesting that Psh1 is also involved in proper plasmid segregation [99]. Metzger et al. initially sought to assess the involvement of the ubiquitin-proteasome system in the turnover of mitochondrial proteins in budding yeast [99]. Then, they found that deletion of a specific ubiquitin ligase (E3), Psh1p, increases the level of a temperature-sensitive mitochondrial protein, mia40-4pHA, when it is expressed from a centromere-containing (*CEN*) plasmid that remains, on average, at one copy per yeast cell. Unexpectedly, they also found deletion of Psh1p elevates the levels of other proteins (not only mitochondrial proteins) expressed from the *CEN* plasmids. The rate of turnover of mia40-4pHA, total protein synthesis, or the protein levels of chromosomal genes is not affected by the loss of Psh1p. On the other hand, *psh1Δ* appears to increase the occurrence of missegregation of centromeric plasmids compared to their normal 1:1 segregation. Their results showed that ongoing missegregation leads to elevated plasmid DNA, mRNA, and protein, all of which they observed in *psh1Δ* cells after generations of growth with selection for the plasmid. Elevation of Cse4p leads to an apparent increase in 1:0 plasmid segregation events, although Cse4p overexpression alone does not phenocopy *psh1Δ* in increasing plasmid DNA and protein levels. Moreover, 2 μm high-copy yeast plasmids also lead to missegregation in *psh1Δ*, but not when Cse4p alone is overexpressed. Their findings demonstrated that Psh1p is required for the faithful inheritance of both centromeric and 2 μm plasmids. In addition, the effects that loss of Psh1p has on plasmid segregation cannot be merely explained by increased levels of Cse4p, arguing two distinct cellular effects by Psh1p and Cse4p.

1.9 Molecular basis for the selective recognition and ubiquitination of Cse4 by Psh1 through Cse4 MIMAS motif

Zhou et al. first solved the structure of the Cse4-binding domain (CBD) of Scm3 in complex with Cse4 and H4 in a single chain model using nuclear magnetic spectroscopy [12]. They suggested that four Cse4-specific residues in the N-terminal region of helix 2 (MIMAS motif; **Table 1**) are sufficient for specific recognition by conserved and functionally important residues in the N-terminal helix of Scm3 through the formation of a hydrophobic cluster.

Scm3 (CBD) also induces major conformational changes and sterically occludes DNA-binding sites in the structure of Cse4 and H4. Furthermore, Zhou et al. showed that Psh1 uses a CBD (residues 1–211) to interact with Cse4-H4 instead of H3-H4, yielding a dissociation constant (K_d) of 27 nM in their isothermal titration calorimetric experiments [100]. They are in vitro pull-down assays revealed that Psh1 interacts with Cse4-specific residues in the L1 loop and α2 helix for Cse4 binding and ubiquitination.

They also mapped the Psh1-binding region of Cse4-H4 and identified a wide range of Cse4 specific residues required for the Psh1-mediated Cse4 recognition and ubiquitination. Consistent with the previous reports of the inhibitory effect of Scm3 on Cse4 ubiquitylation [22], their data showed that the histone chaperone Scm3 prevents Cse4 ubiquitination by abrogating Psh1-Cse4 binding. Their results suggest that Scm3 interacts with the Cse4 MMAS motif (a particular Cse4 region containing residues M181/ M184/ A189/S190 reported previously [12]) to prevent Psh1 from binding to Cse4. Elimination of the Psh1-binding residues outside of the Cse4 MMAS motif promotes the inhibitory effect of Scm3. Thus, the MMAS motif plays a central role in the activation or inhibition of Cse4 ubiquitination as well as yeast cell growth. Taken together, they elucidated a novel Cse4 binding mode distinct from those of known CenH3 chaperones and the mechanism by which Scm3 competes with Psh1 for Cse4 binding.

2. Conclusions

The budding yeast is a powerful organism for centromere-kinetochore research in many aspects. For example, the centromere sequence size of the budding yeast is small and the sequences can be easily mutated to identify the important functional regions [1]. Techniques such as ChIP are also possible, which cannot be easily performed on highly repetitive centromeres in other organisms. Moreover, the centromere can be shifted to other genomic regions, allowing the construction of artificial chromosomes and plasmids as well as tools such as conditional centromeres. As a result, the most common species studied and reported in the past for E3 ligase of CenH3 (Cse4) is budding yeast at present. However, many questions described in this chapter are unanswered even in the budding yeast model. Especially, little has been studied on how each of such multiple E3 ligases of budding yeast selectively recognizes Cse4 substrate and functions specifically. Currently, 4 types of E3 ligases for ubiquitylation and one type of E3 ligases for sumoylation (Slx5/8) have been reported (**Table 1**). In particular, the functions of the 4 types of E3 ligases for ubiquitylation including Psh1 are common, all of which are related to ectopic degradation and/or quality control of soluble or chromatin-bound Cse4, and the functional differences are not clear. It is neither clear why E3 ligases with overlapping functions exist in one species. The simple interpretation is that at least such a number (4–5) of E3 ligases of Cse4 is required as a backup system, so that it can be complemented if one of the E3 functions is defective. As we described the compensatory system of CENP- A PTM (see also the next chapter, Conclusion), compensatory systems and resilience of CENP-A^{Cse4} could be expected as future directions to study the spatiotemporal regulation of E3 ligase of CENP-A^{Cse4}.

No neocentromere has been found in budding yeast. As a simple reason, it seems that there is no or little possibility of ectopic centromere formation, because the kinetochore formation of budding yeast depends on centromeric DNA elements. However, in terms of considering centromeric evolution, it is interesting to question why budding yeast has maintained point centromere which relies on DNA elements, and other species have evolved to regional centromere which allows the system to generate neocentromere? There is also no clear answer as to whether simply introducing centromeric DNA elements into ectopic loci causes neocentromere or it is still eliminated by the specific E3 activity in budding yeast. The building and establishment of artificial chromosomes are facilitated by studying the mechanisms of formation and maintenance of neocentromeres, and these topics of other species are described in the next chapter.

The regulation of budding yeast cenRNA is an essential factor for centromere structure and function as other eukaryotes, but we have little understanding of the causality or feedback between cenRNA transcription and overall transcriptional change after chromosome mis-segregation and CIN. In addition, little is known about the effects of these cenRNAs on the E3 ligase of CENP-A, including how these transcriptional changes and regulation are related to the function of E3 ligase. Although, it is essential to study specific physiological functions of each E3 ligase, the physiological phenotype of budding yeast is limited (e.g., growth, cell death, etc.), thus naturally there is a limit in the discussion of the results in the budding yeast model. Thus, studies of an E3 ligase in CENP-A in higher eukaryotes, mammals, or humans are essential for translational research and informing future therapy, and these topics are described in the next chapter.

Acknowledgements

We thank past and current researchers at Model Animal Research Center, School of Medicine, Nanjing University, Greehey Children's Cancer Research Institute at UT Health Science Center San Antonio, The Research Institute at Nationwide Children's Hospital, and St. Jude Children's Research Hospital for their helpful discussions. Y.N. was supported by Jiangsu Province "Double-First-Class" Construction Fund, Jiangsu Province Natural Science Fund (BK20191252), Jiangsu Province 16th Six Big Talent Peaks Fund (TD-SWYY-001), Jiangsu Province "Foreign Expert Hundred Talents Program" Fund (BX2019082), and National Natural Science Foundation in China (31970665). KK was supported by the National Science Foundation under Grant No.1949653 (KK) and a Mays Cancer Center Pilot Award CCSG P30 CA054174.

Conflict of interest

The authors declare no conflict of interest.

Author details

Yohei Niikura[1,2*] and Katsumi Kitagawa[3*]

1 MOE Key Laboratory of Model Animal for Disease Studies, Model Animal Research Center, School of Medicine, Nanjing University, Nanjing, China

2 Jiangsu Key Laboratory of Molecular Medicine, Nanjing University, Nanjing, China

3 Greehey Children's Cancer Research Institute, Department of Molecular Medicine, School of Medicine, UT Health San Antonio, San Antonio, TX, USA

*Address all correspondence to: niikura@nicemice.cn and kitagawak@uthscsa.edu

References

[1] Biggins S. The composition, functions, and regulation of the budding yeast kinetochore. Genetics. 2013;**194**(4):817-846. DOI: 10.1534/genetics.112.145276

[2] Bloom K, Costanzo V. Centromere structure and function. Progress in Molecular and Subcellular Biology. 2017;**56**:515-539. DOI: 10.1007/978-3-319-58592-5_21

[3] Choy JS, Mishra PK, Au WC, Basrai MA. Insights into assembly and regulation of centromeric chromatin in Saccharomyces cerevisiae. Biochimica et Biophysica Acta. 2012;**1819**(7):776-783. DOI: 10.1016/j.bbagrm.2012.02.008

[4] Pearson CG, Yeh E, Gardner M, Odde D, Salmon ED, Bloom K. Stable kinetochore-microtubule attachment constrains centromere positioning in metaphase. Current Biology. 2004;**14**(21):1962-1967. DOI: 10.1016/j.cub.2004.09.086

[5] Camahort R, Li B, Florens L, Swanson SK, Washburn MP, Gerton JL. Scm3 is essential to recruit the histone h3 variant cse4 to centromeres and to maintain a functional kinetochore. Molecular Cell. 2007;**26**(6):853-865. DOI: 10.1016/j.molcel.2007.05.013

[6] Mizuguchi G, Xiao H, Wisniewski J, Smith MM, Wu C. Nonhistone Scm3 and histones CenH3-H4 assemble the core of centromere-specific nucleosomes. Cell. 2007;**129**(6):1153-1164. DOI: 10.1016/j.cell.2007.04.026

[7] Pidoux AL et al. Fission yeast Scm3: A CENP-A receptor required for integrity of subkinetochore chromatin. Molecular Cell. 2009;**33**(3):299-311. DOI: 10.1016/j.molcel.2009.01.019

[8] Stoler S, Rogers K, Weitze S, Morey L, Fitzgerald-Hayes M, Baker RE. Scm3, an essential Saccharomyces cerevisiae centromere protein required for G2/M progression and Cse4 localization. Proceedings of the National Academy of Sciences of the United States of America. 2007;**104**(25):10571-10576. DOI: 10.1073/pnas.0703178104

[9] Williams JS, Hayashi T, Yanagida M, Russell P. Fission yeast Scm3 mediates stable assembly of Cnp1/CENP-A into centromeric chromatin. Molecular Cell. 2009;**33**(3):287-298. DOI: 10.1016/j.molcel.2009.01.017

[10] Dunleavy EM et al. HJURP is a cell-cycle-dependent maintenance and deposition factor of CENP-A at centromeres. Cell. 2009;**137**(3):485-497. DOI: 10.1016/j.cell.2009.02.040

[11] Foltz DR et al. Centromere-specific assembly of CENP-a nucleosomes is mediated by HJURP. Cell. 2009;**137**(3):472-484. DOI: 10.1016/j.cell.2009.02.039

[12] Zhou Z et al. Structural basis for recognition of centromere histone variant CenH3 by the chaperone Scm3. Nature. 2011;**472**(7342):234-237. DOI: 10.1038/nature09854

[13] Shivaraju M, Camahort R, Mattingly M, Gerton JL. Scm3 is a centromeric nucleosome assembly factor. Journal of Biological Chemistry. 2011;**286**(14):12016-12023. DOI: 10.1074/jbc.M110.183640

[14] Shivaraju M, Gerton JL. The dynamics of the Cse4 chaperone Scm3. Cell Cycle. 2011;**10**(22):3823-3824. DOI: 10.4161/cc.10.22.18232

[15] Xiao H, Mizuguchi G, Wisniewski J, Huang Y, Wei D, Wu C. Nonhistone Scm3

binds to AT-rich DNA to organize atypical centromeric nucleosome of budding yeast. Molecular Cell. 2011;**43**(3):369-380. DOI: 10.1016/j.molcel.2011.07.009

[16] Luconi L, Araki Y, Erlemann S, Schiebel E. The CENP-A chaperone Scm3 becomes enriched at kinetochores in anaphase independently of CENP-A incorporation. Cell Cycle. 2011;**10**(19):3369-3378. DOI: 10.4161/cc.10.19.17663

[17] Black BE, Cleveland DW. Epigenetic centromere propagation and the nature of CENP-a nucleosomes. Cell. 2011;**144**(4):471-479. DOI: 10.1016/j.cell.2011.02.002

[18] Dechassa ML, Wyns K, Li M, Hall MA, Wang MD, Luger K. Structure and Scm3-mediated assembly of budding yeast centromeric nucleosomes. Nature Communications. 2011;**2**:313. DOI: 10.1038/ncomms1320

[19] Cho US, Harrison SC. Recognition of the centromere-specific histone Cse4 by the chaperone Scm3. Proceedings of the National Academy of Sciences of the United States of America. 2011;**108**(23):9367-9371. DOI: 10.1073/pnas.1106389108

[20] Dechassa ML, Wyns K, Luger K. Scm3 deposits a (Cse4-H4)2 tetramer onto DNA through a Cse4-H4 dimer intermediate. Nucleic Acids Research. 2014;**42**(9):5532-5542. DOI: 10.1093/nar/gku205

[21] Ohkuni K et al. Deposition of centromeric histone H3 variant CENP-A/Cse4 into chromatin is facilitated by its C-terminal Sumoylation. Genetics. 2020;**214**(4):839-854. DOI: 10.1534/genetics.120.303090

[22] Hewawasam G et al. Psh1 is an E3 ubiquitin ligase that targets the centromeric histone variant Cse4. Molecular Cell. 2010;**40**(3):444-454. DOI: 10.1016/j.molcel.2010.10.014

[23] Hewawasam GS, Gerton JL. Cse4 gets a kiss-of-death from Psh1. Cell Cycle. 2011;**10**(4):566-567. DOI: 10.4161/cc.10.4.14770

[24] Mishra PK et al. Pat1 protects centromere-specific histone H3 variant Cse4 from Psh1-mediated ubiquitination. Molecular Biology of the Cell. 2015;**26**(11):2067-2079. DOI: 10.1091/mbc.E14-08-1335

[25] Shahnejat-Bushehri S, Ehrenhofer-Murray AE. The ATAD2/ANCCA homolog Yta7 cooperates with Scm3(HJURP) to deposit Cse4(CENP-A) at the centromere in yeast. Proceedings of the National Academy of Sciences of the United States of America. 2020;**117**(10):5386-5393. DOI: 10.1073/pnas.1917814117

[26] Hewawasam GS, Dhatchinamoorthy K, Mattingly M, Seidel C, Gerton JL. Chromatin assembly factor-1 (CAF-1) chaperone regulates Cse4 deposition into chromatin in budding yeast. Nucleic Acids Research. 2018;**46**(9):4440-4455. DOI: 10.1093/nar/gky169

[27] Eisenstatt JR et al. Reduced gene dosage of histone H4 prevents CENP-A mislocalization and chromosomal instability in *Saccharomyces cerevisiae*. Genetics. 2021;**218**(1). DOI: 10.1093/genetics/iyab033

[28] Ohkuni K et al. SUMO-targeted ubiquitin ligase (STUbL) Slx5 regulates proteolysis of centromeric histone H3 variant cse4 and prevents its mislocalization to euchromatin. Molecular Biology of the Cell. 2016;**27**(9):1500-1510. DOI: 10.1091/mbc.E15-12-0827

[29] Hewawasam GS et al. Phosphorylation by casein kinase 2 facilitates Psh1 protein-assisted degradation of Cse4 protein. Journal of Biological Chemistry. 2014;**289**(42):29297-29309. DOI: 10.1074/jbc.M114.580589

[30] Ciftci-Yilmaz S et al. A genome-wide screen reveals a role for the HIR histone chaperone complex in preventing mislocalization of budding yeast CENP-A. Genetics. 2018;**210**(1):203-218. DOI: 10.1534/genetics.118.301305

[31] Deyter GM, Hildebrand EM, Barber AD, Biggins S. Histone H4 facilitates the proteolysis of the budding yeast CENP-ACse4 Centromeric histone variant. Genetics. 2017;**205**(1):113-124. DOI: 10.1534/genetics.116.194027

[32] Ohkuni K, Abdulle R, Kitagawa K. Degradation of centromeric histone H3 variant Cse4 requires the Fpr3 peptidyl-prolyl cis-trans isomerase. Genetics. 2014;**196**(4):1041-1045. DOI: 10.1534/genetics.114.161224

[33] Cheng H, Bao X, Gan X, Luo S, Rao H. Multiple E3s promote the degradation of histone H3 variant Cse4. Scientific Reports. 2017;**7**(1):8565. DOI: 10.1038/s41598-017-08923-w

[34] Au WC et al. Skp, Cullin, F-box (SCF)-Met30 and SCF-Cdc4-mediated proteolysis of CENP-A prevents Mislocalization of CENP-A for chromosomal stability in budding yeast. PLoS Genetics. 2020;**16**(2):e1008597. DOI: 10.1371/journal.pgen.1008597

[35] Arunkumar G, Melters DP. Centromeric transcription: A conserved Swiss-Army knife. Genes (Basel). 2020;**11**(8):1-22. DOI: 10.3390/genes11080911

[36] Leclerc S, Kitagawa K. The role of human Centromeric RNA in chromosome stability. Frontiers in Molecular Biosciences. 2021;**8**:642732. DOI: 10.3389/fmolb.2021.642732

[37] Rosic S, Erhardt S. No longer a nuisance: Long non-coding RNAs join CENP-A in epigenetic centromere regulation. Cellular and Molecular Life Sciences. 2016;**73**(7):1387-1398. DOI: 10.1007/s00018-015-2124-7

[38] Ling YH, Yuen KWY. Point centromere activity requires an optimal level of centromeric noncoding RNA. Proceedings of the National Academy of Sciences of the United States of America. 2019;**116**(13):6270-6279. DOI: 10.1073/pnas.1821384116.

[39] Ling YH, Yuen KWY. Centromeric non-coding RNA as a hidden epigenetic factor of the point centromere. Current Genetics. 2019;**65**(5):1165-1171. DOI: 10.1007/s00294-019-00988-6

[40] Chen CF, Pohl TJ, Chan A, Slocum JS, Zakian VA. Saccharomyces cerevisiae centromere RNA is negatively regulated by Cbf1 and its unscheduled synthesis impacts CenH3 binding. Genetics. 2019;**213**(2):465-479. DOI: 10.1534/genetics.119.302528

[41] Maehara K, Takahashi K, Saitoh S. CENP-A reduction induces a p53-dependent cellular senescence response to protect cells from executing defective mitoses. Molecular and Cellular Biology. 2010;**30**(9):2090-2104. DOI: 10.1128/MCB.01318-09

[42] Lomonte P, Sullivan KF, Everett RD. Degradation of nucleosome-associated centromeric histone H3-like protein CENP-A induced by herpes simplex virus type 1 protein ICP0. Journal of Biological Chemistry. 2001;**276**(8):5829-5835. DOI: 10.1074/jbc.M008547200

[43] Collins KA, Furuyama S, Biggins S. Proteolysis contributes to the exclusive centromere localization of the yeast Cse4/CENP-A histone H3 variant. Current Biology. 2004;**14**(21):1968-1972. DOI: 10.1016/j.cub.2004.10.024

[44] Krogan NJ et al. RNA polymerase II elongation factors of Saccharomyces cerevisiae: A targeted proteomics

approach. Molecular and Cellular Biology. 2002;**22**(20):6979-6992. DOI: 10.1128/MCB.22.20.6979-6992.2002

[45] Ranjitkar P, Press MO, Yi X, Baker R, MacCoss MJ, Biggins S. An E3 ubiquitin ligase prevents ectopic localization of the centromeric histone H3 variant via the centromere targeting domain. Molecular Cell. 2010;**40**(3):455-464. DOI: 10.1016/j.molcel.2010.09.025

[46] Gkikopoulos T et al. The SWI/SNF complex acts to constrain distribution of the centromeric histone variant Cse4. The EMBO Journal. 2011;**30**(10):1919-1927. DOI: 10.1038/emboj.2011.112

[47] Au WC, Dawson AR, Rawson DW, Taylor SB, Baker RE, Basrai MA. A novel role of the N terminus of budding yeast histone H3 variant Cse4 in ubiquitin-mediated proteolysis. Genetics. 2013;**194**(2):513-518. DOI: 10.1534/genetics.113.149898

[48] Folco HD et al. The CENP-A N-tail confers epigenetic stability to centromeres via the CENP-T branch of the CCAN in fission yeast. Current Biology. 2015;**25**(3):348-356. DOI: 10.1016/j.cub.2014.11.060

[49] Gonzalez M, He H, Dong Q, Sun S, Li F. Ectopic centromere nucleation by CENP--a in fission yeast. Genetics. 2014;**198**(4):1433-1446. DOI: 10.1534/genetics.114.171173

[50] Le Goff S et al. The H3 histone chaperone NASP(SIM3) escorts CenH3 in Arabidopsis. The Plant Journal. 2020;**101**(1):71-86. DOI: 10.1111/tpj.14518

[51] Logsdon GA et al. Both tails and the centromere targeting domain of CENP-A are required for centromere establishment. Journal of Cell Biology. 2015;**208**(5):521-531. DOI: 10.1083/jcb.201412011

[52] Ohkuni K et al. N-terminal sumoylation of centromeric histone H3 variant Cse4 regulates its proteolysis to prevent mislocalization to non-centromeric chromatin. G3 (Bethesda). 2018;**8**(4):1215-1223. DOI: 10.1534/g3.117.300419

[53] Tan HL et al. Prolyl isomerization of the CENP-A N-terminus regulates centromeric integrity in fission yeast. Nucleic Acids Research. 2018;**46**(3):1167-1179. DOI: 10.1093/nar/gkx1180

[54] Tan HL, Zeng YB, Chen ES. N-terminus does not govern protein turnover of *Schizosaccharomyces pombe* CENP-A. International Journal of Molecular Sciences. 2020;**21**(17):1-13. DOI: 10.3390/ijms21176175

[55] Yang J et al. Heterochromatin and RNAi regulate centromeres by protecting CENP-A from ubiquitin-mediated degradation. PLoS Genetics. 2018;**14**(8):e1007572. DOI: 10.1371/journal.pgen.1007572

[56] Hochwagen A, Tham WH, Brar GA, Amon A. The FK506 binding protein Fpr3 counteracts protein phosphatase 1 to maintain meiotic recombination checkpoint activity. Cell. 2005;**122**(6):861-873. DOI: 10.1016/j.cell.2005.07.010

[57] Shan X, Xue Z, Melese T. Yeast NPI46 encodes a novel prolyl cis-trans isomerase that is located in the nucleolus. The Journal of Cell Biology. 1994;**126**(4):853-862. DOI: 10.1083/jcb.126.4.853

[58] Deyter GM, Biggins S. The FACT complex interacts with the E3 ubiquitin ligase Psh1 to prevent ectopic localization of CENP-A. Genes & Development. 2014;**28**(16):1815-1826. DOI: 10.1101/gad.243113.114

[59] Hildebrand EM, Biggins S. Regulation of budding yeast CENP-A levels prevents Misincorporation at promoter nucleosomes and transcriptional defects. PLoS Genetics. 2016;**12**(3):e1005930. DOI: 10.1371/journal.pgen.1005930

[60] Castillo AG et al. Plasticity of fission yeast CENP-A chromatin driven by relative levels of histone H3 and H4. PLoS Genetics. 2007;**3**(7):e121. DOI: 10.1371/journal.pgen.0030121

[61] Au WC, Crisp MJ, DeLuca SZ, Rando OJ, Basrai MA. Altered dosage and mislocalization of histone H3 and Cse4p lead to chromosome loss in *Saccharomyces cerevisiae*. Genetics. 2008;**179**(1):263-275. DOI: 10.1534/genetics.108.088518

[62] Partridge JF. Centromeric chromatin in fission yeast. Frontiers in Bioscience. 2008;**13**:3896-3905. DOI: 10.2741/2977

[63] Fujita Y et al. Priming of centromere for CENP-A recruitment by human hMis18alpha, hMis18beta, and M18BP1. Developmental Cell. 2007;**12**(1):17-30. DOI: 10.1016/j.devcel.2006.11.002

[64] Hayashi T, Fujita Y, Iwasaki O, Adachi Y, Takahashi K, Yanagida M. Mis16 and Mis18 are required for CENP-A loading and histone deacetylation at centromeres. Cell. 2004;**118**(6):715-729. DOI: 10.1016/j.cell.2004.09.002

[65] Loyola A, Almouzni G. Histone chaperones, a supporting role in the limelight. Biochimica et Biophysica Acta. 2004;**1677**(1-3):3-11. DOI: 10.1016/j.bbaexp.2003.09.012

[66] Shang WH et al. Chromosome engineering allows the efficient isolation of vertebrate neocentromeres. Developmental Cell. 2013;**24**(6):635-648. DOI: 10.1016/j.devcel.2013.02.009

[67] Boltengagen M et al. A novel role for the histone acetyltransferase Hat1 in the CENP-A/CID assembly pathway in *Drosophila melanogaster*. Nucleic Acids Research. 2016;**44**(5):2145-2159. DOI: 10.1093/nar/gkv1235

[68] Kim IS et al. Roles of Mis18alpha in epigenetic regulation of centromeric chromatin and CENP-A loading. Molecular Cell. 2012;**46**(3):260-273. DOI: 10.1016/j.molcel.2012.03.021

[69] Zasadzinska E, Foltz DR. Orchestrating the specific assembly of centromeric nucleosomes. Progress in Molecular and Subcellular Biology. 2017;**56**:165-192. DOI: 10.1007/978-3-319-58592-5_7

[70] Mitra S, Srinivasan B, Jansen LET. Stable inheritance of CENP-A chromatin: Inner strength versus dynamic control. Journal of Cell Biology. 2020;**219**(10): 1-17. DOI: 10.1083/jcb.202005099

[71] Flotho A, Melchior F. Sumoylation: A regulatory protein modification in health and disease. Annual Review of Biochemistry. 2013;**82**:357-385. DOI: 10.1146/annurev-biochem-061909-093311

[72] Jentsch S, Psakhye I. Control of nuclear activities by substrate-selective and protein-group SUMOylation. Annual Review of Genetics. 2013;**47**:167-186. DOI: 10.1146/annurev-genet-111212-133453

[73] Mukhopadhyay D, Arnaoutov A, Dasso M. The SUMO protease SENP6 is essential for inner kinetochore assembly. Journal of Cell Biology. 2010;**188**(5): 681-692. DOI: 10.1083/jcb.200909008

[74] Liebelt F et al. The poly-SUMO2/3 protease SENP6 enables assembly of the constitutive centromere-associated network by group deSUMOylation.

Nature Communications. 2019;**10**(1):3987. DOI: 10.1038/s41467-019-11773-x

[75] Mitra S et al. Genetic screening identifies a SUMO protease dynamically maintaining centromeric chromatin. Nature Communications. 2020;**11**(1):501. DOI: 10.1038/s41467-019-14276-x

[76] Samel A, Nguyen TKL, Ehrenhofer-Murray AE. Defects in methylation of arginine 37 on CENP-A/Cse4 are compensated by the ubiquitin ligase complex Ubr2/Mub1. FEMS Yeast Research. 2017;**17**(1):1-8. DOI: 10.1093/femsyr/fox009

[77] Akiyoshi B, Nelson CR, Duggan N, Ceto S, Ranish JA, Biggins S. The Mub1/Ubr2 ubiquitin ligase complex regulates the conserved Dsn1 kinetochore protein. PLoS Genetics. 2013;**9**(2):e1003216. DOI: 10.1371/journal.pgen.1003216

[78] Herrero E, Thorpe PH. Synergistic control of kinetochore protein levels by Psh1 and Ubr2. PLoS Genetics. 2016;**12**(2):e1005855. DOI: 10.1371/journal.pgen.1005855

[79] Niikura Y, Kitagawa R, Ogi H, Abdulle R, Pagala V, Kitagawa K. CENP-A K124 ubiquitylation is required for CENP-A deposition at the centromere. Developmental Cell. 2015;**32**(5):589-603. DOI: 10.1016/j.devcel.2015.01.024

[80] Sharp JA, Franco AA, Osley MA, Kaufman PD. Chromatin assembly factor I and Hir proteins contribute to building functional kinetochores in *S. cerevisiae*. Genes & Development. 2002;**16**(1):85-100. DOI: 10.1101/gad.925302

[81] Lopes da Rosa J, Holik J, Green EM, Rando OJ, Kaufman PD. Overlapping regulation of CenH3 localization and histone H3 turnover by CAF-1 and HIR proteins in *Saccharomyces cerevisiae*. Genetics. 2011;**187**(1):9-19. DOI: 10.1534/genetics.110.123117

[82] Lacoste N et al. Mislocalization of the centromeric histone variant CenH3/CENP-A in human cells depends on the chaperone DAXX. Molecular Cell. 2014;**53**(4):631-644. DOI: 10.1016/j.molcel.2014.01.018

[83] Athwal RK et al. CENP-A nucleosomes localize to transcription factor hotspots and subtelomeric sites in human cancer cells. Epigenetics & Chromatin. 2015;**8**:2. DOI: 10.1186/1756-8935-8-2

[84] Mishra PK et al. Misregulation of Scm3p/HJURP causes chromosome instability in *Saccharomyces cerevisiae* and human cells. PLoS Genetics. 2011;**7**(9):e1002303. DOI: 10.1371/journal.pgen.1002303

[85] Zhang F et al. HJURP is a prognostic biomarker for clear cell renal cell carcinoma and is linked to immune infiltration. International Immunopharmacology. 2021;**99**:107899. DOI: 10.1016/j.intimp.2021.107899

[86] Li Y et al. Hypomethylation-driven overexpression of HJURP promotes progression of hepatocellular carcinoma and is associated with poor prognosis. Biochemical and Biophysical Research Communications. 2021;**566**:67-74. DOI: 10.1016/j.bbrc.2021.05.102

[87] Kang DH et al. Prognostic relevance of HJURP expression in patients with surgically resected colorectal cancer. International Journal of Molecular Sciences. 2020;**21**(21). DOI: 10.3390/ijms21217928

[88] Wei Y et al. Knockdown of HJURP inhibits non-small cell lung cancer cell proliferation, migration, and

invasion by repressing Wnt/beta-catenin signaling. European Review for Medical and Pharmacological Sciences. 2019;**23**(9):3847-3856. DOI: 10.26355/eurrev_201905_17812

[89] Chen T et al. HJURP promotes hepatocellular carcinoma proliferation by destabilizing p21 via the MAPK/ERK1/2 and AKT/GSK3beta signaling pathways. Journal of Experimental & Clinical Cancer Research. 2018;**37**(1):193. DOI: 10.1186/s13046-018-0866-4

[90] Hu Z et al. The expression level of HJURP has an independent prognostic impact and predicts the sensitivity to radiotherapy in breast cancer. Breast Cancer Research. 2010;**12**(2):R18. DOI: 10.1186/bcr2487

[91] Hori T et al. Essentiality of CENP-A depends on its binding mode to HJURP. Cell Reports. 2020;**33**(7):108388. DOI: 10.1016/j.celrep.2020.108388

[92] Serafim RB et al. HJURP knockdown disrupts clonogenic capacity and increases radiation-induced cell death of glioblastoma cells. Cancer Gene Therapy. 2020;**27**(5):319-329. DOI: 10.1038/s41417-019-0103-0

[93] Chen T et al. HJURP promotes epithelial-to-mesenchymal transition via upregulating SPHK1 in hepatocellular carcinoma. International Journal of Biological Sciences. 2019;**15**(6):1139-1147. DOI: 10.7150/ijbs.30904

[94] Hu B, Wang Q, Wang Y, Chen J, Li P, Han M. Holliday junction-recognizing protein promotes cell proliferation and correlates with unfavorable clinical outcome of hepatocellular carcinoma. Oncotargets and Therapy. 2017;**10**:2601-2607. DOI: 10.2147/OTT.S127738

[95] Eisenstatt JR et al. Dbf4-dependent kinase (DDK)-mediated proteolysis of CENP-A prevents mislocalization of CENP-A in *Saccharomyces cerevisiae*. G3 (Bethesda). 2020;**10**(6):2057-2068. DOI: 10.1534/g3.120.401131

[96] Canzonetta C, Leo M, Guarino SR, Montanari A, Francisci S, Filetici P. SAGA complex and Gcn5 are necessary for respiration in budding yeast. Biochimica et Biophysica Acta. 2016;**1863**(12):3160-3168. DOI: 10.1016/j.bbamcr.2016.10.002

[97] Henry KW et al. Transcriptional activation via sequential histone H2B ubiquitylation and deubiquitylation, mediated by SAGA-associated Ubp8. Genes & Development. 2003;**17**(21):2648-2663. DOI: 10.1101/gad.1144003

[98] Daniel JA et al. Deubiquitination of histone H2B by a yeast acetyltransferase complex regulates transcription. Journal of Biological Chemistry. 2004;**279**(3):1867-1871. DOI: 10.1074/jbc.C300494200

[99] Metzger MB, Scales JL, Dunklebarger MF, Weissman AM. The ubiquitin ligase (E3) Psh1p is required for proper segregation of both centromeric and two-micron plasmids in *Saccharomyces cerevisiae*. G3 (Bethesda). 2017;**7**(11):3731-3743. DOI: 10.1534/g3.117.300227

[100] Zhou N, Shi L, Shan S, Zhou Z. Molecular basis for the selective recognition and ubiquitination of centromeric histone H3 by yeast E3 ligase Psh1. Journal of Genetics and Genomics. 2021;**48**(6):463-472. DOI: 10.1016/j.jgg.2021.04.007

Chapter 10

E3 Ligase for CENP-A (Part 2)

Yohei Niikura and Katsumi Kitagawa

Abstract

Centromeric CENP-A, a variant of histone H3, plays a central role in proper chromosome segregation and its function is highly conserved among different species. In most species with regional centromeres, an active centromere relies not on defined DNA sequences, but on the presence of CENP-A proteins in centromeric nucleosomes. CENP-A is proposed to be the non-DNA indicator (epigenetic mark) that defines proper centromere assembly and function. Recently, many post-translational modifications (PTMs) of CENP-A and their functions have been reported. They revealed the importance of the functions of CENP-A PTMs in CENP-A deposition at centromeres, proteolysis/protein stability, and recruitment of other centromere-kinetochore proteins. Ubiquitylation and sumoylation by E3 ligases regulate multiple functions, including proteolysis and signaling, and play important roles in the cell cycle and mitotic control. Recently, the function of E3 ligase that ubiquitylates/sumoylates and controls CENP-A protein has emerged as an important regulatory paradigm in different species. Many have reported the importance of CENP-A ubiquitylation and sumoylation in CENP-A deposition at centromeres and for protein stability, which is regulated by specific E3 ligases. Therefore, here we summarize what is known about the E3 ligases for CENP-A ubiquitylation and sumoylation and their biological functions and significance in different species.

Keywords: CENP-A, Cse4, Cnp1, CID, E3 ligase, centromere, kinetochore, ubiquitylation, sumoylation, epigenetics, Psh1, Siz1 and Siz2, Slx5 and Slx8, CUL3/RDX, SCF, APC, CUL4A/RBX1/COPS8, DAXX (fruit fly DLP), SGT1-HSP90, Scm3, CAF-1 complex, CAL1, HJURP, Mis18 (human Mis18α and Mis18β) and Mis16 (human RbAp46 and RbAp48)

1. Introduction

During cell division, proper chromosome segregation must be achieved to avoid unequal distribution of chromosomes to daughter cells. Spindle microtubules must attach to a single region of each chromosome, termed the "centromere," in most eukaryotes. The kinetochore is a complex of proteins that are located at the centromere. Defects in the centromere-kinetochore and spindle check point functions lead to aneuploidy and cancer and are often associated with a poor prognosis. Therefore, it is highly important to study the spatiotemporal regulation and the structures of centromere and kinetochore proteins to understand chromosome instability (CIN) during development and cancer progression. The key question is how the chromosomal location and function of a centromere (i.e., centromere identity) are determined and thus participate in accurate chromosome segregation. In most species with regional centromeres (see the previous chapter for an exception of the budding yeast *Saccharomyces cerevisiae* that has genetically defined point centromeres), centromere identity relies not on a defined DNA sequence, but on the presence of a special nucleosome that contains a specific histone-like protein called CENP-A. CENP-A is proposed to be the non-DNA indicator (epigenetic mark) of centromere identity. CENP-A partially replaces histone H3 in the centromeric regions. CENP-A-containing nucleosomes are the basis for kinetochore formation and are the most important marker for centromere function in eukaryotes [1].

The structure of CENP-A-containing nucleosomes is more compact than H3-containing nucleosomes [2–4]. Although it is commonly reported that CENP-A-containing nucleosomes are formed with the canonical histones H2A, H2B, and H4 at the active centromeres, their structure remains controversial among different research groups [5]. CENP-A is at the top of a hierarchy of the pathway that determines the assembly of kinetochore components [6], and how CENP-A defines the position of the centromere in humans is the fundamental question. While the function of CENP-A protein is highly conserved among most eukaryotes, its protein sequence has apparently undergone both convergent and divergent evolution [7], and the centromere DNA repeats with which the CENPA-containing nucleosome interacts are also highly diverged. The architectures of CENP-A chromatin with quantified numbers of CENP-A (CenH3) molecules (e.g., ~400 molecules of human CENP-A/kinetochore) have been reported using fluorescence microscope assays among different species [8–11]. CENP-A is also called CenH3 (centromere-specific histone H3). Its homologs in different species are summarized in **Table 1**.

CENP-A contains a short centromere targeting domain (CATD) within the histone fold region [2] in the C-terminus. Replacement of the corresponding region of histone H3 with the CATD is sufficient to direct histone H3 to the centromere [2], and this chimeric histone can rescue the viability of CENP-A-depleted cells [2, 12]. The CENP-A C-terminus contains another tail domain that recruits CENP-C to promote centromere and kinetochore assembly [13, 14]. CENP-N was also identified as the first protein to selectively bind CENP-A nucleosomes but not H3 nucleosomes during centromere assembly [15].

Meanwhile, the functions of the N-terminal CENP-A are also reported for some species [16–27] (see also previous chapter, Sections 2.1, 2.2.2, 2.4.1, 2.9 and this chapter, Sections 2.1, 2.3, 2.4, 3.1, 4.1, 4.6, and 5.1). Loading of CENP-A at centromeres and its incorporation/deposition and maintenance in centromeric chromatin is cell cycle-regulated. In cells overexpressing CENP-A, the ectopic protein incorporates throughout the chromatin in interphase [28]. By the next G1, however, mis-incorporated CENP-A seems to have been cleared from chromatin by a mechanism that likely involves ubiquitin-mediated proteolysis, as suggested by experiments in yeast and *Drosophila* cells [28]. Importantly, the timing of deposition of newly synthesized CENP-A within the cell cycle may be variable not only among different species but also different developmental stages within the same species. Yeast suppressor of chromosome missegregation protein 3 (Scm3) [29] (previous chapter, **Figure 1**; **Table 1**) is a distant counterpart of human Holliday junction recognition protein (HJURP) (**Figures 2** and **3**; **Table 1**), and they are CENP-A (CenH3)-specific chromatin assembly factors [29, 41–43]. The incorporation of newly synthesized CenH3 (CENP-A) into centromeric nucleosomes depends on Scm3/HJURP [41–43] in budding, fission yeasts, and humans. In addition, other components and factors that contribute to CENP-A deposition, maintenance, and inheritance in centromeric nucleosomes have been reported [28, 44].

Species	CENP-A homolog	E3 ligase (ubiquitylation or sumoylation)	Function	Preceding PTMs before ubiquitylation or sumoylation	Another proposed factor relevant to the E3 function
Saccharomyces pombe	Cnp1/ SpCENP-A	N.D.	Proteasomal degradation to remove non-centromeric Cnp1	N.D.	N-terminal domain of Cnp1, Overexpression of H3/H4
Drosophila melanogaster	CID/Cid	CUL3/RDX (ubiquitylation)	Interacts with CAL3 and promotes CAL3 function, loading and stabilizing (maintenance) of CID protein at centromeres (proteasomal independent mechanism)	N.D.	N.D.
		SCF^{Ppa} (ubiquitylation)	Prevents the promiscuous incorporation of CID across chromatin during replication, (targeting CID that is not in complex with CAL1)	S20 phosphorylation	S20 phosphorylation
		APC/C^{Cdh1} (ubiquitylation)	Degradation of the CAL1-CID complex (likely regulates centromeric CID deposition)	N.D.	N.D.
Homo sapiens	CENP-A	CUL4A/RBX1/COPS8	Facilitate interaction of CENP-A with HJURP through CENP-A ubiquitylation, CENP-A deposition at the centromere (proteasomal independent mechanism)	N.D.	COPS8 as an adaptor, heterodimerization of CENP-A, SUGT1-HSP90
Arabidopsis thaliana (CENH3 was expressed in Nicotiana tabacum)	AtCENH3	N.D. (VHHGFP4-human SPOP as synthetic E3 ligase expressed in *Nicotiana tabacum*)	Proteasomal degradation of AtCENH3	N.D.	N.D.

Note: E3s of some species (e.g., Caenorhabditis elegans, Xenopus laevis, zebrafish Danio rerio, chicken Gallus domesticus DT40 cells, Mus musculus, etc.) are not discovered, and so are not described in this table. N.D. = not determined; PTMs = post-translational modifications.

Table 1.
E3 ligases for CENP-A in species with regional centromeres.

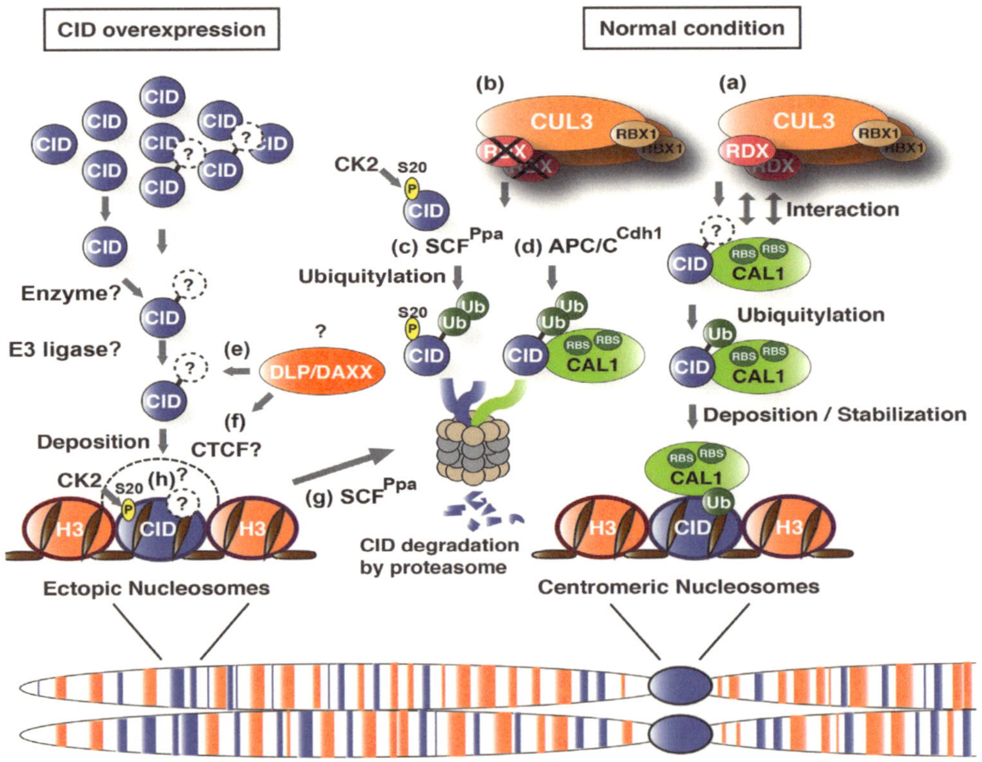

Figure 1.
Mechanistic scheme for Drosophila melanogaster CENP-A^{CID} pathways. (Right) (a) Bade et al. proposed the role of the E3 ligase CUL3/RDX in CUL1-dependent ubiquitylation of CID [30, 31]. CUL3/RDX activity, which is presumably in dimer form [30], leads to the monoubiquitylation of CID through an interaction with the RDX-binding sites (RBS) of CAL1. This monoubiquitylation of CID is proteasomal-independent but is required for stable localization of CID and CAL1 to the centromere. The ubiquitin-conjugating enzyme (E2) bound on RBX1 is omitted for simplicity. (b) In the absence of RDX, CENP-A is not monoubiquitylated by CUL3, and both CENP-A and CAL1 are subjected to proteasome-dependent degradation, but presumably ubiquitylated by (c) SCF^{Ppa} or (d) APC/C^{Cdh1} as Moreno-Moreno proposed [32] (see below). Furthermore, the absence of RDX results in cell death and severe chromosomal aberration (e.g., chromosome fragmentation), some of which may be attributed to the loss of CENP-A and CAL1 from centromeric regions (not shown in this cartoon). Moreno-Moreno et al. suggest that (c) whereas SCF^{Ppa} targets the fraction of CID that is not in complex with CAL1, (d) APC/C^{Cdh1} contributes to the degradation of the CAL1-CID complex and, thus, likely regulates centromeric CID deposition [32] as previously proposed [30]. Huang et al. proposed that phosphorylation of CID of serine 20 (S20) regulates both protein turn-over and centromere-specific loading [33] (see also left). The CID S20 phosphorylation renders CID a substrate for ubiquitylation by SCF^{Ppa}, thereby regulating the abundance of free pre-nucleosomal CID through subsequent proteasomal degradation (see also left (g)). (Left) (e) The role of DLP/DAXX in CID deposition into ectopic nucleosomes through CID ubiquitylation as proposed in a human cell model [34] has yet to be confirmed experimentally. (f) CTCF occlusion by the aberrant nucleosome of heterotypic tetramer consisting of CENP-A-H4 with H3.3-H4 as proposed in human cell models [34] has not been confirmed in D. melanogaster. (g) Huang et al. observed that CID S20 is phosphorylated by casein kinase II (CK2) not only insoluble but also chromatin-bound CID, and this phosphorylation also facilitates removal of CID from ectopic but not from centromeric sites in chromatin [33]. (h) Factors/components that stabilize ectopically incorporated CID and are required for neocentromere formation and its maintenance are not yet clear. The status of overall PTMs, including polyubiquitylation of CID, especially in the ectopic nucleosome, has yet to be elucidated.

Recently, many post-translational modifications of CENP-A and their functions have been reported [45]. They revealed the importance of these changes in CENP-A deposition at centromeres, proteolysis/protein stability, and recruitment of the CCAN (constitutive centromere-associated network) proteins [45]. Thus, here we focus on E3

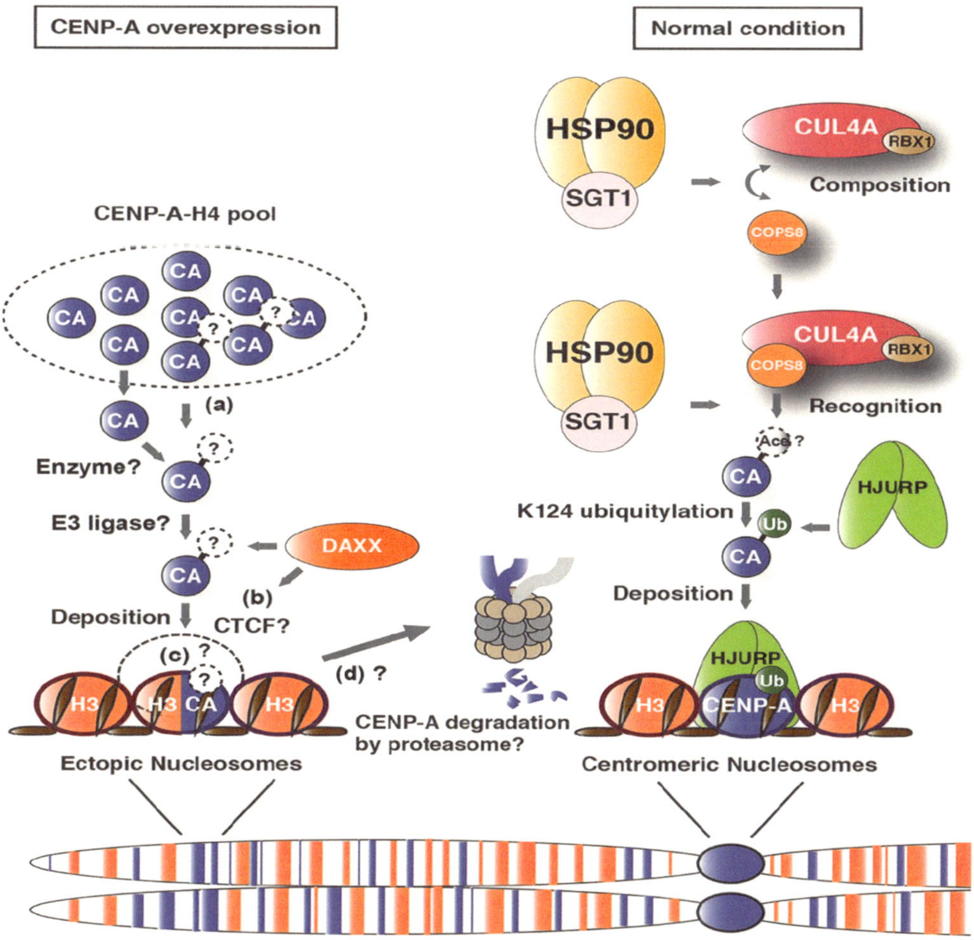

Figure 2.
Mechanistic scheme for human CENP-A pathways. (Right) In normal conditions, CUL4A-RBX1-COPS8 E3 ligase activity is required for CENP-A mono- or di-ubiquitylation on lysine 124 (K124) and CENP-A centromere localization [35]. CENP-A K124 mono- or di-ubiquitylation is required for CENP-A's interaction with the chromatin assembly factor HJURP and CENP-A deposition at the centromere. The CUL4A complex targets CENP-A through the adaptor COPS8/CSN8 that has WD40 motifs. In non-canonical CRL4 machinery, CUL4/RBX1/COPS8 may dimerize as a CUL4/DACAF1 complex [36, 37], but the dimerization unit remains unknown [31]. Here only the CUL4/RBX1/COPS8 monomer is shown for simplicity. Upstream, the SGT1-HSP90 complex is required for the composition of the CUL4A complex and recognition of COPS8 to target CENP-A. Therefore, the SGT1-HSP90 complex is also required for CENP-A ubiquitylation and localization of CENP-A to centromeres. "CA" refers to the CENP-A monomer. "Ace" refers to the putative acetylated lysine 124 (K124) previously reported by Bui et al. that is concurrent with the structural transitions of CENP-A-containing nucleosomes through the cell cycle [38]. Their computational modeling suggests that acetylation of K124 causes tightening of the histone core and hampers accessibility to its C-terminus, which in turn reduces CENP-C interaction [39] (not shown in this cartoon). However, its precise function and relationship with K124 ubiquitylation remain to be studied. (Left) When human CENP-A is overexpressed, CENP-A is incorporated into ectopic nucleosomes consisting of a heterotypic tetramer that contains CENP-A-H4 with H3.3-H4 [34]. This ectopic localization of this particle (aberrant nucleosome) depends on the H3.3 chaperone DAXX rather than the centromeric CENP-A-specific chaperone HJURP. (a) Post-translational modifications of human CENP-A, especially before recognition by DAXX and after incorporation into the ectopic nucleosome, have yet to be elucidated. (b) CTCF occlusion by the aberrant nucleosome of a heterotypic tetramer consisting of CENP-A-H4 with H3.3-H4 was also proposed in a human cell model [34], but specific DAXX localization on these CTCF sites under CENP-A overexpression has not been confirmed experimentally. (c) Factors/components that stabilize ectopically incorporated CENP-A and are required for neocentromere formation and its maintenance are not yet clear. In addition, the status of overall post-translational modifications, including polyubiquitylation of CENP-A, especially in the ectopic nucleosome, is unknown. (d) Currently, the proteolysis mechanism for mis-incorporated human CENP-A and its E3 ligase is not yet clear. Note that histone H4 and phosphorylation of human CENP-A are omitted for simplicity.

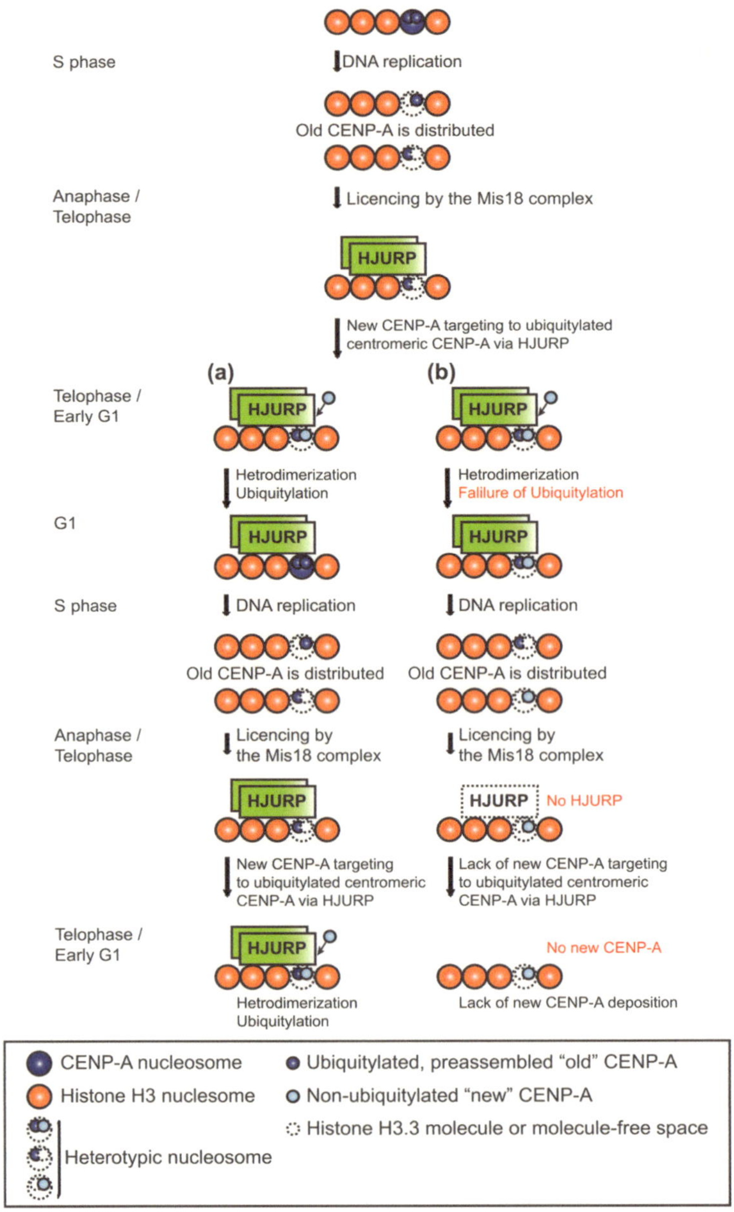

Figure 3.
Models of epigenetic inheritance of CENP-A ubiquitylation through heterodimerization. In the octamer model, two CENP-A dimers in one nucleosome are split/diluted between the two daughter centromere-DNA sequences, and one CENP-A molecule replaces with one H3 molecule or leaves a molecule-free space during the replication/S phase. HJURP (Holliday junction recognition protein) predominantly interacts with ubiquitylated, preassembled "old" CENP-A, which resides mostly in nucleosomes. A non-ubiquitylated newly synthesized ("new") CENP-A monomer targets ubiquitylated centromeric CENP-A through preassembled HJURP. Note that histone H4 is omitted for simplicity. (a) New CENP-A is appropriately ubiquitylated in a heterodimerization-dependent manner (i.e., dimers of old CENP-A with new CENP-A). In this way, both ubiquitylation and the location of the centromere are inherited epigenetically. (b) If K124 ubiquitylation does not occur on new CENP-A, the non-ubiquitylated CENP-A nucleosome distributed during the S phase does not recruit HJURP to the centromere because the affinity of non-ubiquitylated new CENP-A to HJURP is low. Subsequently, this loss of localization of HJURP at the centromere leads to the lack of new CENP-A targeting to ubiquitylated centromeric CENP-A through HJURP, and eventually to the lack of new CENP-A deposition. This figure is partly adapted from Niikura et al. [40].

ligase activities (i.e., on ubiquitylation and sumoylation) of CENP-A and summarize these functions for each species with regional centromeres in the following sections.

2. E3 ligase for fission yeast (*Schizosaccharomyces pombe*) CENP-A^{Cnp1} and its function

2.1 Overview of CENP-A^{Cnp1}

Fission yeast (*Schizosaccharomyces pombe*) centromeres consist of large (40–100 kb) inverted repeats that display heterochromatic features. Therefore, fission yeast provides a good model for higher eukaryotic centromeres. The mechanistic processes to establish centromeric chromatin of fission yeast and its structures have been reviewed [46, 47]. This section focuses on the E3 ligase(s) for fission yeast CENP-A^{Cnp1} and its function through ubiquitylation, although endogenous E3 ligase for *S. pombe* CENP-A^{Cnp1} is not yet identified and its specific regulation is still unclear.

In fission yeast, the recruitment of the CENP-A-specific chaperone to the centromere is an essential step in epigenetic inheritance. The fission yeast Scm3 could be functionally homologous to HJURP. It interacts with CENP-A, localizes to centromeres during most of the cell cycle (except in mitosis), and is required for CENP-A deposition [48, 49]. Sequence analysis revealed a shared common domain in Scm3 and HJURP proteins [29]. Dunleavy et al. identified another chaperone known as Sim3 (start independent of mitosis 3) in fission yeast [50, 51]. Sim3 is homologous to known histone binding proteins NASP (human) and N1/N2 (xenopus) and aligns with Hif1 (*S. cerevisiae*), defining the SHNi-TPR family [51]. Sim3 is distributed throughout the nucleoplasm, yet it associates with CENP-A^{Cnp1} and also binds H3. It interacts also with non-chromosomal CENP-A and is required for its incorporation in *S. pombe*. These results are consistent with those in *Arabidopsis thaliana* [19] (see also Section 5.1). Sim3 also has been proposed to share some common roles with the histone chaperone Asf1, mutations in which cause a defect in overall chromatin structure [52, 53]. It has been suggested that Sim3 could function as an escort chaperone, handing off CENP-A to Scm3, a role that human HJURP may accomplish by itself [43, 48, 50].

Mis16 (human homologs of Mis16 are RbAp46 and RbAp48) and Mis18 (human homologs of Mis18 are Mis18α and Mis18β) are required for loading of newly synthesized Cnp1/CENP-A into centromeric chromatin [54, 55], but are absent from organisms with point centromeres [44] (see also previous chapter, Section 2.3.3 and this chapter, Sections 3.1 and 4.1). Mis16 and Mis18 are also required for the maintenance of the hypoacetylation of histone H4 specifically within the central domain of the centromere [55], and Mis16 homologs are components of several histone chaperon complexes [56]. Moreover, acetylation of histone H4 lysine 5 and 12 (H4K5ac and H4K12ac) within the pre-nucleosomal CENP-A-H4-HJURP complex mediated by the RbAp46/48-Hat1 complex is required for CENP-A deposition into centromeres in chicken and humans [57], consistent with Hat1's role in *Drosophila melanogaster* [58] (see also Sections 3.1 and 4.1). In mouse studies, Mis18α interacts with DNMT3A/3B, and this interaction is required to maintain DNA methylation [59]. *Mis18α* deficiency leads to not only the reduction of DNA methylation, but altered histone H3 modifications, and uncontrolled noncoding transcripts in the centromere region (see also Section 4.1). It is an interesting model that Mis16 and Mis18 complexes "prime centromeres" affect post-translational modifications of histone H3/H4 proteins and centromeric DNA in advance of

CENP-A incorporation. How such chromatin structures feedback with the regulation of E3 ligases of CENP-A has not yet been reported, which could be important.

In *S. pombe*, spMis16, and spMis18 mutants eliminate Cnp1 incorporation to centromeres and Mis18 directly interacts with Scm3 *in vitro*, suggesting they cooperate to assemble Cnp1 into centromeric chromatin [48]. *S. pombe* lacks the vertebrate Mis18BP1 ortholog, and the Mis18BP1 function in *S. pombe* is replaced by the Eic1 protein (a.k.a Mis19) [44, 60, 61]. While Eic2 (a.k.a Mis20) is dispensable for the recruitment of Cnp1 to the centromere, Eic1 is required for the recruitment of the Mis18, Mis16, and Scm3 proteins to the centromere and Cnp1 incorporation. Both of the Eic1 and Eic2 proteins co-purify with the spMis18 and exhibit a similar centromeric localization throughout the cell cycle [60, 61]. Taken together, these data suggest that Eic1 is functionally analogous to the Mis18BP1 subunit [60, 61]. However, Eic1 is evolutionarily distinct and no homolog of Mis19 has been found in the human genome, and Eic1 does not share any apparent sequence homology to Mis18BP1 [60, 61]. Centromere localization and function of Mis18 require Yippee-like domain-mediated oligomerization [62]. Furthermore, there are at least two mechanisms to restrict the assembly of CENP-A nucleosomes in G1—disruption of Mis18 multimerization by HJURP-Mis18 interaction, and ubiquitylation and degradation of Mis18β through $SCF^{\beta TrCP}$ E3 ligase [44].

Domain-specific function, such as the N-terminal function, of fission yeast Cnp1/CENP-A is also reported as budding yeast Cse4 [24, 25] (see also previous chapter, Section 2.4). Folco et al. demonstrated that alteration of the Cnp1 N-tail does not affect Cnp1 loading at centromeres, outer kinetochore recruitment, or spindle checkpoint signaling but significantly increases chromosome loss [17]. On the other hand, their N-tail mutants exhibit centromere inactivation enhanced by an altered centromere. The N-tail mutants specifically reduced localization of the CCAN proteins CENP-T^{Cnp20} and CENP-I^{Mis6}, but not CENP-C^{Cnp3}. Therefore, these authors suggest that the Cnp1 N-tail maintains the epigenetic stability of centromeres in fission yeast, at least in part via assembly of the CENP-T branch of the CCAN. Tan et al. identified a proline-rich "GRANT" (Genomic stability Regulating site within CENP-A N-Terminus) motif that is essential for Cnp1 centromeric targeting [24]. They showed that especially GRANT proline-15 (P15) undergoes cis-trans isomerization to drive proper chromosome segregation. This cis-trans isomerization appears to be carried out by two FK506-binding protein (FKBP) family prolyl cis-trans isomerases. In addition, they identified Sim3 as a Cnp1 NTD interacting protein that is dependent on GRANT proline residues. Together, they suggest cis-trans proline isomerization of Cnp1 is required for precise propagation of centromeric integrity in fission yeast, presumably via targeting Cnp1 to the centromere. Thus, the requirement of cis-trans proline isomerization of $CenH3^{Cnp1}$ in fission yeast studies appears to be consistent with the one of $CenH3^{Cse4}$ proposed in budding yeast studies [63] (see also previous chapter, Section 2.2.3). However, they suggest that the GRANT-prolines of Cnp1 do not coordinate proteolysis of the SpCENP-A protein as do proline residues in the budding yeast Cse4 NTD. In addition, Tan et al. showed that sequential truncation of the NTD did not improve the stability of the protein, suggesting that the NTD of Cnp1 does not regulate the turnover of the protein [25]. Instead, they proposed that heterochromatin integrity may contribute to Cnp1 stability and promote its chromatin incorporation.

Compared to the studies of budding yeast and some of the other species, currently, there are few studies on post-translational modifications and domain-specific functions of fission yeast CenH3/Cnp1. Further research is required on the relationships among Cnp1 post-translational modifications, structural change, interaction with its chaperones (e.g., Scm3 and Sim3), and surrounding heterochromatin regulation.

2.2 Dos1/2-Cdc20 complex

In *S. cerevisiae*, all pre-existing CENP-A is replaced by newly synthesized CENP-A during the S phase [64], whereas in *S. pombe*, two pathways of CENP-A deposition exist at the S and G2 phases of the cell cycle [50, 65]. Parental CENP-A is deposited at centromeres during the S phase, whereas newly synthesized CENP-A is deposited during later stages of the cell cycle [66]. The mechanisms involved in the deposition of CENP-A at centromeres during the S phase remain poorly understood [66]. In *S. pombe*, the GATA-like transcription factor Ams2, a key factor in CENP-A deposition during the S phase, appears to work, at least in part, through the regulation of transcription of core histones [65].

Li et al. reported that the DNA polymerase (Pol) epsilon catalytic subunit A (pol2), Cdc20, interacts with the Dos1-Dos2 silencing complex to facilitate heterochromatin assembly and inheritance of H3K9 methylation during the S phase [67]. We note that fission yeast *S. pombe* Cdc20 (UniProtKB—P87154) is not the ortholog of human CDC20 (cell division cycle protein 20 homolog, UniProtKB—Q12834), but of human POLE (UniProtKB—Q07864). Gonzalez et al. showed that the Dos1/2-Cdc20 complex is also required for localization of Cnp1 at centromeres at this stage [66]. Disruption of Dos1 (also known as Raf1/Clr8/Cmc1), Dos2 (also known as Raf2/Clr7/Cmc2), or Cdc20, a DNA polymerase epsilon subunit, leads to delocalization of CENP-A from centromeres and mislocalization of the protein to ectopic (non-centromeric) sites. All three mutants of Dos1, Dos2, and Cdc20 exhibit spindle disorganization and mitotic defects. Inactivation of Dos1 or Cdc20 also results in the accumulation of noncoding RNA transcripts from centromeric cores, a feature common to mutants affecting kinetochore integrity. These authors found that Dos1 physically associates with Ams2 and contributes to the interaction of Ams2 with centromeric cores during the S phase. They further showed that Dos2 associates with centromeric cores during the S phase and that its recruitment to centromeric cores depends on Cdc20. Together, this study identifies a physical link between DNA replication and the CENP-A assembly machinery and provides mechanistic insight into how CENP-A is faithfully inherited during the S phase.

It is important to clarify how exactly the Dos1-Dos2-Cdc20 complex contributes to the inheritance of preexisting Cnp1 during centromere replication [66]. Interestingly, Rik1 is a component of silencing factors. The heterochromatic methylation of histone H3-K9 by Clr4 is promoted by silencing factors: Dos1-Dos2-Rik1-Lid2 [67]. Horn et al. reported that subunits of a cullin-dependent E3 ubiquitin ligase interact with Rik1 and Clr4, and Rik1-TAP preparations exhibit robust E3 ubiquitin ligase activity [68]. They also demonstrated that the expression of a dominant-negative allele of the Pcu4 cullin subunit (the human Cullin-4 homolog) disrupts the regulation of K4 methylation within heterochromatin. Hong et al. also reported a novel complex that associates with the Clr4 methyltransferase, termed the CLRC (CLr4-Rik1-Cul4) complex using affinity purification of Rik1, and found that Rik1 interacts with the fission yeast Cullin4 (Cul4, encoded by $cul4^{+}$), the ubiquitin-like protein, Ned8, and two previously uncharacterized proteins, designated Cmc1 and Cmc2 [69]. They also demonstrated a defect in the processing of noncoding RNA to small RNA caused by the defective Clr4-Rik1-Cul4 complex, suggesting that the components of the Clr4-Rik1-Cul4 complex collaborate at an early step in heterochromatin formation. Unlike the studies of CUL3/RDX in fruit flies (**Figure 1**, right; see also Section 3), the function of Cul4 E3 ligase targeting non-centromeric CENP-A^{Cnp1} and the mechanism of its proteolysis are not yet studied in fission yeast. In fission yeast, there is no report about the involvement of Cul4 E3 ligase in CENP-A^{Cnp1} deposition at the centromere, unlike in humans (**Figure 2**, right; see also Section 4). On the other hand, it would be interesting to test if the Cul4 E3 function for heterochromatin assembly is conserved in other species, including humans.

2.3 Assembly of Cnp1 at non-centromeric chromatin

Consistent with the results in budding yeast Cse4 [23, 70, 71] (see also the previous chapter, Section 2.1), Gonzalez et al. reported that the overexpression of fission yeast Cnp1 results in the assembly of Cnp1 at non-centromeric chromatin during mitosis and meiosis [18]. The non-centromeric Cnp1 is preferentially recruited near heterochromatin and is able to recruit kinetochore components, and Cnp1 overexpression leads to severe chromosome missegregation and spindle microtubule disorganization. Moreover, ectopic Cnp1-containing chromatin is inherited over multiple generations using pulse induction of Cnp1 overexpression. Interestingly, ectopic assembly of Cnp1 is suppressed by overexpression of histone H3 or H4 (**Table 1**), as other groups suggest that the balance between histones H3 and H4 and CENP-A is important for centromeric chromatin assembly [72, 73]. Further, Gonzalez et al. demonstrated that deletion of the N-terminal domain of Cnp1 results in an increase in the number of ectopic CENP-A sites, suggesting that the N-terminal domain of CENP-A prevents CENP-A assembly at ectopic loci via the ubiquitin-dependent proteolysis [18].

However, it is not yet clear by which E3 ligase the exogenous Cse4 expressed in the fission yeast *S. pombe* is targeted, and a budding yeast Psh1 homolog is not yet identified in fission yeast. Further study is required to elucidate how the activity of a specific E3 ligase targeting endogenous Cnp1 is regulated in fission yeast.

2.4 Heterochromatin and RNAi regulate centromeres by protecting Cnp1 from ubiquitin-mediated degradation

In most eukaryotes, the centromere is flanked and bordered by the epigenetically distinct heterochromatin domain. The establishment of centromeric heterochromatin profoundly correlates to centromere function, but the precise role of heterochromatin in centromere specification and activation is not yet clear. The transition between point centromeres (e.g., budding yeast *S. cerevisiae*) and regional centromeres (e.g., fission yeast *S. pombe*) is considered one of the most substantial centromere evolutionary events.

Yang et al. demonstrated that budding yeast Cse4 can localize to centromeres in fission yeast and partially substitute fission yeast Cnp1, however, overexpressed Cse4 localizes to heterochromatin regions [26]. Cse4 undergoes efficient ubiquitin-dependent degradation in *S. pombe*, and its N-terminal domain contributes to its centromere distribution via ubiquitination. Importantly, their results showed that GFP-Cse4 fails to localize at centromeres without heterochromatin and RNA interference (RNAi) using Clr4 mutant (*clr4Δ*) and dicer mutant (*dcr1Δ*), respectively. Therefore, they showed that RNAi-dependent heterochromatin is required for centromeric localization of Cse4 and protects Cse4 from ubiquitin-dependent degradation. Heterochromatin is also required for the deposition of native Cnp1 at the centromere via the same mechanism. Together, they suggest that protection of CENP-A from degradation by heterochromatin is a conserved mechanism used for centromere assembly and provided novel insights into centromere evolution from point centromere to regional centromere.

However, E3 ligase targets endogenous Cnp1 is still unclear, and its degradation mechanism through heterochromatin and RNAi machinery in fission yeast is still elusive. Further study is required to elucidate how E3 ligase activity is involved in RNAi-dependent heterochromatin formation and maintenance in fission yeast.

3. E3 ligase for fruit fly (*Drosophila melanogaster*) CENP-A^{CID} and its function

3.1 Overview of CENP-A^{CID}

Fruit fly (*Drosophila melanogaster*) centromeres extend for 200–420 kb and contain repetitive DNA that is interspersed with transposable elements (TEs) [74]. TEs are sequences that have the capacity to move other chromosomal locations and are a component of the "interspersed repeat" fraction of most genomes [75]. In fruit flies and other species (e.g., plants, wallabies, humans), the significance and function of these TEs in centromeric DNA remain to be studied. In plants, Jiang et al. suggest that the retention of active transcriptional machinery within the long terminal repeat may promote demarcation of the active centromere [76] (see also Section 5). The importance of centromeric long noncoding RNA (cenRNA) for centromere integrity has been suggested in various species [77–79]. In humans, a cenRNA is required for targeting CENP-A to the centromere [80] (see also Section 6). Arunkumar and Melters hypothesize that loading of both CENP-A and CENP-C could be one major function of centromeric transcripts, and RNA-DNA triplexes (e.g., R-loops) could be involved in loading both proteins; thereby, one may elucidate the role of RNA-DNA triplexes in both CENP-A and CENP-C loading [77].

The mechanism of heterochromatin silencing in fruit flies has been reported [81], including the position-effect variegation [82], histone modification [83], and the RNAi machinery [84]. Recently, a PIWI-interacting RNAs (piRNAs) system has been implicated in heterochromatin formation [85–88], and the mechanism of heterochromatic piRNA production is being elucidated in *Drosophila* [89]. Ectopic CID/the *Drosophila* CENP-A homolog is prone to localize at euchromatin-heterochromatin boundaries, and this observation suggests that CID chromatin is likely to localize right next to a heterochromatin domain [90]. Kwenda et al. showed that RNA polymerase I transcription is required for efficient CID assembly in meiosis, as well as centromere tethering to nucleoli [91]. Recent work in mammalian and fruit fly cell lines showed that chemical inhibition of activated RNA polymerase 2 (RNAP2) resulted in the loss of centromeric CID chromatin [80, 92]; and the elongation factor Spt6 facilitates maintenance of centromeric CID [93]. These reports strongly suggest that transcription and RNA production are involved in CID incorporation.

The timing of CID incorporation occurs during metaphase/anaphase in *D. melanogaster* [74]. In human cells, the incorporation of newly synthesized CENP-A occurs in telophase/early G1 [94, 95]. Similar to humans, in the fast cycles of *Drosophila* syncytial embryos, CID incorporates in anaphase [96]. However, in *Drosophila* Kc cells, GFP-tagged CID was detected in metaphase cells 2 h after induction of its expression, implying that incorporation occurred at some point between the preceding G2 and metaphase [97]. In S2 cells, newly synthesized CAL1 is deposited at centromeres in prophase, preceding CID loading in metaphase [98]. Based on this observation, CAL1, like the Mis18 complex in humans, was suggested to prime the centromere before assisting in CID loading [98, 99]. In somatic tissues of *Drosophila* larvae, centromeric CID deposition initiates at late telophase and continues during G1 when APC/C^{Cdh1} is active [32, 100].

In *Drosophila*, Erhardt et al. performed an RNAi-based genome-wide search and identified CAL1 and CENP-C for CID localization determinants [101]. CID, CAL1, and CENP-C co-immunoprecipitate and are mutually dependent for centromere targeting and function. However, the molecular mechanism underlying these dependencies remains to be clarified. No homologs for CAL1 have been reported in other organisms. They also proposed that the mitotic cyclin A (CYCA) localizes at the centromere, and CYCA and RCA1/Emi1 couple centromere assembly to the cell cycle through regulation of the fizzy-related/CDH1 subunit of the APC [101], while Moreno-Moreno et al. proposed that APC/C^{Cdh1} contributes to the degradation of the CAL1-CID complex [32] (see also Section 3.3). Consistent with the role of histone H4 acetylation in chickens and humans [57], Boltengagen et al. showed that the histone acetyltransferase Hat1 contributes to the CID/CENP-A assembly pathway in *D. melanogaster* [58] (see also previous chapter, Section 2.3.3 and this chapter, Sections 2.1 and 4.1).

Recently, there have been more reports published on the mechanism of how these three proteins (CID, CAL1, and CENP-C) work in CID incorporation. Chen et al. showed that the constitutive centromere protein CENP-C is required for recruitment of the *Drosophila melanogaster (mel)* CAL1 protein to existing centromeres [102]. Rosin and Mellone showed that exogenously expressed CAL1 from two different *Drosophila* species was efficiently recruited to *D. melanogaster* endogenous centromeres [103]. The CENP-C interaction with CAL1 is conserved across the *Drosophila* phylogeny. Whereas the coordinated evolutionary changes between CAL1 and CID prevent the recruitment of *Drosophila species bipectinata* (*bip*) CID to melanogaster centromeres, the CAL1 proteins showed no species specificity in their recruitment. The importance of the CENP-C protein for recruiting the CENP-A deposition machinery is shared in the fly and human centromere assembly pathways. However, in humans, the Mis18 complex, which is absent in *Drosophila*, interacts with CENP-C to recruit HJURP and CENP-A to existing centromeres [104] (see also Section 4.1).

CENP-A is maintained to mark paternal centromeres, whereas most histones are removed from mature sperm. In *Drosophila* males, Kwenda et al. showed that the centromere assembly factors CAL1 and CENP-C are required for meiotic chromosome segregation, CID assembly and maintenance on sperm, and fertility [91]. They showed that CID accumulates with CAL1 in nucleoli in meiosis, and CENP-C normally limits the release of CAL1 and CID from nucleoli for proper centromere assembly in meiotic prophase I. Pauleau et al. found that overexpression of CAL1 is associated with increased CID levels at centromeres and uncouples CID loading from mitosis [105]. CID levels inversely correlate with mitosis, and mitosis length is influenced by the spindle assembly checkpoint. They found that CAL1 interacts with the SAC protein and RZZ complex component Zw10 and thus constitutes the anchor for the recruitment of RZZ. Demirdizen et al. showed that the N-terminus of CID contributes to nuclear localization and protein stability [106]. While co-expression of mutant CID with RbAp48 leads to exclusive non-centromeric CID incorporation, co-expression with CAL1 leads to exclusive centromere loading of CID, suggesting that CID-associated proteins, rather than CID itself, determine its localization. Their further analysis revealed that NuRD is required for ectopic CID incorporation. The interaction of the NuRD complex with CENP-A is mediated by RbAp48 and MTA1-like (i.e., a subunit of NuRD complex), which binds specifically to the N-terminal region of CENP-A. Roure et al. showed a positive feedback loop between CID, CENP-C, and CAL1 [107], and

Medina-Pritchard et al. showed that CAL1 binds both CID and CENP-C without the requirement for the Mis18 complex, using X-ray crystallography [108].

Studies of the neocentromere have also been performed in *Drosophila* [109–111], and the requirements, mechanism, and transmission for the neocentromere are actively under study. Two groups independently reported overexpressed CID mis-localization and ectopic incorporation into non-centromeric chromatin [112, 113]. Heun et al. demonstrated that overexpressed CID is mislocalized into normally non-centromeric regions in *Drosophila* tissue culture cells (S2 cells) and animals and induces severe mitotic defects [113]. These CID mis-incorporated regions display the presence of microtubule motors and binding proteins, and spindle attachments. Moreno-Moreno et al. showed that centromeric localization of transiently expressed CID is impaired in the presence of the proteasome inhibitor MG132 in Kc cells, and mislocalization of CID affects cell cycle progression with strong mitotic defects [112]. Recently, Palladino et al. used a LacO/LacI ectopic centromeric chromatin assembly system and showed that multiple genomic locations can acquire centromere activity. In addition, they demonstrated that these de novo centromeres can be transmitted and maintained epigenetically in mitotic tissues [114]. Together, their data suggest that proteolysis-mediated regulation of ectopic $CenH3^{CID}$ is also present in fruit flies as in other species. Further mechanisms of CID protein degradation, including the identification of E3 ligase, are described in the following Sections 3.2 and 3.3.

3.2 CUL3/RDX E3 Ligase

In *Drosophila*, $CENP\text{-}A^{CID}$ deposition to centromeres depends on a specialized loading factor that is called CAL1 [30]. Bade et al. showed that CAL1 directly interacts with RDX, an adaptor for CUL3/RDX-mediated ubiquitylation, through the two conserved RDX-binding sites (RBSs) of CAL1 [30] (**Figure 1a**; **Table 1**). However, CAL1 is not a substrate of the CUL3/RDX ligase but functions as an additional substrate-specifying factor for the CUL3/RDX-mediated ubiquitylation of CID. It is noteworthy that this fly CID ubiquitylation is proteasomal independent—ubiquitylation of CID by CUL3/RDX does not trigger its degradation but stabilizes CID and CAL1. Loss of RDX leads to rapid degradation of CAL1 and CID and to massive chromosome segregation defects during development (**Figure 1b**). Therefore, they suggest a proteolysis-independent role of ubiquitin conjugation in centromere regulation that is essential for the maintenance of the centromere-defining protein CID and its loading factor CAL1. Bade et al. proposed that this CID ubiquitylation event induces a conformational change within the CAL1/CID complex, or alternatively, increases the affinity toward centromeric chromatin, where it is protected from proteasomal degradation. The data of Bade et al. support a dual role of CAL1 in both loading and stabilizing CID protein (**Figure 1a**). Interestingly, their proposed "proteasomal-independent" mechanism of CUL3/RDX-mediated fly CID ubiquitylation is consistent with one of CUL4-mediated human CENP-A ubiquitylation found independently by our group [35, 115–117] (**Figure 2**, right; **Table 1;** see also Section 4.2). In humans, our group speculates that CENP-A mono- or di-ubiquitylation might sterically affect the overall conformational change, L112 residue, or C-terminal portion of the CATD on which HJURP recognition is mainly dependent (see also Section 4.2).

In humans, ectopic localization of CID depends on the H3.3 chaperone DAXX rather than the centromeric CENP-A specific chaperone HJURP [34] (**Figure 2**, left). This human CENP-A-containing ectopic nucleosome involves a heterotypic tetramer that contains CENP-A-H4 with H3.3-H4 [34] (**Figure 2**, left). Cells overexpressing human CENP-A are more tolerant of DNA damage induced by camptothecin or ionizing radiation, and both the survival advantage and CTCF occlusion by the aberrant nucleosome of heterotypic tetramer in these human cells are dependent on DAXX [34] (**Figure 2**, left). Although *D. melanogaster* has a DAXX ortholog, Daxx-like protein (DLP), the role of DLP/DAXX in CID deposition into ectopic nucleosomes through CID ubiquitylation (**Figure 1e**) and the CTCF occlusion by the aberrant nucleosome (**Figure 1f**) must be confirmed experimentally in *D. melanogaster*.

3.3 The E3-ligases SCF^{Ppa} and APC/C^{Cdh1} co-operate to regulate CID expression across the cell cycle

Moreno-Moreno et al. reported that the F box protein partner of paired (Ppa), which is a variable component of an SCF E3-ubiquitin ligase complex, controls CenH3^{CID} stability in *Drosophila* [44, 118] (**Figure 1b**; **Table 1**). They showed that Ppa depletion results in increased CenH3^{CID} levels, and Ppa physically interacts with CenH3^{CID} through the CATD^{CID} and regulates CenH3^{CID} stability in *Drosophila* [44, 118]. Their results showed that most known SCF complexes are inactive at mitosis when newly synthesized CenH3^{CID} is deposited at centromeres. Therefore, they suggest that CenH3^{CID} deposition and proteolysis are synchronized events in *Drosophila*. They further reported that, in *Drosophila*, CID expression levels are regulated throughout the cell cycle by the combined action of SCF^{Ppa} and APC/C^{Cdh1} [32] (**Table 1**). They showed that SCF^{Ppa} regulates CID expression in G1. Importantly, in S phase SCF^{Ppa} prevents the promiscuous incorporation of CID across chromatin during replication. In the G1 phase, CID expression is also controlled by APC/C^{Cdh1}. They also showed that CAL1, the specific chaperone that deposits CENP-A^{CID} at centromeres, protects CID from SCF^{Ppa}-mediated degradation but not from APC/C^{Cdh1}-mediated degradation. Together, they suggest that, whereas SCF^{Ppa} targets the fraction of CID that is not in complex with CAL1 (**Figure 1c**; **Table 1**), APC/C^{Cdh1} contributes to the degradation of the CAL1-CID complex and, thus, likely regulates centromeric CID deposition (**Figure 1d**; **Table 1**).

3.4 Phosphorylation of *Drosophila* CID on serine 20 regulates protein turnover and centromere-specific loading

Huang et al. showed that CID is phosphorylated at serine 20 (S20) by casein kinase II (CK2) and that the phosphorylated form is enriched on chromatin during mitosis [33] (**Figure 1c** and **g**; **Table 1**). Their results revealed that S20 phosphorylation regulates the turnover of prenucleosomal CID through the SCF^{Ppa}-proteasome pathway (**Figure 1c**; **Table 1**) and that phosphorylation facilitates removal of CID from ectopic but not from centromeric sites in chromatin (**Figure 1g** and **h**; **Table 1**). They provided multiple lines of evidence for an essential role of S20 phosphorylation in regulating restricted incorporation of CID into centromeric chromatin, suggesting that modulation of the phosphorylation state of S20 may lead to fine-tuned control of CID levels to prevent malignant incorporation into non-centromeric chromatin.

On the other hand, factors/components that stabilize ectopically incorporated CID and are required for neocentromere formation and its maintenance are not clear in *D. melanogaster* (**Figure 1h**). The status of overall post-translational modifications, including polyubiquitylation of CID, especially in ectopic nucleosomes, remains to be elucidated.

4. E3 ligase for human CENP-A its function

4.1 Overview of human CENP-A

In most eukaryotes, including humans, the centromere has no defined DNA sequence but is associated with large arrays of repetitive DNA; in humans, this sequence is a 171-bp alpha-satellite DNA, although several other sequence types are found in this region. CENP-A-containing nucleosomes are formed with canonical histones H2A, H2B, and H4 at the active centromeres [5]. CENP-A nucleosomes localize to the inner plate of mammalian kinetochores [119] and bind to the 171-bp alpha-satellite DNA. Recently, the importance of centromeric cis-element, transcription, and centromeric long noncoding RNA (cenRNA) for centromere integrity has been suggested in various species, including humans [77–79] (see also Sections 3 and 5). Interestingly, when the CENP-B box DNA sequence is located proximal to the CENP-A nucleosome, CENP-B forms a more stable complex with the CENP-A nucleosome through specific interactions with CENP-A [120]. In humans, a centromeric long noncoding RNA (cenRNA) is required for targeting CENP-A to the centromere [80].

Currently, it is commonly reported that CENP-A-containing nucleosomes are formed with canonical histones H2A, H2B, and H4 at the active centromeres, however, their structure remains controversial among different research groups [5]. Bui et al. suggest that CENP-A nucleosomes alter from tetramers to octamers before replication and revert to tetramers after replication, using combinatory methods, including atomic force microscopy [38]. It is noteworthy that reversible chaperone binding, chromatin fiber folding changes, and CENP-A K124 acetylation (K124ac) and H4 K79 acetylation (K79ac) are concurrent with these structural transitions. Further computational modeling suggests that acetylation of K124 causes tightening of the histone core and hampers accessibility to its C-terminus, which in turn reduces CENP-C interaction [39] (see also the following paragraph about the function of histone H4 acetylation). Further study, including the solution of real-time post-translational modifications or the 3D structure of free Cse4 complexes, is required to determine how different chaperons recognize Cse4/CENP-A-H4 for incorporation into different locations of chromatin.

CENP-A contains a short centromere targeting domain (CATD) within the histone fold region [2]. Replacement of the corresponding region of H3 with the CATD is sufficient to direct H3 to the centromere [2], and this chimeric histone can rescue the viability of CENP-A-depleted cells [2, 12]. On the other hand, Logsdon et al. found contributions from small portions of the N-terminal tail and the CATD in the initial recruitment of CENP-C and CENP-T, using a LacO/LacI ectopic centromeric chromatin assembly system [20]. Jing et al. reported that deletion of the first 53 but not the first 29 residues of CENP-A from the N-terminus, resulted in its cytoplasmic localization [121]. They identified two motifs for CENP-A nuclear accumulation and one motif involved in the centromeric accumulation of CENP-A, as well as the interaction of CENP-A with core histone H4 and CENP-B.

Early studies in human cells showed that CENP-A mRNA and protein start to accumulate in the mid-S phase and peak in G2 [122, 123], however, further cell type-specific regulation of human CENP-A mRNA and protein remains to be studied.

In human cells, the incorporation of newly synthesized CENP-A occurs in telophase/early G1 [94, 95]. The incorporation of newly synthesized CENP-A into centromeric nucleosomes depends on the HJURP, which is a CENP-A-specific chromatin assembly factor [41–43]. Like CENP-A, HJURP is also assembled during early

G1 to centromeres [42, 43, 94, 96]. The primary structural analysis demonstrated that human HJURP is a distant counterpart of Scm3, which is required to deposit centromeric nucleosomes in yeast [29]. CENP-A interacts with HJURP as a soluble pre-nucleosomal complex, and the unique structural dynamics of HJURP together with CENP-A/H4 heterodimer/tetramer (pre-nucleosomal CENP-A-H4-HJURP complex) have been reported [3, 124–132]. HJURP recruitment to centromeres depends on the activity of the Mis18 complex [41, 104], which affects the histone modification and DNA methylation status of centromeres [54, 59]. The human proteins hMis18 and M18BP1/KNL2 are recruited to the centromere at telophase/G1, suggesting that the hMis18 complex and RbAp46/48 (homologs of Mis16) prime the centromere for CENP-A localization [54, 133]. Moreover, acetylation of histone H4 lysine 5 and 12 (H4K5ac and H4K12ac) within pre-nucleosomal CENP-A-H4-HJURP complex mediated by the RbAp46/48-Hat1 complex is required for CENP-A deposition into centromeres in chickens and humans [57], consistent with the role of Hat1 shown in *D. melanogaster* [58] (see also Section 3.1). In mouse studies, Mis18α interacts with DNMT3A/3B, and this interaction is required to maintain DNA methylation [59]. *Mis18α* deficiency leads to not only the reduction of DNA methylation, but altered histone H3 modifications, and uncontrolled noncoding transcripts in the centromere region. Faithful CENP-A deposition requires integrated signals from Plk1 and cyclin-dependent kinase (CDK), with Plk1 promoting the localization of the Mis18 complex, and CDK inhibiting Mis18 complex assembly [134]. Moreover, the remodeling and spacing factor complex is required for the assembly of CENP-A chromatin [135], and the CENP-A licensing factor M18BP1/KNL2 and the small GTPases-activating protein MgcRacGAP cooperate to maintain the stability of newly loaded CENP-A at centromeres [136, 137].

Currently, the proteolysis mechanism for mis-incorporated human CENP-A and its E3 ligase is not yet clear (**Figure 2d**), and there are no reports to date on proteasome-mediated degradation of human CENP-A [138]. We reported that mono- or di-ubiquitylation of CENP-A K124 is required for CENP-A deposition at the centromere [35] (**Figure 2**, right). However, the stability of endogenous CENP-A is not affected by CUL4A or RBX1 depletion, and the stability of exogenous CENP-A K124R is the same as in wild-type cells. Rather, overexpression of a monoubiquitin-fused CENP-A mutant induces neocentromere formation, suggesting that signaling CENP-A mono- or di-ubiquitylation determines centromere location and activity [115] (see also Sections 4.2 and 4.3). Future studies are required to reveal how ectopic CENP-A is degraded and removed from the non-centromeric chromosome, and/or how the neocentromere established through CENP-A ubiquitylation is deactivated in humans (**Figure 2c** and **d**). This proteolysis could be initiated on chromatin and the machinery involved could be specifically excluded from centromeric regions. Alternatively, mis-incorporated CENP-A nucleosomes may dissociate more easily than those properly localized and be subsequently degraded in the nucleoplasm [139]. Obuse et al. performed chromatin immunoprecipitation with an anti-CENP-A monoclonal antibody using HeLa interphase nuclei and systematic identification of its interactors by mass spectrometric analyses [140]. They identified UV-damaged DNA binding protein 1 (DDB1) as a component of the CEN complex and BMI-1 that is transiently co-localized with the centromeric region in interphase.

RbAp46 forms a complex with the CRL4 ubiquitin ligase and DDB1 protein (where DDB1 mediates the association of CUL4 with its substrate-specific receptor—RbAP46) [141, 142]. RbAp46 is required for stabilizing CENP-A protein levels and the CRL4-RbAp46 complex activity promotes efficient new CENP-A deposition in humans [142]. This is in contrast to studies in yeast and fruit flies, where the association of CENP-A with the SCF E3-ubiquitin ligase complex leads to CENP-A degradation. However, our group showed that CUL4A-RBX1-COPS8 E3 ligase activity is required for CENP-A mono- or di-ubiquitylation on lysine 124 (K124) and CENP-A centromere localization, although our results suggest that DDB1 is not required for CENP-A recruitment to centromeres [35] (**Figure 2**, right; see also Sections 4.2–4.5). In humans, soluble CENP-A is associated with the centromeric CENP-A specific chaperone HJURP (see also Introduction). Depletion of HJURP leads to a significant decrease in CENP-A levels, suggesting that HJURP protects the fraction of CENP-A that will be incorporated at the centromere in G1 while remaining "free" CENP-A will be degraded to prevent its incorporation into non-centromeric chromatin [42, 43]. Our results also support this model, because CENP-A ubiquitylation enhances the affinity between HJURP with ubiquitylated CENP-A [35] (see also Sections 4.2–4.5).

One question is also generated about the function of H3.3 histone chaperone proteins, HIRA and DAXX, which were previously reported to promote ectopic CENP-A deposition in human cancer cells [34, 143]. Lacoste et al. found that CENP-A overexpression in human cells leads to ectopic enrichment at sites of active histone turnover involving a heterotypic tetramer that contains CENP-A-H4 with H3.3-H4 [34] (**Figure 2**, left). Ectopic localization of this particle (aberrant nucleosome) depends on the H3.3 chaperone DAXX rather than the centromeric CENP-A specific chaperone HJURP (**Figure 2**, left). Cells overexpressing CENP-A are more tolerant of DNA damage induced by camptothecin or ionizing radiation, and both the survival advantage and CTCF occlusion by the aberrant nucleosome of heterotypic tetramer in these cells are dependent on DAXX (**Figure 2**, left). However, post-translational modifications of human CENP-A, especially before recognition by DAXX and after incorporation into the ectopic nucleosome, must be elucidated (**Figure 2a**), and specific DAXX localization on these CTCF sites under CENP-A overexpression has to be confirmed experimentally (**Figure 2b**).

Shrestha et al. showed that mislocalization of CENP-A to chromosome arms is one of the major contributors to CIN, as depletion of histone chaperone DAXX prevents CENP-A mislocalization and rescues the reduced interkinetochore distance and CIN phenotype in CENP-A-overexpressing cells [144]. Nye et al. reported that in human colon cancer cells, the H3.3 chaperones HIRA and DAXX promote ectopic CENP-A incorporation [143]. They found that a correct balance between levels of the centromeric chaperone HJURP and CENP-A is required to prevent ectopic assembly by H3.3 chaperones. Their results also suggest that CENP-A occupancy at the 8q24 locus is significantly correlated with amplification and overexpression of the MYC gene within that locus. Together, CENP-A mislocalization into non-centromeric regions resulting from its overexpression leads to chromosomal segregation aberrations and genome instability [145]. Overexpression of CENP-A is a feature of many cancers and is likely associated with malignant progression and poor outcomes [146–148]. CENP-A overexpression is often accompanied by overexpression of its chaperone HJURP, leading to "epigenetic addiction" in which increased levels of HJURP and CENP-A become necessary to support rapidly dividing p53-deficient cancer cells [149]. In addition, the functional roles of DAXX and HIRA in the development of cancer and other diseases have been described [150–153]. Elucidation of the proper mechanism of H3.3 incorporation into chromatin through DAXX and HIRA may also lead to proper CENP-A incorporation at centromeres as well as an effective disease (e.g., cancer) therapy.

Recently, the importance of the site-specific posttranslational modifications of human CENP-A and their biological functions has been reported [44, 45]. The functional roles of phosphorylation at CENP-A-Ser68 are still under active investigation [124, 125, 154–156]. How the defects of CENP-A PTMs and the dysfunction of centromere contribute to the generation and the development of cancer is an unsolved question. Takada et al. demonstrated that CENP-A Ser18 hyperphosphorylation by cyclin E1/CDK2 occurred upon loss of FBW7, a tumor suppressor whose inactivation leads to CIN [157]. This CENP-A Ser18 hyperphosphorylation reduced the CENP-A centromeric localization, increased CIN, and promoted anchorage-independent growth and xenograft tumor formation. Defects of CENP-A PTMs are significantly associated with chromosome segregation errors and CIN [149].

4.2 CENP-A K124 ubiquitylation is required for CENP-A deposition at the centromere

In budding yeast, Scm3 and Psh1 might compete for binding to Cse4. Cse4 that is not associated with Scm3 may be targeted by Psh1 for proteolysis, but Cse4 in a complex with Scm3 may be protected [71] (see also previous chapter, Section 2.1). On the other hand, in *D. melanogaster* it was proposed that CENP-A^{CID} ubiquitylation induces a conformational change within the CAL1/CENP-A complex, or alternatively, increases the affinity toward centromeric chromatin, where it is protected from proteasomal degradation [30] (see also Section 3.2).

In humans, our group found that CUL4A-RBX1-COPS8 E3 ligase activity is required for CENP-A mono- or di-ubiquitylation on lysine 124 (K124) and CENP-A centromere localization [35] (**Figure 2**, right). CUL4A complex targets CENP-A through the adaptor COPS8/CSN8 that has WD40 motifs in non-canonical CRL4 machinery (**Figure 2**, right). A mutation of CENP-A, K124R, reduces interaction with HJURP and abrogates localization of CENP-A to the centromere. The addition of monoubiquitin is sufficient to restore CENP-A K124R to centromeres and the interaction with HJURP, indicating that "signaling" ubiquitylation is required for CENP-A loading at centromeres (**Figure 2**, right).

However, one question remains—how does such mono- or di-ubiquitylation of CENP-A facilitate the interaction of CENP-A with HJURP? The CENP-A K124 site and its proximal residues might not directly affect CENP-A-HJURP interaction in the crystal structure of the HJURP-CENP-A-histone H4 complex, since we did not detect defects in CENP-A dimerization of K124R mutant (**Figure 3**; see also Section 4.3) or any ubiquitin interacting motif in HJURP. Therefore, we speculate that CENP-A mono- or di-ubiquitylation might sterically affect the overall conformational change, L112 residue (the closest CENP-A's residue to K124 out of the seven residues reported to be important for appropriate interaction with HJURP), or C-terminal portion of the CATD on which HJURP recognition is mainly dependent. In addition, acetylated lysine 124 (K124) was previously reported by Bui et al. [38], but the functional role of K124 acetylation and its relationship with K124 ubiquitylation remains to be studied (**Figure 2**, right). Moreover, currently, the proteolysis mechanism for mis-incorporated human CENP-A and its E3 ligase is not clear, and there are no reports to date regarding proteasome-mediated degradation of human CENP-A [138] (**Figure 2d**). Future studies are required to reveal how ectopic CENP-A is degraded and removed from the non-centromeric chromosome (**Figure 2c** and **d**).

4.3 CENP-A ubiquitylation is inherited through dimerization between cell divisions

The mechanism by which centromere inheritance occurs is largely unknown. Gassmann et al. suggested that in *Caenorhabditis elegans*, pre-existing CENP-A^{HCP-3} nucleosomes are not necessary to guide the recruitment of new CENP-A nucleosomes [158]. In contrast, in *Drosophila melanogaster*, CENP-A^{CID} is present in mature sperm, and the amount of CID that is loaded during each cell cycle appears to be determined primarily by the pre-existing centromeric CID, a finding that is consistent with a "template-governed" mechanism [159]. However, in humans, it is unclear how CENP-A works as the epigenetic mark at the molecular level.

Our group showed that pre-existing ubiquitylated CENP-A is necessary for the recruitment of newly synthesized CENP-A to the centromere and that CENP-A ubiquitylation is inherited between cell divisions (**Figure 3**). *In vivo* and *in vitro* analyses using dimerization mutants and dimerization domain fusion mutants revealed that the inheritance of CENP-A ubiquitylation requires CENP-A dimerization. Therefore, we propose models in which CENP-A ubiquitylation is inherited and centromere location is determined through dimerization (**Figure 3**).

Numerous studies have found that CENP-A can be experimentally mistargeted to non-centromeric regions of chromatin and that this mistargeting leads to the formation of ectopic centromeres in model organisms [160]. Chromosome engineering has allowed the efficient isolation of neocentromeres on a wide range of both transcriptionally active and inactive sequences in chicken DT40 cells [57]. More than 100 neocentromeres in human clinical samples have been described [161]. They form on diverse DNA sequences and are associated with CENP-A localization, but not with alpha-satellite arrays; thus, these findings provide strong evidence that human centromeres result from sequence-independent epigenetic mechanisms. However, neocentromeres have not yet been created experimentally in humans; overexpression of CENP-A induces mislocalization of CENP-A, but not the formation of functional neocentromeres [162].

Our group demonstrated that overexpression of a monoubiquitin-fused CENP-A mutant induces neocentromeres at non-centromeric regions of chromosomes, and this result further supports our model in which CENP-A ubiquitylation is inherited and determines centromere location through dimerization (**Figure 3**). Our assay using the LacO/LacI ectopic centromeric chromatin assembly system clearly revealed that CENP-A ubiquitylation contributes to the recruitment of CENP-A chaperones (HJURP and DAXX) and outer kinetochore components (HEC1 and SKA1). It is possible that ubiquitylation of CENP-A contributes to maintain and stabilize ectopic neocentromeres in humans (**Figure 2c**).

However, it remains unclear how the neocentromere established through CENP-A ubiquitylation is deactivated. Future studies are required to reveal the mechanism of site-specific (centromeric and/or non-centromeric) deubiquitylation CENP-A and subsequent proteolysis in humans (**Figure 2c** and **d**). In this context, it would be interesting to test if the Ubp8-driven deubiquitylation mechanism in budding yeast [163] (see also previous chapter, Section 2.7) is conserved in humans.

4.4 SGT1-HSP90 complex is required for CENP-A loading at centromeres

The mechanism that controls the E3 ligase activity of the CUL4A-RBX1-COPS8 complex remains obscure. Our group found that the SGT1-HSP90 complex is required for recognition of CENP-A by COPS8 [164] (**Figure 2**, right). SGT1/SUGT1, a co-chaperone of HSP90, is involved in multiple cellular activities, including cullin E3 ubiquitin ligase activity [165]. The *SGT1* gene was originally isolated as a dosage suppressor of the *skp1–4* mutant in yeast *S. cerevisiae*, which causes defects in yeast kinetochore function, but also as a novel subunit of the Skp1-Cullin-F-box (SCF) ubiquitin ligase complex [166]. In both yeast and humans, the interaction between SGT1 and heat shock protein 90 (HSP90) is also required for kinetochore assembly [167–169]. In humans, cancer cells utilize Hsp90 as a chaperone to promote the folding and function of mutated or overexpressed oncoproteins, because aberrant oncoproteins are unstable [170, 171]. SGT1 contributes to cancer development by stabilizing oncoproteins, and the SGT1-HSP90 complex is a potential target for therapies aimed at cancer, brain, and heart disease [165].

Our group initially applied RNA interference (RNAi)-mediated SGT1 and/or HSP90 depletion in HeLa cells and found that the SGT1-HSP90 complex is required for CENP-A ubiquitylation *in vivo* and CENP-A deposition at centromeres [164] (**Figure 2**, right). Moreover, our group and others demonstrated *in vivo* interactions of SGT1A with CUL4A [164] and HSP90 with CUL4 [172], respectively (**Figure 2**, right). Previously, we had also reported that the CUL4A complex targets CENP-A through the adaptor COPS8/CSN8 that has WD40 motifs in non-canonical CRL4 machinery [35] (**Figure 2**, right; see also Section 4.2). Therefore, we hypothesized that depletion of SGT1 or HSP90 protein promotes loss of interaction among components of the CUL4A complex. Indeed, SGT1 or HSP90 siRNA disrupted interactions of COPS8 with CENP-A and CUL4A. These results suggest that the SGT1-HSP90 complex is required for the composition of the CUL4A complex as well as recognition of CENP-A by COPS8 (**Figure 2**, right). Thus, we clarified how the SGT1-HSP90 complex contributes to the E3 ligase activity of the CUL4A complex in CENP-A ubiquitylation (**Figure 2**, right).

In our study, SKP1 siRNA treatment did not lead to any signal reduction of CENP-A at centromeres [164]. Therefore, we proposed that the SGT1-HSP90 complex is involved in CENP-A deposition at centromeres in an SKP1-independent and/or SCF-independent manner. This conclusion is consistent with our previous report that the CUL4A-RBX1 complex, which does not require SKP1 to function, contributes to CENP-A deposition at centromeres [35]. Because our results suggest that SKP1 is not required for the recruitment of CENP-A to centromeres, it is unlikely that SKP1 activity affects the CENP-A loading pathway. Because CENP-A is at the top of a hierarchy of the pathway that determines the assembly of kinetochore components [6], destabilization of the MIS12 complex at the kinetochore was observed by Davies et al. [173] could be partially due to the defect in CENP-A recruitment. This idea is supported by our results demonstrating that SGT1 siRNA treatment did not significantly change the recruitment of endogenous MIS12, HEC1, and SKA1 proteins in LacO arrays after ectopic loci were determined through LacO-LacI-CENP-A interactions. Collectively, these data suggest that the losses of immunofluorescence signals of the central-outer kinetochore proteins at the kinetochore caused by SGT1 siRNA defects, including ones reported previously [174], are explained by CENP-A mislocalization caused by SGT1 siRNA defects.

4.5 CENP-A ubiquitylation is indispensable to cell viability

Our group reported that CENP-A K124 ubiquitylation, mediated by the CUL4A-RBX1-COPS8 complex, is essential for CENP-A deposition at the centromere [35] (**Figure 2**, right; see also Section 4.2). On the other hand, Fachinetti et al. reported that CENP-A K124R mutants show no defects in centromere localization and cell viability [156]. However, there are substantive problems with their experiments that yielded these results. We reported our response describing potential issues with the results and their conclusions [117]. A major caveat is that they used a fusion protein much larger molecular size than CENP-A. In their RPE-1 CENP-A$^{-/F}$ knockout system, the enhanced yellow fluorescent protein (EYFP) is approximately 30 kDa, and endogenous CENP-A is about 16 kDa. Fachinetti et al. also used SNAP-tags, and they found that SNAP-CENP-A K124R showed no defects in centromere deposition. Because the SNAP-tag (20 kDa) is also a larger tag than CENP-A (approximately 16 kDa) and has 10 lysines, SNAP-CENP-A K124R, presumably, is ubiquitylated at a site different than K124. One possibility is that the tagging of a large protein may endogenously lead to ubiquitylation at an amino acid other than K124 in the CENP-A K124R mutant protein, and this ubiquitylation at another site could suppress the mutant phenotype as a compensatory mechanism. Therefore, our group hypothesized that the presence of a large fusion protein promotes ubiquitylation at a different lysine in the CENP-A K124R mutant protein.

Indeed, our group found that EYFP tagging induces additional ubiquitylation of EYFP-CENP-A K124R, which allows the mutant protein to bind to HJURP [116]. Our immunoprecipitation mass spectrometry analysis showed that lysine 306 (K306) in the EYFP-CENP-A K124R mutant is ubiquitylated *in vivo*. This site corresponds to lysine 56 (K56) in CENP-A. These data suggest that once EYFP is tagged to a K124R mutant, another ubiquitylation occurs at a different site than K124 as endogenous compensatory machinery. Using a previously developed conditional CENP-A knockout system and our CENP-A K124R knockin mutant created by the CRISPR-Cas9 system, we show that the small size Flag-tagged or untagged CENP-A K124R mutant is lethal. This lethality is rescued by monoubiquitin fusion, indicating that CENP-A ubiquitylation is essential for viability. Therefore, our group suggests a caveat in the use of GFP/EYFP as a tool to analyze the function of a protein, and our data still support that the CENP-A ubiquitylation is indispensable to cell viability.

4.6 Hypothetical regulation of human CENP-A through sumoylation

In budding yeast, CENP-A^{Cse4} is sumoylated on its N-terminal tail by Siz1/Siz2 SUMO E3 ligases [22] (previous chapter, **Figure 1a** and **b**) (see also previous chapter, Section 2.4.1). Cse4 is poly-sumoylated at K65 in its N-terminal domain, which recruits the yeast SUMO-targeted ubiquitin ligase (STUbl) Slx5, leading to the polyubiquitination of poly-sumoylated Cse4 and its subsequent degradation [21]. Cse4 K215/216 sumoylation in C-terminus also controls its interaction with the histone chaperones Scm3 and CAF-1, facilitating the deposition of overexpressed Cse4 into CEN and non-CEN regions, respectively [175] (previous chapter, **Figure 1**) (see also previous chapter, Section 2.4.2).

In humans, depletion of the human Slx5 homolog ring finger protein 4 (RNF4) contributes to SUMOylation-dependent degradation of the CCAN protein CENP-I, while SENP6 stabilizes CENP-I by antagonizing RNF4 [176]. SENP6 is a desumoylation enzyme as well as a member of a large family of Sentrin-specific protease enzymes (SENP1–7) [138, 177]. In budding yeast, two SUMO proteases are known, ubiquitin-like protease 1 and 2 (Ulp1 and 2); in mammalian cells, these have diverged into the SENP family. SENP1–5 is evolutionarily conserved to Ulp1, while the more divergent

SENP6 and SENP7 belong to the Ulp2 group. Depletion of SENP6 in HeLa cells leads to the loss of the CENP-H/I/K complex from the centromeres, but not an apparent r eduction in centromeric CENP-A/B/C levels recognized by CREST sera [176].

Liebelt et al. identified a protein group de-modification by SENP6, including the constitutive centromere-associated network (CCAN), the CENP-A loading factors Mis18BP1 and Mis18A, and DNA damage response factors [178]. SENP6-deficient cells are severely compromised for proliferation, accumulate in the G2/M phases, and frequently form micronuclei. Centromeric assembly of CENP-T, CENP-W, and CENP-A is impaired in the absence of SENP6. However, the increase of SUMO chains is not required for ubiquitin-dependent proteasomal degradation of the CCAN subunits. Therefore, their results indicated that SUMO polymers can act in a proteolysis-independent manner and consequently, have a more diverse signaling function than previously expected. On the other hand, Mitra et al. identified the SUMO-protease SENP6 as a key factor, not only controlling CENP-A stability but virtually the entire centromere and kinetochore using a genetic screen coupled to pulse-chase labeling [179]. Loss of SENP6 results in hyper-sumoylation of CENP-C and CENP-I, but not CENP-A itself. SENP6 activity is required throughout the cell cycle, suggesting that a dynamic SUMO cycle underlies continuous surveillance of the centromere complex that in turn ensures stable transmission of CENP-A chromatin. Mitra et al. and other groups did not detect sumoylation of CENP-A, suggesting that CENP-A is not a direct substrate of SENP6 [138, 179]. However, the effect of SENP6 depletion on CENP-A stability is much greater than observed on depletion of CENP-C or -B alone [179]. This result suggests that there may be other components required for the SENP6-mediated stabilization of centromeric chromatin [138].

5. E3 ligases for plant CENP-A (CENH3) its function

5.1 Overview of plant CENP-A (CENH3)

Studies of E3 ligases at plant centromeres-kinetochores are not as advanced as those in model animal species. The structure and organization of plant centromeric DNA have been described, and satellite repeats associated with centromeres have been reported in many plant species [76]. Plant centromeres also have mega-base-sized arrays of tandem repetitive DNA sequences, as in centromeres of humans and other mammals, and transposable elements are abundant in centromeric and para-centromeric regions [76, 180]. In early studies, Jiang et al. suggest that the retention of active transcriptional machinery within the long terminal repeat may promote demarcation of an active centromere [76]. A Ty3/gypsy class of centromere-specific retrotransposons, the centromeric retrotransposon (CR) family, was discovered in the grass species. Highly conserved motifs were found in the long terminal repeat of the CR elements from rice, maize, and barley [181]. The CR elements are highly enriched in chromatin domains associated with CENH3/CENP-A, the centromere-specific histone H3 variant. CR elements as well as their flanking centromeric satellite DNA are actively transcribed in maize. These data suggest that the deposition of centromeric histones might be a transcription-coupled event. The importance of centromeric transcription and centromeric long noncoding RNA (cenRNA) for centromere integrity has been suggested in various species, including plants [77–79] (see also Sections 3 and 4). Moreover, in maize, CENP-C binding to centromeric DNA is associated with small RNA [182], whereas in humans CENP-A loading is linked to lncRNAs [80]. It is not yet known whether the same transcript can recruit and stabilize both CENP-A and CENP-C at centromeric chromatin [77].

Plant CENH3/CENP-A and other centromere-kinetochore proteins have been reported showing high conservation among species. On the other hand, DNA sequences of plant centromeres, of which loci are determined epigenetically by centromeric histone 3 (CENH3), have revealed high structural diversity, ranging from the canonical monocentric form seen in vertebrates, to polycentric and holocentric forms [183, 184]. Plant centromeres can change position over evolutionary time or upon genomic stress, such as in McClintock's genome shock [185] or physically damaged or broken chromosomes [183]. Jiang et al. suggested that the centromeric state is reinforced and maintained by the tension applied during spindle attachment [76]. The chromatin damaged by such mechano-force could then be marked for repair by the replication-independent mechanism similar to the one originally incorporated in CENH3. Indeed, human centromere-kinetochore proteins, including CENP-A, are involved in DNA damage/repair [186], and the incorporation of newly synthesized CENP-A occurs "right after mitosis" (i.e., telophase/early G1) [94, 95]. However, the model of CenH3 (CENP-A) incorporation upon mechano-force-induced DNA damage/repair is not yet experimentally demonstrated, and its precise mechanism needs to be elucidated. Meanwhile, there is evidence of divergent evolution originating in CenH3 in plants [187, 188] and *Drosophila* [189]. The CenH3 (CENP-A) has apparently undergone both convergent and divergent evolution [7]. Nagaki et el. and others described that the centromere DNA repeats with which CENH3-containing nucleosome interacts are also highly diverged, proposing an "arms race" hypothesis where centromere DNA repeats are changing and expanding to increase their segregation properties, while CENH3 is changing to curb this process and keep segregation frequencies equal to avoid fixing traits [180, 184, 189, 190].

Plant studies of dicentric centromeres and neocentromeres have been described along with those of other eukaryotes [180, 183]. The active state of one of the two centromeres on the wheat dicentric chromosome can be epigenetically silenced [180], as in the human dicentric chromosome [191]. Neocentromeres have been described extensively in human and fruit fly chromosomes as well as in some plant species, such as barley, maize, and rice [114, 184, 192]. In *D. melanogaster*, Palladino et al. showed that multiple genomic locations can acquire centromere activity, using a LacO/LacI ectopic centromeric chromatin assembly system. In addition, they demonstrated that these de novo centromeres can be transmitted and maintained epigenetically in mitotic tissues [114]. Although studies of human neocentromeres have indicated that they are generated at new positions in a single step; the barley neocentromere appears to have shifted several times along the chromosomal arm region during the deletion steps to finally reach the observed position [180]. The emergence of new centromeres was also observed in hybrid conditions [183, 184], and Wang et al. described the proposed model for hybrids between maize and oat [193]. The "centromere repositioning" then generates neocentromeres; the establishment of a new centromere does not require specific DNA composition in the target loci [76, 194]. Most new centromeres have no satellite DNA [195]. However, most mature centromeres are overwhelmingly composed of repetitive DNA, especially satellite DNA [76, 194]. One hypothesis to explain this apparent contradiction was described by Oliveira et al. as "satellite DNA invasion mechanism"—a new satellite repeat or one already present in other centromeres may invade and occupy the CenH3 domain of the new centromere [184]. The satellite DNA invasion mechanism is still elusive, and the retrotransposons would be the main source for the origin of new repeats [184, 196].

Plant studies of minichromosomes and artificial chromosomes also have been reported, as in other eukaryotes [180, 183]. The main issues of these studies are what are the size and factors required for the maintenance and stability of such special chromosomes during cell division. Harrington et al. constructed human artificial minichromosomes [197], and Ananiev et al. artificially generated minichromosomes in maize by introducing the DNA molecule containing native centromere segment, ori, and telomere repeats [198]. These studies suggested that repetitive DNA may play an important but unknown role in centromere function. The repetitive centromeric DNA may be still important, although it is not essential for centromeric function, since plant centromeric DNA does not generate functional centromeres when reintroduced into plant cells [199] and new centromeres are functional even if located in loci with non-centromeric DNA [161].

In terms of the plant CENH3 recruitment mechanism to centromeres, most CENP-A is loaded in G2 by a replication-independent mechanism in *Arabidopsis thaliana* [200]. However, in plants as in other species, the timing of deposition of newly synthesized CENP-A within the cell cycle may be variable—not only among different plant species but also different developmental stages within the same species. Le Goff et al. reported that the H3 histone chaperone $NASP^{SIM3}$ escorts CENH3 in *Arabidopsis* [19]. They showed that the *Arabidopsis* ortholog of the mammalian nuclear autoantigenic sperm protein (NASP) and *S. pombe* histone chaperone Sim3 is a soluble nuclear protein that interacts with CENH3 and influences its abundance at the centromeres [19]. $NASP^{SIM3}$ is co-expressed with *Arabidopsis* CENH3 in dividing cells and binds directly to both the N-terminal tail and the C-terminal histone fold domain of non-nucleosomal CENH3. Reduced $NASP^{SIM3}$ expression by $NASP^{SIM3}$ knockdown impairs CenH3 deposition. Thus, they identified $NASP^{SIM3}$ as a CenH3 histone chaperone as demonstrated in fission yeast (see also Section 2.1).

5.2 Engineered degradation of EYFP-tagged CENH3 in plants

Currently, an endogenous E3 ligase for plant CENP-A (CENH3) is not yet identified. Sorge et al. developed a synthetic biology approach to degrade plant CENP-A using E3-ligase adapter protein SPOP (Speckle-type POZ adapter protein) with a specific anti-GFP nanobody (VHHGFP4) [201] (**Table 1**). To determine the function of proteins, CRISPR/Cas9-based methods and antisense/RNAi strategies are commonly used to remove the selected protein from all organs in a cell- and tissue-specific manner. However, CRISPR/Cas9 and antisense/RNAi strategies are still error-prone and can show off-target effects [202]. Classical genetic strategies to knock out/down protein function in plants still have problems, such as the time-consuming process of generating homozygous transgenic lines or the risk of lethal phenotypes at early developmental stages.

Sorge et al. attempted to solve these problems by utilizing the synthetic E3 ligase activity in protein ubiquitylation and degradation pathway. They successfully recruited the 26S proteasome pathway to directly degrade CENP-A of *A. thaliana* via replacement of the interaction domain of the E3 ligase adaptor protein SPOP (Speckle-type POZ adapter protein) with a specific anti-GFP nanobody (VHHGFP4). They proved that the target protein CENH3 of *A. thaliana* fused to EYFP was subjected to nanobody-guided proteasomal degradation *in planta*. Thus, their results show the potential of the modified E3-ligase adapter protein VHHGFP4-SPOP to induce nucleus-specific protein degradation in plants. However, further studies are required to identify endogenous E3 ligase for plant CENP-A

(CENH3) and determine the function of the plant CENP-A (CENH3) proteolysis or deposition at centromeres.

6. Conclusions

Each species reviewed in our articles, including the previous chapter has advantages and disadvantages for research. For example, the centromere sequence size of the budding yeast is small and the sequences can be easily mutated to identify the important functional regions [203]. Techniques such as ChIP are also possible, which cannot be easily performed on highly repetitive centromeres in other organisms. Moreover, the centromere can be shifted to other genomic regions, allowing the construction of artificial chromosomes and plasmids as well as tools, such as conditional centromeres. Fission yeast and fruit fly models have progressed more than others in studies of heterochromatin regulation and gene silencing. Plant models have advanced more in evolutionary studies of centromeric DNA structures, including CR family comparisons among different plant species.

On the other hand, in fission yeast and plant species, the E3 ligase of CENP-A (CenH3) and its specific regulation and/or function are not yet identified. The E3 ligase of CENP-A is unknown in multiple species (e.g., *Caenorhabditis elegans*, *Xenopus laevis*, zebrafish *Danio rerio*, chicken *Gallus domesticus* DT40 cells, *Mus musculus*, etc.) or this research is sparse in these species compared with others. At present, the most common species studied and reported in the past for E3 ligase of CenH3 (Cse4) is budding yeast. However, in other species, much is not understood, particularly about control of the balance between E3 ligase and deubiquitylase and the balance among SUMO E3 ligase, the desumoylation enzyme, SUMO proteases (e.g., SENP6). Future research into the E3 ligases of CENP-A will elucidate the regulation and mechanisms of these subtle balances in each species and human diseases.

Studying the mechanisms of formation and maintenance of neocentromeres will deepen our understanding of the centromere-kinetochore formation and promote the building and establishment of artificial chromosomes. Such studies will lead to the construction of artificial cells and tissues that can be controlled by DNA levels through chromosome dynamics. As a result, the function of E3 ligase can be artificially adjusted, which will increase the effectiveness of future gene therapies. Minichromosomes generated to date suggest that the repetitive centromeric DNA may be still important, although perhaps, it is not essential for centromeric function. In addition, it is unclear whether there is causality or feedback between cenRNA transcription and overall transcriptional change after chromosome missegregation and CIN. As of now, we have little understanding of the effects of these cenRNAs on the E3 ligase of CENP-A, including how these transcriptional changes and regulation are related to the function of E3 ligase.

Although our group showed that ubiquitylation occurs at a different site than CENP-A K124 as endogenous compensatory machinery, the compensatory machinery of post-translational modifications in endogenous conditions is poorly understood. This machinery can be incorporated in a process of disease progress or development. For example, suppose a post-translational modification is required for host cancer cell development but its activity can be blocked by cancer drugs. However, another site's post-translational modification could compensate for that change, so that host cancer cells can survive, proliferate, and eventually metastasize. For cell proliferation and differentiation in general, such compensatory machinery could be a versatile backup system. However, such backup systems may not have been detected experimentally due to our limited technology or brief experimental periods. Thus, many E3 ligases may work in similar signal pathways (see also the previous chapter, Conclusion), or the function of a post-translational modification in one site may be compensated for or complemented by another site, but it is currently unknown how likely such complementary machineries would be. Research to predict such compensatory systems and resilience could be expected as future directions to study the spatiotemporal regulation of E3 ligase of CENP-A.

Ultimately, studies of E3 ligase in CENP-A in higher mammals or humans are essential for translational research and informing future therapy. Overexpression and mislocalization of human CENP-A are presumably features of cancer development, however, the detailed mechanisms for cancer development and possible therapies still remain unclear. In addition to cancer, translational studies of CENP-A and its E3 ligase could be beneficial for CREST autoimmune diseases and other diseases. Centromere proteins, including CENP-A, have been identified as antigens from CREST patients [204, 205], but the mechanism that causes CREST syndrome and how CENP-A and other centromere-kinetochore proteins are involved is unknown. Observations of neocentromeres were also reported in patients with other developmental diseases [206], but research has been limited, in part because of the relatively smaller number of patients.

Defects in ubiquitin E3 ligases promote the pathogenesis of several human diseases, including cancer, and CRL4 [207], a well-defined E3 ligase, has been reported to be upregulated and is proposed to be a potential drug target in cancers [208]. However, the biological functions of CRL4 and the underlying mechanism regulating cancer chemoresistance are still largely elusive. In humans, proteolysis activity of CRL4 ubiquitin ligase targeting CENP-A has not been observed so far, and other E3 ligases that function in CENP-A proteolysis are unidentified (**Figure 2d**). It is also important to determine if ubiquitylation or sumoylation-related enzymes, including E3 ligases, can be druggable targets.

Tumors develop in complex tissue microenvironments, where they depend on for sustained growth, invasion, and metastasis [209]. We could be at a turning point to fill the gap between the detailed intracellular mechanisms of CENP-A function studied in the past and its mechanism in complex tissue microenvironments. Thus, cell type and/or tissue-specific CENP-A function involved in different types of cancer in different organs is a likely focus for future research. There are many unknowns about whether the function of E3 ligase of CENP-A represents a cell or tissue-specific difference, or whether the cell or tissue completely replaces E3 ligase itself. The utilization and application of organoid, spheroid, and coculture systems may reduce the effort, time, and cost that is required to answer these questions and ultimately yield better therapies.

Acknowledgements

We thank past and current researchers at Model Animal Research Center, School of Medicine, Nanjing University, Greehey Children's Cancer Research Institute at UT Health Science Center San Antonio, the Research Institute at Nationwide Children's Hospital, and St. Jude Children's Research Hospital for their helpful discussions. Y.N. was supported by Jiangsu Province "Double-First-Class" Construction Fund, Jiangsu Province Natural Science Fund (BK20191252), Jiangsu Province 16th Six Big Talent Peaks Fund (TD-SWYY-001), Jiangsu Province

"Foreign Expert Hundred Talents Program" Fund (BX2019082), and National Natural Science Foundation in China (31970665). KK was supported by the National Science Foundation under Grant No.1949653 (KK) and a Mays Cancer Center Pilot Award CCSG P30 CA054174.

Conflict of interest

The authors declare no conflict of interest.

Author details

Yohei Niikura[1,2*] and Katsumi Kitagawa[3*]

1 MOE Key Laboratory of Model Animal for Disease Studies, Model Animal Research Center, School of Medicine, Nanjing University, Nanjing, China

2 Jiangsu Key Laboratory of Molecular Medicine, Nanjing University, Nanjing, China

3 Department of Molecular Medicine, Greehey Children's Cancer Research Institute, UT Health San Antonio, San Antonio, TX, USA

*Address all correspondence to: niikura@nicemice.cn and kitagawak@uthscsa.edu

References

[1] Allshire RC, Karpen GH. Epigenetic regulation of centromeric chromatin: Old dogs, new tricks? Nature Reviews. Genetics. 2008;**9**(12):923-937. DOI: 10.1038/nrg2466

[2] Black BE, Foltz DR, Chakravarthy S, Luger K, Woods VL Jr, Cleveland DW. Structural determinants for generating centromeric chromatin. Nature. 2004;**430**(6999):578-582. DOI: 10.1038/nature02766

[3] Sekulic N, Bassett EA, Rogers DJ, Black BE. The structure of (CENP-A-H4)(2) reveals physical features that mark centromeres. Nature. 2010;**467**(7313):347-351. DOI: 10.1038/nature09323

[4] Tachiwana H et al. Crystal structure of the human centromeric nucleosome containing CENP-A. Nature. 2011;**476**(7359):232-235. DOI: 10.1038/nature10258

[5] Black BE, Cleveland DW. Epigenetic centromere propagation and the nature of CENP-a nucleosomes (in eng). Cell. 2011;**144**(4):471-479. DOI: 10.1016/j.cell.2011.02.002

[6] Liu ST, Rattner JB, Jablonski SA, Yen TJ. Mapping the assembly pathways that specify formation of the trilaminar kinetochore plates in human cells. The Journal of Cell Biology. 2006;**175**(1):41-53. DOI: 10.1083/jcb.200606020

[7] Ekwall K. Epigenetic control of centromere behavior. Annual Review of Genetics. 2007;**41**:63-81. DOI: 10.1146/annurev.genet.41.110306.130127

[8] Aravamudhan P, Felzer-Kim I, Joglekar AP. The budding yeast point centromere associates with two Cse4 molecules during mitosis. Current Biology. 2013;**23**(9):770-774. DOI: 10.1016/j.cub.2013.03.042

[9] Bodor DL et al. The quantitative architecture of centromeric chromatin. eLife. 2014;**3**:e02137. DOI: 10.7554/eLife.02137

[10] Coffman VC, Wu P, Parthun MR, Wu JQ. CENP-A exceeds microtubule attachment sites in centromere clusters of both budding and fission yeast. The Journal of Cell Biology. 2011;**195**(4):563-572. DOI: 10.1083/jcb.201106078

[11] Lawrimore J, Bloom KS, Salmon ED. Point centromeres contain more than a single centromere-specific Cse4 (CENP-A) nucleosome. The Journal of Cell Biology. 2011;**195**(4):573-582. DOI: 10.1083/jcb.201106036

[12] Black BE et al. Centromere identity maintained by nucleosomes assembled with histone H3 containing the CENP-A targeting domain. Molecular Cell. 2007;**25**(2):309-322. DOI: 10.1016/j.molcel.2006.12.018

[13] Carroll CW, Milks KJ, Straight AF. Dual recognition of CENP-A nucleosomes is required for centromere assembly. The Journal of Cell Biology. 2010;**189**(7):1143-1155. DOI: 10.1083/jcb.201001013

[14] Guse A, Carroll CW, Moree B, Fuller CJ, Straight AF. In vitro centromere and kinetochore assembly on defined chromatin templates. Nature. 2011;**477**(7364):354-358. DOI: 10.1038/nature10379

[15] Carroll CW, Silva MC, Godek KM, Jansen LE, Straight AF. Centromere assembly requires the direct recognition

of CENP-A nucleosomes by CENP-N. Nature Cell Biology. 2009;**11**(7):896-902. DOI: 10.1038/ncb1899

[16] Au WC, Dawson AR, Rawson DW, Taylor SB, Baker RE, Basrai MA. A novel role of the N terminus of budding yeast histone H3 variant Cse4 in ubiquitin-mediated proteolysis. Genetics. 2013;**194**(2):513-518. DOI: 10.1534/genetics.113.149898

[17] Folco HD et al. The CENP-A N-tail confers epigenetic stability to centromeres via the CENP-T branch of the CCAN in fission yeast. Current Biology. 2015;**25**(3):348-356. DOI: 10.1016/j.cub.2014.11.060

[18] Gonzalez M, He H, Dong Q, Sun S, Li F. Ectopic centromere nucleation by CENP--a in fission yeast. Genetics. 2014;**198**(4):1433-1446. DOI: 10.1534/genetics.114.171173

[19] Le Goff S et al. The H3 histone chaperone NASP(SIM3) escorts CenH3 in Arabidopsis. The Plant Journal. 2020;**101**(1):71-86. DOI: 10.1111/tpj.14518

[20] Logsdon GA et al. Both tails and the centromere targeting domain of CENP-A are required for centromere establishment. The Journal of Cell Biology. 2015;**208**(5):521-531. DOI: 10.1083/jcb.201412011

[21] Ohkuni K et al. N-terminal sumoylation of centromeric histone H3 variant Cse4 regulates its proteolysis to prevent mislocalization to non-centromeric chromatin. G3 (Bethesda). 2018;**8**(4):1215-1223. DOI: 10.1534/g3.117.300419

[22] Ohkuni K et al. SUMO-targeted ubiquitin ligase (STUbL) Slx5 regulates proteolysis of centromeric histone H3 variant Cse4 and prevents its mislocalization to euchromatin. Molecular Biology of the Cell. 2016;27(9):1500-1510. DOI: 10.1091/mbc.E15-12-0827

[23] Ranjitkar P, Press MO, Yi X, Baker R, MacCoss MJ, Biggins S. An E3 ubiquitin ligase prevents ectopic localization of the centromeric histone H3 variant via the centromere targeting domain. Molecular Cell. 2010;**40**(3):455-464. DOI: 10.1016/j.molcel.2010.09.025

[24] Tan HL et al. Prolyl isomerization of the CENP-A N-terminus regulates centromeric integrity in fission yeast. Nucleic Acids Research. 2018;**46**(3):1167-1179. DOI: 10.1093/nar/gkx1180

[25] Tan HL, Zeng YB, Chen ES. N-terminus does not govern protein turnover of *Schizosaccharomyces pombe* CENP-A. International Journal of Molecular Sciences. 2020;**21**(17):1-13. DOI: 10.3390/ijms21176175

[26] Yang J et al. Heterochromatin and RNAi regulate centromeres by protecting CENP-A from ubiquitin-mediated degradation. PLoS Genetics. 2018;**14**(8):e1007572. DOI: 10.1371/journal.pgen.1007572

[27] Zhou Z et al. Structural basis for recognition of centromere histone variant CenH3 by the chaperone Scm3. Nature. 2011;**472**(7342):234-237. DOI: 10.1038/nature09854

[28] Bernad R, Sanchez P, Losada A. Epigenetic specification of centromeres by CENP-A. Experimental Cell Research. 2009;**315**(19):3233-3241. DOI: 10.1016/j.yexcr.2009.07.023

[29] Sanchez-Pulido L, Pidoux AL, Ponting CP, Allshire RC. Common ancestry of the CENP-A chaperones Scm3 and HJURP. Cell. 2009;**137**(7):1173-1174. DOI: 10.1016/j.cell.2009.06.010

[30] Bade D, Pauleau AL, Wendler A, Erhardt S. The E3 ligase CUL3/RDX controls centromere maintenance by ubiquitylating and stabilizing CENP-A in a CAL1-dependent manner. Developmental Cell. 2014;**28**(5):508-519. DOI: 10.1016/j.devcel.2014.01.031

[31] Merlet J, Burger J, Gomes JE, Pintard L. Regulation of cullin-RING E3 ubiquitin-ligases by neddylation and dimerization. Cellular and Molecular Life Sciences. 2009;**66**(11-12):1924-1938. DOI: 10.1007/s00018-009-8712-7

[32] Moreno-Moreno O, Torras-Llort M, Azorin F. The E3-ligases SCFPpa and APC/CCdh1 co-operate to regulate CENP-ACID expression across the cell cycle. Nucleic Acids Research. 2019;**47**(7):3395-3406. DOI: 10.1093/nar/gkz060

[33] Huang A et al. Phosphorylation of Drosophila CENP-A on serine 20 regulates protein turn-over and centromere-specific loading. Nucleic Acids Research. 2019;**47**(20):10754-10770. DOI: 10.1093/nar/gkz809

[34] Lacoste N et al. Mislocalization of the centromeric histone variant CenH3/CENP-A in human cells depends on the chaperone DAXX. Molecular Cell. 2014;**53**(4):631-644. DOI: 10.1016/j.molcel.2014.01.018

[35] Niikura Y, Kitagawa R, Ogi H, Abdulle R, Pagala V, Kitagawa K. CENP-A K124 ubiquitylation is required for CENP-A deposition at the centromere. Developmental Cell. 2015;**32**(5):589-603. DOI: 10.1016/j.devcel.2015.01.024

[36] Ahn J et al. The cullin-RING E3 ubiquitin ligase CRL4-DCAF1 complex dimerizes via a short helical region in DCAF1. Biochemistry. 2011;**50**(8):1359-1367. DOI: 10.1021/bi101749s

[37] Chew EH, Poobalasingam T, Hawkey CJ, Hagen T. Characterization of cullin-based E3 ubiquitin ligases in intact mammalian cells--evidence for cullin dimerization. Cellular Signalling. 2007;**19**(5):1071-1080. DOI: 10.1016/j.cellsig.2006.12.002

[38] Bui M et al. Cell-cycle-dependent structural transitions in the human CENP-A nucleosome in vivo. Cell. 2012;**150**(2):317-326. DOI: 10.1016/j.cell.2012.05.035

[39] Bui M et al. Internal modifications in the CENP-A nucleosome modulate centromeric dynamics. Epigenetics & Chromatin. 2017;**10**:17. DOI: 10.1186/s13072-017-0124-6

[40] Niikura Y, Kitagawa R, Kitagawa K. The inheritance of centromere identity. Molecular & Cellular Oncology. 2016;**3**(4):e1188226. DOI: 10.1080/23723556.2016.1188226

[41] Barnhart MC et al. HJURP is a CENP-A chromatin assembly factor sufficient to form a functional de novo kinetochore. The Journal of Cell Biology. 2011;**194**(2):229-243. DOI: 10.1083/jcb.201012017

[42] Dunleavy EM et al. HJURP is a cell-cycle-dependent maintenance and deposition factor of CENP-A at centromeres. Cell. 2009;**137**(3):485-497. DOI: 10.1016/j.cell.2009.02.040

[43] Foltz DR et al. Centromere-specific assembly of CENP-a nucleosomes is mediated by HJURP (in eng). Cell. 2009;**137**(3):472-484. DOI: 10.1016/j.cell.2009.02.039

[44] Zasadzinska E, Foltz DR. Orchestrating the specific assembly of centromeric nucleosomes. Progress in Molecular and Subcellular Biology.

2017;**56**:165-192. DOI: 10.1007/978-3-319-58592-5_7

[45] Srivastava S, Foltz DR. Posttranslational modifications of CENP-A: Marks of distinction. Chromosoma. 2018;**127**(3):279-290. DOI: 10.1007/s00412-018-0665-x

[46] Partridge JF. Centromeric chromatin in fission yeast. Frontiers in Bioscience. 2008;**13**:3896-3905. DOI: 10.2741/2977

[47] Creamer KM, Partridge JF. RITS-connecting transcription, RNA interference, and heterochromatin assembly in fission yeast. Wiley Interdisciplinary Reviews: RNA. 2011;**2**(5):632-646. DOI: 10.1002/wrna.80

[48] Pidoux AL et al. Fission yeast Scm3: A CENP-A receptor required for integrity of subkinetochore chromatin. Molecular Cell. 2009;**33**(3):299-311. DOI: 10.1016/j.molcel.2009.01.019

[49] Williams JS, Hayashi T, Yanagida M, Russell P. Fission yeast Scm3 mediates stable assembly of Cnp1/CENP-A into centromeric chromatin. Molecular Cell. 2009;**33**(3):287-298. DOI: 10.1016/j.molcel.2009.01.017

[50] Dunleavy EM et al. A NASP (N1/N2)-related protein, Sim3, binds CENP-A and is required for its deposition at fission yeast centromeres. Molecular Cell. 2007;**28**(6):1029-1044. DOI: 10.1016/j.molcel.2007.10.010

[51] Liu H et al. Structural insights into yeast histone chaperone Hif1: A scaffold protein recruiting protein complexes to core histones. The Biochemical Journal. 2014;**462**(3):465-473. DOI: 10.1042/BJ20131640

[52] Tanae K, Horiuchi T, Yamakawa T, Matsuo Y, Kawamukai M. Sim3 shares some common roles with the histone chaperone Asf1 in fission yeast. FEBS Letters. 2012;**586**(23):4190-4196. DOI: 10.1016/j.febslet.2012.10.020

[53] Tanae K, Horiuchi T, Matsuo Y, Katayama S, Kawamukai M. Histone chaperone Asf1 plays an essential role in maintaining genomic stability in fission yeast. PLoS One. 2012;**7**(1):e30472. DOI: 10.1371/journal.pone.0030472

[54] Fujita Y et al. Priming of centromere for CENP-A recruitment by human hMis18alpha, hMis18beta, and M18BP1. Developmental Cell. 2007;**12**(1):17-30. DOI: 10.1016/j.devcel.2006.11.002

[55] Hayashi T, Fujita Y, Iwasaki O, Adachi Y, Takahashi K, Yanagida M. Mis16 and Mis18 are required for CENP-A loading and histone deacetylation at centromeres. Cell. 2004;**118**(6):715-729. DOI: 10.1016/j.cell.2004.09.002

[56] Loyola A, Almouzni G. Histone chaperones, a supporting role in the limelight. Biochimica et Biophysica Acta. 2004;**1677**(1-3):3-11. DOI: 10.1016/j.bbaexp.2003.09.012

[57] Shang WH et al. Chromosome engineering allows the efficient isolation of vertebrate neocentromeres. Developmental Cell. 2013;**24**(6):635-648. DOI: 10.1016/j.devcel.2013.02.009

[58] Boltengagen M et al. A novel role for the histone acetyltransferase Hat1 in the CENP-A/CID assembly pathway in *Drosophila melanogaster*. Nucleic Acids Research. 2016;**44**(5):2145-2159. DOI: 10.1093/nar/gkv1235

[59] Kim IS et al. Roles of Mis18alpha in epigenetic regulation of centromeric chromatin and CENP-A loading. Molecular Cell. 2012;**46**(3):260-273. DOI: 10.1016/j.molcel.2012.03.021

[60] Hayashi T, Ebe M, Nagao K, Kokubu A, Sajiki K, Yanagida M. *Schizosaccharomyces pombe* centromere protein Mis19 links Mis16 and Mis18 to recruit CENP-A through interacting with NMD factors and the SWI/SNF complex. Genes to Cells. 2014;**19**(7):541-554. DOI: 10.1111/gtc.12152

[61] Subramanian L, Toda NR, Rappsilber J, Allshire RC. Eic1 links Mis18 with the CCAN/Mis6/Ctf19 complex to promote CENP-A assembly. Open Biology. 2014;**4**:140043. DOI: 10.1098/rsob.140043

[62] Subramanian L et al. Centromere localization and function of Mis18 requires Yippee-like domain-mediated oligomerization. EMBO Reports. 2016;**17**(4):496-507. DOI: 10.15252/embr.201541520

[63] Ohkuni K, Abdulle R, Kitagawa K. Degradation of centromeric histone H3 variant Cse4 requires the Fpr3 peptidyl-prolyl Cis-Trans isomerase. Genetics. 2014;**196**(4):1041-1045. DOI: 10.1534/genetics.114.161224

[64] Pearson CG, Yeh E, Gardner M, Odde D, Salmon ED, Bloom K. Stable kinetochore-microtubule attachment constrains centromere positioning in metaphase. Current Biology. 2004;**14**(21):1962-1967. DOI: 10.1016/j.cub.2004.09.086

[65] Takayama Y, Sato H, Saitoh S, Ogiyama Y, Masuda F, Takahashi K. Biphasic incorporation of centromeric histone CENP-A in fission yeast. Molecular Biology of the Cell. 2008;**19**(2):682-690. DOI: 10.1091/mbc.e07-05-0504

[66] Gonzalez M, He H, Sun S, Li C, Li F. Cell cycle-dependent deposition of CENP-A requires the Dos1/2-Cdc20 complex. Proceedings of the National Academy of Sciences of the United States of America. 2013;**110**(2):606-611. DOI: 10.1073/pnas.1214874110

[67] Li F, Martienssen R, Cande WZ. Coordination of DNA replication and histone modification by the Rik1-Dos2 complex. Nature. 2011;**475**(7355): 244-248. DOI: 10.1038/nature10161

[68] Horn PJ, Bastie JN, Peterson CL. A Rik1-associated, cullin-dependent E3 ubiquitin ligase is essential for heterochromatin formation. Genes & Development. 2005;**19**(14):1705-1714. DOI: 10.1101/gad.1328005

[69] Hong EJ, Villen J, Gerace EL, Gygi SP, Moazed D. A cullin E3 ubiquitin ligase complex associates with Rik1 and the Clr4 histone H3-K9 methyltransferase and is required for RNAi-mediated heterochromatin formation. RNA Biology. 2005;**2**(3):106-111. DOI: 10.4161/rna.2.3.2131

[70] Hewawasam G et al. Psh1 is an E3 ubiquitin ligase that targets the centromeric histone variant Cse4. Molecular Cell. 2010;**40**(3):444-454. DOI: 10.1016/j.molcel.2010.10.014

[71] Hewawasam GS, Gerton JL. Cse4 gets a kiss-of-death from Psh1. Cell Cycle. 2011;**10**(4):566-567. DOI: 10.4161/cc.10.4.14770

[72] Castillo AG et al. Plasticity of fission yeast CENP-A chromatin driven by relative levels of histone H3 and H4. PLoS Genetics. 2007;**3**(7):e121. DOI: 10.1371/journal.pgen.0030121

[73] Au WC, Crisp MJ, DeLuca SZ, Rando OJ, Basrai MA. Altered dosage and mislocalization of histone H3 and Cse4p lead to chromosome loss in *Saccharomyces cerevisiae*. Genetics. 2008;**179**(1):263-275. DOI: 10.1534/genetics.108.088518

[74] Sullivan BA. Cell architecture and function/centromeres. In: Encyclopedia of Biological Chemistry II. 2nd ed. Elsevier Inc; 2013. pp. 500-502. DOI: 10.1016/B978-0-12-374984-0.00219-9

[75] Brown JD, O'Neill RJ. The Evolution of Centromeric DNA Sequences. Wiley; 2014. DOI: 10.1002/9780470015902.a0020827.pub2. https://onlinelibrary.wiley.com/doi/10.1002/9780470015902.a0020827.pub2

[76] Jiang J, Birchler JA, Parrott WA, Dawe RK. A molecular view of plant centromeres. Trends in Plant Science. 2003;**8**(12):570-575. DOI: 10.1016/j.tplants.2003.10.011

[77] Arunkumar G, Melters DP. Centromeric transcription: A conserved swiss-army knife. Genes (Basel). 2020;**11**(8):1-22. DOI: 10.3390/genes11080911

[78] Leclerc S, Kitagawa K. The role of human centromeric RNA in chromosome stability. Frontiers in Molecular Biosciences. 2021;**8**:642732. DOI: 10.3389/fmolb.2021.642732

[79] Rosic S, Erhardt S. No longer a nuisance: Long non-coding RNAs join CENP-A in epigenetic centromere regulation. Cellular and Molecular Life Sciences. 2016;**73**(7):1387-1398. DOI: 10.1007/s00018-015-2124-7

[80] Quenet D, Dalal Y. A long non-coding RNA is required for targeting centromeric protein A to the human centromere. eLife. 2014;**3**:e03254. DOI: 10.7554/eLife.03254

[81] Elgin SC. Heterochromatin and gene regulation in Drosophila. Current Opinion in Genetics & Development. 1996;**6**(2):193-202. DOI: 10.1016/s0959-437x(96)80050-5

[82] Elgin SC, Reuter G. Position-effect variegation, heterochromatin formation, and gene silencing in Drosophila. Cold Spring Harbor Perspectives in Biology. 2013;**5**(8):a017780. DOI: 10.1101/cshperspect.a017780

[83] Walther M et al. Heterochromatin formation in Drosophila requires genome-wide histone deacetylation in cleavage chromatin before mid-blastula transition in early embryogenesis. Chromosoma. 2020;**129**(1):83-98. DOI: 10.1007/s00412-020-00732-x

[84] Riddle NC, Elgin SC. A role for RNAi in heterochromatin formation in Drosophila. Current Topics in Microbiology and Immunology. 2008;**320**:185-209. DOI: 10.1007/978-3-540-75157-1_9

[85] Gu T, Elgin SC. Maternal depletion of Piwi, a component of the RNAi system, impacts heterochromatin formation in Drosophila. PLoS Genetics. 2013;**9**(9):e1003780. DOI: 10.1371/journal.pgen.1003780

[86] Klenov MS et al. Impact of nuclear Piwi elimination on chromatin state in *Drosophila melanogaster* ovaries. Nucleic Acids Research. 2014;**42**(10):6208-6218. DOI: 10.1093/nar/gku268

[87] Lin H, Yin H. A novel epigenetic mechanism in Drosophila somatic cells mediated by Piwi and piRNAs. Cold Spring Harbor Symposia on Quantitative Biology. 2008;**73**:273-281. DOI: 10.1101/sqb.2008.73.056

[88] Sentmanat M, Wang SH, Elgin SC. Targeting heterochromatin formation to transposable elements in Drosophila: Potential roles of the piRNA system. Biochemistry (Moscow). 2013;**78**(6):562-571. DOI: 10.1134/S0006297913060023

[89] ElMaghraby MF et al. A heterochromatin-specific RNA export

pathway facilitates piRNA production. Cell. 2019;**178**(4):964-979 e20. DOI: 10.1016/j.cell.2019.07.007

[90] Olszak AM et al. Heterochromatin boundaries are hotspots for de novo kinetochore formation. Nature Cell Biology. 2011;**13**(7):799-808. DOI: 10.1038/ncb2272

[91] Kwenda L, Collins CM, Dattoli AA, Dunleavy EM. Nucleolar activity and CENP-C regulate CENP-A and CAL1 availability for centromere assembly in meiosis. Development. 2016;**143**(8):1400-1412. DOI: 10.1242/dev.130625

[92] Bobkov GOM, Gilbert N, Heun P. Centromere transcription allows CENP-A to transit from chromatin association to stable incorporation. The Journal of Cell Biology. 2018;**217**(6):1957-1972. DOI: 10.1083/jcb.201611087

[93] Bobkov GOM et al. Spt6 is a maintenance factor for centromeric CENP-A. Nature Communications. 2020;**11**(1):2919. DOI: 10.1038/s41467-020-16695-7

[94] Jansen LE, Black BE, Foltz DR, Cleveland DW. Propagation of centromeric chromatin requires exit from mitosis. The Journal of Cell Biology. 2007;**176**(6):795-805. DOI: 10.1083/jcb.200701066

[95] Hemmerich P, Weidtkamp-Peters S, Hoischen C, Schmiedeberg L, Erliandri I, Diekmann S. Dynamics of inner kinetochore assembly and maintenance in living cells (in eng). The Journal of Cell Biology. 2008;**180**(6):1101-1114. DOI: 10.1083/jcb.200710052

[96] Schuh M, Lehner CF, Heidmann S. Incorporation of Drosophila CID/CENP-A and CENP-C into centromeres during early embryonic anaphase. Current Biology. 2007;**17**(3):237-243. DOI: 10.1016/j.cub.2006.11.051

[97] Ahmad K, Henikoff S. Centromeres are specialized replication domains in heterochromatin. The Journal of Cell Biology. 2001;**153**(1):101-110. DOI: 10.1083/jcb.153.1.101

[98] Mellone BG, Grive KJ, Shteyn V, Bowers SR, Oderberg I, Karpen GH. Assembly of Drosophila centromeric chromatin proteins during mitosis. PLoS Genetics. 2011;**7**(5):e1002068. DOI: 10.1371/journal.pgen.1002068

[99] Pauleau AL, Erhardt S. Centromere regulation: New players, new rules, new questions. European Journal of Cell Biology. 2011;**90**(10):805-810. DOI: 10.1016/j.ejcb.2011.04.016

[100] Dunleavy EM, Beier NL, Gorgescu W, Tang J, Costes SV, Karpen GH. The cell cycle timing of centromeric chromatin assembly in Drosophila meiosis is distinct from mitosis yet requires CAL1 and CENP-C. PLoS Biology. 2012;**10**(12):e1001460. DOI: 10.1371/journal.pbio.1001460.

[101] Erhardt S, Mellone BG, Betts CM, Zhang W, Karpen GH, Straight AF. Genome-wide analysis reveals a cell cycle-dependent mechanism controlling centromere propagation. The Journal of Cell Biology. 2008;**183**(5):805-818. DOI: 10.1083/jcb.200806038

[102] Chen CC et al. CAL1 is the Drosophila CENP-A assembly factor. The Journal of Cell Biology. 2014;**204**(3): 313-329. DOI: 10.1083/jcb.201305036

[103] Rosin L, Mellone BG. Co-evolving CENP-A and CAL1 domains mediate centromeric CENP-A deposition across Drosophila species. Developmental Cell. 2016;**37**(2):136-147. DOI: 10.1016/j.devcel.2016.03.021

[104] Moree B, Meyer CB, Fuller CJ, Straight AF. CENP-C recruits M18BP1

to centromeres to promote CENP-A chromatin assembly. The Journal of Cell Biology. 2011;**194**(6):855-871. DOI: 10.1083/jcb.201106079

[105] Pauleau AL, Bergner A, Kajtez J, Erhardt S. The checkpoint protein Zw10 connects CAL1-dependent CENP-A centromeric loading and mitosis duration in Drosophila cells. PLoS Genetics. 2019;**15**(9):e1008380. DOI: 10.1371/journal.pgen.1008380

[106] Demirdizen E et al. Localization of Drosophila CENP-A to non-centromeric sites depends on the NuRD complex. Nucleic Acids Research. 2019;**47**(22):11589-11608. DOI: 10.1093/nar/gkz962

[107] Roure V et al. Reconstituting Drosophila centromere identity in human cells. Cell Reports. 2019;**29**(2):464-479 e5. DOI: 10.1016/j.celrep.2019.08.067

[108] Medina-Pritchard B et al. Structural basis for centromere maintenance by Drosophila CENP-A chaperone CAL1. The EMBO Journal. 2020;**39**(7):e103234. DOI: 10.15252/embj.2019103234

[109] Williams BC, Murphy TD, Goldberg ML, Karpen GH. Neocentromere activity of structurally acentric mini-chromosomes in Drosophila. Nature Genetics. 1998;**18**(1):30-37. DOI: 10.1038/ng0198-30

[110] Wong LH, Saffery R, Choo KH. Construction of neocentromere-based human minichromosomes for gene delivery and centromere studies. Gene Therapy. 2002;**9**(11):724-726. DOI: 10.1038/sj.gt.3301756

[111] Choo KH. Domain organization at the centromere and neocentromere. Developmental Cell. 2001;**1**(2):165-177. DOI: 10.1016/s1534-5807(01)00028-4

[112] Moreno-Moreno O, Torras-Llort M, Azorin F. Proteolysis restricts localization of CID, the centromere-specific histone H3 variant of Drosophila, to centromeres. Nucleic Acids Research. 2006;**34**(21):6247-6255. DOI: 10.1093/nar/gkl902

[113] Heun P, Erhardt S, Blower MD, Weiss S, Skora AD, Karpen GH. Mislocalization of the Drosophila centromere-specific histone CID promotes formation of functional ectopic kinetochores. Developmental Cell. 2006;**10**(3):303-315. DOI: 10.1016/j.devcel.2006.01.014

[114] Palladino J, Chavan A, Sposato A, Mason TD, Mellone BG. Targeted de novo centromere formation in Drosophila reveals plasticity and maintenance potential of CENP-A chromatin. Developmental Cell. 2020;**52**(3):379-394 e7. DOI: 10.1016/j.devcel.2020.01.005

[115] Niikura Y, Kitagawa R, Kitagawa K. CENP-A ubiquitylation Is Inherited through dimerization between cell divisions. Cell Reports. 2016;**15**(1):61-76. DOI: 10.1016/j.celrep.2016.03.010

[116] Niikura Y, Kitagawa R, Fang L, Kitagawa K. CENP-A ubiquitylation is indispensable to cell viability. Developmental Cell. 2019;**50**(6):683-689 e6. DOI: 10.1016/j.devcel.2019.07.015

[117] Niikura Y, Kitagawa R, Kitagawa K. CENP-A ubiquitylation is required for CENP-A deposition at the centromere. Developmental Cell. 2017;**40**(1):7-8. DOI: 10.1016/j.devcel.2016.12.020

[118] Moreno-Moreno O, Medina-Giro S, Torras-Llort M, Azorin F. The F box protein partner of paired regulates stability of Drosophila centromeric histone H3, CenH3(CID). Current Biology. 2011;**21**(17):1488-1493. DOI: 10.1016/j.cub.2011.07.041

[119] Warburton PE et al. Immunolocalization of CENP-A suggests a distinct nucleosome structure at the inner kinetochore plate of active centromeres. Current Biology. 1997;7(11):901-904. DOI: 10.1016/s0960-9822(06)00382-4

[120] Fujita R et al. Stable complex formation of CENP-B with the CENP-A nucleosome. Nucleic Acids Research. 2015;**43**(10):4909-4922. DOI: 10.1093/nar/gkv405

[121] Jing R et al. Motifs in the amino-terminus of CENP-A are required for its accumulation within the nucleus and at the centromere. Oncotarget. 2017;**8**(25):40654-40667. DOI: 10.18632/oncotarget.17204

[122] Shelby RD, Vafa O, Sullivan KF. Assembly of CENP-A into centromeric chromatin requires a cooperative array of nucleosomal DNA contact sites. The Journal of Cell Biology. 1997;**136**(3): 501-513. DOI: 10.1083/jcb.136.3.501

[123] Shelby RD, Monier K, Sullivan KF. Chromatin assembly at kinetochores is uncoupled from DNA replication. The Journal of Cell Biology. 2000;**151**(5): 1113-1118. DOI: 10.1083/jcb.151.5.1113

[124] Bassett EA et al. HJURP uses distinct CENP-A surfaces to recognize and to stabilize CENP-A/histone H4 for centromere assembly. Developmental Cell. 2012;**22**(4):749-762. DOI: 10.1016/j.devcel.2012.02.001

[125] Hu H et al. Structure of a CENP-A-histone H4 heterodimer in complex with chaperone HJURP. Genes & Development. 2011;**25**(9):901-906. DOI: 10.1101/gad.2045111

[126] Yoda K, Tomonaga T. Centromere identity originates in the structure of CENP-A/H4 tetramer itself: A mechanism for aneuploidy. Lancet. 2004;**364**(9439):1022-1024. DOI: 10.1016/S0140-6736(04)17077-3

[127] Zhao H, Winogradoff D, Bui M, Dalal Y, Papoian GA. Promiscuous histone mis-assembly is actively prevented by chaperones. Journal of the American Chemical Society. 2016;**138**(40):13207-13218. DOI: 10.1021/jacs.6b05355

[128] Zhao H, Winogradoff D, Dalal Y, Papoian GA. The oligomerization landscape of histones. Biophysical Journal. 2019;**116**(10):1845-1855. DOI: 10.1016/j.bpj.2019.03.021

[129] Shang WH et al. Acetylation of histone H4 lysine 5 and 12 is required for CENP-A deposition into centromeres. Nature Communications. 2016;7:13465. DOI: 10.1038/ncomms13465

[130] Pan D, Walstein K, Take A, Bier D, Kaiser N, Musacchio A. Mechanism of centromere recruitment of the CENP-A chaperone HJURP and its implications for centromere licensing. Nature Communications. 2019;**10**(1):4046. DOI: 10.1038/s41467-019-12019-6

[131] Shuaib M, Ouararhni K, Dimitrov S, Hamiche A. HJURP binds CENP-A via a highly conserved N-terminal domain and mediates its deposition at centromeres. Proceedings of the National Academy of Sciences of the United States of America. 2010;**107**(4):1349-1354. DOI: 10.1073/pnas.0913709107

[132] Bodor DL, Valente LP, Mata JF, Black BE, Jansen LE. Assembly in G1 phase and long-term stability are unique intrinsic features of CENP-A nucleosomes. Molecular Biology of the Cell. 2013;**24**(7):923-932. DOI: 10.1091/mbc.E13-01-0034

[133] Maddox PS, Hyndman F, Monen J, Oegema K, Desai A. Functional genomics

identifies a Myb domain-containing protein family required for assembly of CENP-A chromatin. The Journal of Cell Biology. 2007;**176**(6):757-763. DOI: 10.1083/jcb.200701065

[134] McKinley KL, Cheeseman IM. Polo-like kinase 1 licenses CENP-A deposition at centromeres. Cell. 2014;**158**(2):397-411. DOI: 10.1016/j.cell.2014.06.016

[135] Perpelescu M, Nozaki N, Obuse C, Yang H, Yoda K. Active establishment of centromeric CENP-A chromatin by RSF complex. The Journal of Cell Biology. 2009;**185**(3):397-407. DOI: 10.1083/jcb.200903088

[136] Lagana A, Dorn JF, De Rop V, Ladouceur AM, Maddox AS, Maddox PS. A small GTPase molecular switch regulates epigenetic centromere maintenance by stabilizing newly incorporated CENP-A. Nature Cell Biology. 2010;**12**(12):1186-1193. DOI: 10.1038/ncb2129

[137] Prendergast L, Sullivan KF. A GTPase switch maintains CENP-A at centromeric chromatin. Nature Cell Biology. 2010;**12**(12):1128-1130. DOI: 10.1038/ncb1210-1128

[138] Mitra S, Srinivasan B, Jansen LET. Stable inheritance of CENP-A chromatin: Inner strength versus dynamic control. The Journal of Cell Biology. 2020;**219**(10). DOI: 10.1083/jcb.202005099

[139] Conde e Silva N, Black BE, Sivolob A, Filipski J, Cleveland DW, Prunell A. CENP-A-containing nucleosomes: Easier disassembly versus exclusive centromeric localization. Journal of Molecular Biology. 2007;**370**(3):555-573. DOI: 10.1016/j.jmb.2007.04.064

[140] Obuse C, Yang H, Nozaki N, Goto S, Okazaki T, Yoda K. Proteomics analysis of the centromere complex from HeLa interphase cells: UV-damaged DNA binding protein 1 (DDB-1) is a component of the CEN-complex, while BMI-1 is transiently co-localized with the centromeric region in interphase. Genes to Cells. 2004;**9**(2):105-120. DOI: 10.1111/j.1365-2443.2004.00705.x

[141] Lee J, Zhou P. DCAFs, the missing link of the CUL4-DDB1 ubiquitin ligase. Molecular Cell. 2007;**26**(6):775-780. DOI: 10.1016/j.molcel.2007.06.001

[142] Mouysset J et al. CRL4(RBBP7) is required for efficient CENP-A deposition at centromeres. Journal of Cell Science. 2015;**128**(9):1732-1745. DOI: 10.1242/jcs.162305

[143] Nye J, Sturgill D, Athwal R, Dalal Y. HJURP antagonizes CENP-A mislocalization driven by the H3.3 chaperones HIRA and DAXX. PLoS One. 2018;**13**(10):e0205948. DOI: 10.1371/journal.pone.0205948

[144] Shrestha RL et al. Mislocalization of centromeric histone H3 variant CENP-A contributes to chromosomal instability (CIN) in human cells. Oncotarget. 2017;**8**(29):46781-46800. DOI: 10.18632/oncotarget.18108

[145] Sharma AB, Dimitrov S, Hamiche A, Van Dyck E. Centromeric and ectopic assembly of CENP-A chromatin in health and cancer: Old marks and new tracks. Nucleic Acids Research. 2019;**47**(3):1051-1069. DOI: 10.1093/nar/gky1298

[146] Tomonaga T et al. Overexpression and mistargeting of centromere protein-A in human primary colorectal cancer. Cancer Research. 2003;**63**(13):3511-3516. Available from: https://www.ncbi.nlm.nih.gov/pubmed/12839935

[147] Valdivia MM, Hamdouch K, Ortiz M, Astola A. CENPA a genomic

marker for centromere activity and human diseases. Current Genomics. 2009;**10**(5):326-335. DOI: 10.2174/138920209788920985

[148] Li Y et al. ShRNA-targeted centromere protein A inhibits hepatocellular carcinoma growth. PLoS One. 2011;**6**(3):e17794. DOI: 10.1371/journal.pone.0017794

[149] Mahlke MA, Nechemia-Arbely Y. Guarding the genome: CENP-A-chromatin in health and cancer. Genes (Basel). 2020;**11**(7). DOI: 10.3390/genes11070810

[150] Burgess RJ, Zhang Z. Histone chaperones in nucleosome assembly and human disease. Nature Structural & Molecular Biology. 2013;**20**(1):14-22. DOI: 10.1038/nsmb.2461

[151] Huang X, Liu J, Ma Q. Prohibitin participates in the HIRA complex to promote cell metastasis in breast cancer cell lines. FEBS Open Bio. 2020;**10**(10):2182-2190. DOI: 10.1002/2211-5463.12966

[152] Mahmud I, Liao D. DAXX in cancer: Phenomena, processes, mechanisms and regulation. Nucleic Acids Research. 2019;**47**(15):7734-7752. DOI: 10.1093/nar/gkz634

[153] Ricketts MD, Marmorstein R. A molecular prospective for HIRA complex assembly and H3.3-specific histone chaperone function. Journal of Molecular Biology. 2017;**429**(13):1924-1933. DOI: 10.1016/j.jmb.2016.11.010

[154] Wang K, Yu Z, Liu Y, Li G. Ser68 phosphorylation ensures accurate cell-cycle-dependent CENP-A deposition at centromeres. Developmental Cell. 2017;**40**(1):5-6. DOI: 10.1016/j.devcel.2016.12.015

[155] Yu Z et al. Dynamic phosphorylation of CENP-A at Ser68 orchestrates its cell-cycle-dependent deposition at centromeres. Developmental Cell. 2015;**32**(1):68-81. DOI: 10.1016/j.devcel.2014.11.030

[156] Fachinetti D, Logsdon GA, Abdullah A, Selzer EB, Cleveland DW, Black BE. CENP-A modifications on Ser68 and Lys124 are dispensable for establishment, maintenance, and long-term function of human centromeres. Developmental Cell. 2017;**40**(1):104-113. DOI: 10.1016/j.devcel.2016.12.014

[157] Takada M et al. FBW7 loss promotes chromosomal instability and tumorigenesis via cyclin E1/CDK2-mediated phosphorylation of CENP-A. Cancer Research. 2017;**77**(18):4881-4893. DOI: 10.1158/0008-5472.CAN-17-1240

[158] Gassmann R et al. An inverse relationship to germline transcription defines centromeric chromatin in *C. elegans*. Nature. 2012;**484**(7395):534-537. DOI: 10.1038/nature10973

[159] Raychaudhuri N, Dubruille R, Orsi GA, Bagheri HC, Loppin B, Lehner CF. Transgenerational propagation and quantitative maintenance of paternal centromeres depends on Cid/Cenp-A presence in Drosophila sperm. PLoS Biology. 2012;**10**(12):e1001434. DOI: 10.1371/journal.pbio.1001434

[160] Fukagawa T, Earnshaw WC. The centromere: Chromatin foundation for the kinetochore machinery. Developmental Cell. 2014;**30**(5):496-508. DOI: 10.1016/j.devcel.2014.08.016

[161] Marshall OJ, Chueh AC, Wong LH, Choo KH. Neocentromeres: New insights into centromere structure, disease development, and karyotype evolution. American Journal of Human Genetics.

2008;**82**(2):261-282. DOI: 10.1016/j.ajhg.2007.11.009

[162] Van Hooser AA et al. Specification of kinetochore-forming chromatin by the histone H3 variant CENP-A. Journal of Cell Science. 2001;**114**(Pt 19):3529-3542. Available from: https://www.ncbi.nlm.nih.gov/pubmed/11682612

[163] Canzonetta C, Leo M, Guarino SR, Montanari A, Francisci S, Filetici P. SAGA complex and Gcn5 are necessary for respiration in budding yeast. Biochimica et Biophysica Acta. 2016;**1863**(12):3160-3168. DOI: 10.1016/j.bbamcr.2016.10.002

[164] Niikura Y, Kitagawa R, Ogi H, Kitagawa K. SGT1-HSP90 complex is required for CENP-A deposition at centromeres. Cell Cycle. 2017;**16**(18):1683-1694. DOI: 10.1080/15384101.2017.1325039

[165] Niikura Y, Kitagawa K. Functions of SGT1, a Co-chaperone. In: Kaur P, Asea AAA, editors. Heat Shock Protein 90 in Human Diseases and Disorders (Asea AA, Calderwood SK, editors. Heat Shock Proteins). Vol. 19. Switzerland AG: Springer Nature; 2019. ch. 16. pp. 317-370

[166] Kitagawa K, Skowyra D, Elledge SJ, Harper JW, Hieter P. SGT1 encodes an essential component of the yeast kinetochore assembly pathway and a novel subunit of the SCF ubiquitin ligase complex. Molecular Cell. 1999;**4**(1):21-33. DOI: 10.1016/s1097-2765(00)80184-7

[167] Niikura Y, Ohta S, Vandenbeldt KJ, Abdulle R, McEwen BF, Kitagawa K. 17-AAG, an Hsp90 inhibitor, causes kinetochore defects: A novel mechanism by which 17-AAG inhibits cell proliferation. Oncogene. 2006;**25**(30):4133-4146. DOI: 10.1038/sj.onc.1209461

[168] Bansal PK, Abdulle R, Kitagawa K. Sgt1 associates with Hsp90: An initial step of assembly of the core kinetochore complex. Molecular and Cellular Biology. 2004;**24**(18):8069-8079. DOI: 10.1128/MCB.24.18.8069-8079.2004

[169] Lingelbach LB, Kaplan KB. The interaction between Sgt1p and Skp1p is regulated by HSP90 chaperones and is required for proper CBF3 assembly. Molecular and Cellular Biology. 2004;**24**(20):8938-8950. DOI: 10.1128/MCB.24.20.8938-8950.2004

[170] Trepel J, Mollapour M, Giaccone G, Neckers L. Targeting the dynamic HSP90 complex in cancer. Nature Reviews. Cancer. 2010;**10**(8):537-549. DOI: 10.1038/nrc2887

[171] Ogi H et al. The oncogenic role of the cochaperone Sgt1. Oncogene. 2015;**4**:e149. DOI: 10.1038/oncsis.2015.12

[172] Dias DC, Dolios G, Wang R, Pan ZQ. CUL7: A DOC domain-containing cullin selectively binds Skp1.Fbx29 to form an SCF-like complex. Proceedings of the National Academy of Sciences of the United States of America. 2002;**99**(26):16601-16606. DOI: 10.1073/pnas.252646399

[173] Davies AE, Kaplan KB. Hsp90-Sgt1 and Skp1 target human Mis12 complexes to ensure efficient formation of kinetochore-microtubule binding sites. The Journal of Cell Biology. 2010;**189**(2):261-274. DOI: 10.1083/jcb.200910036

[174] Steensgaard P et al. Sgt1 is required for human kinetochore assembly. EMBO Reports. 2004;**5**(6):626-631. DOI: 10.1038/sj.embor.7400154

[175] Ohkuni K et al. Deposition of centromeric histone H3 variant CENP-A/Cse4 into chromatin is facilitated by

its C-terminal sumoylation. Genetics. 2020;**214**(4):839-854. DOI: 10.1534/genetics.120.303090

[176] Mukhopadhyay D, Arnaoutov A, Dasso M. The SUMO protease SENP6 is essential for inner kinetochore assembly. The Journal of Cell Biology. 2010;**188**(5):681-692. DOI: 10.1083/jcb.200909008

[177] Jentsch S, Psakhye I. Control of nuclear activities by substrate-selective and protein-group SUMOylation. Annual Review of Genetics. 2013;**47**: 167-186. DOI: 10.1146/annurev-genet-111212-133453

[178] Liebelt F et al. The poly-SUMO2/3 protease SENP6 enables assembly of the constitutive centromere-associated network by group deSUMOylation. Nature Communications. 2019;**10**(1): 3987. DOI: 10.1038/s41467-019-11773-x

[179] Mitra S et al. Genetic screening identifies a SUMO protease dynamically maintaining centromeric chromatin. Nature Communications. 2020;**11**(1):501. DOI: 10.1038/s41467-019-14276-x

[180] Kiyotaka N, Jason W, Cory H, Jiming J, Minoru M. Structure and evolution of plant centromeres. In: Ugarkovic D, editor. Centromere (Progress in Molecular and Subcellular Biology). Springer-Verlag Berlin Heidelbarg; 2009. pp. 153-179 ch. 6. DOI: 10.1007/978-3-642-00182-6_6

[181] Nagaki K et al. Molecular and cytological analyses of large tracks of centromeric DNA reveal the structure and evolutionary dynamics of maize centromeres. Genetics. 2003;**163**(2):759-770. DOI: 10.1093/genetics/163.2.759

[182] Du Y, Topp CN, Dawe RK. DNA binding of centromere protein C (CENPC) is stabilized by single-stranded RNA. PLoS Genetics. 2010;**6**(2):e1000835. DOI: 10.1371/journal.pgen.1000835

[183] Comai L, Maheshwari S, Marimuthu MPA. Plant centromeres. Current Opinion in Plant Biology. 2017;**36**:158-167. DOI: 10.1016/j.pbi.2017.03.003

[184] Oliveira LC, Torres GA. Plant centromeres: Genetics, epigenetics and evolution. Molecular Biology Reports. 2018;**45**(5):1491-1497. DOI: 10.1007/s11033-018-4284-7

[185] McClintock B. The significance of responses of the genome to challenge. Science. 1984;**226**(4676):792-801. DOI: 10.1126/science.15739260

[186] Zeitlin SG et al. Double-strand DNA breaks recruit the centromeric histone CENP-A. Proceedings of the National Academy of Sciences of the United States of America. 2009;**106**(37):15762-15767. DOI: 10.1073/pnas.0908233106

[187] Cooper JL, Henikoff S. Adaptive evolution of the histone fold domain in centromeric histones. Molecular Biology and Evolution. 2004;**21**(9):1712-1718. DOI: 10.1093/molbev/msh179

[188] Talbert PB, Bryson TD, Henikoff S. Adaptive evolution of centromere proteins in plants and animals. Journal of Biology. 2004;**3**(4):18. DOI: 10.1186/jbiol11

[189] Malik HS, Henikoff S. Adaptive evolution of Cid, a centromere-specific histone in Drosophila. Genetics. 2001;**157**(3):1293-1298. DOI: 10.1093/genetics/157.3.1293

[190] Talbert PB, Masuelli R, Tyagi AP, Comai L, Henikoff S. Centromeric localization and adaptive evolution of an Arabidopsis histone H3 variant. Plant Cell. 2002;**14**(5):1053-1066. DOI: 10.1105/tpc.010425

[191] Sullivan BA, Schwartz S. Identification of centromeric antigens in dicentric Robertsonian translocations: CENP-C and CENP-E are necessary components of functional centromeres. Human Molecular Genetics. 1995;**4**(12): 2189-2197. DOI: 10.1093/hmg/4.12.2189

[192] Murillo-Pineda M, Jansen LET. Genetics, epigenetics and back again: Lessons learned from neocentromeres. Experimental Cell Research. 2020;**389**(2):111909. DOI: 10.1016/j.yexcr.2020.111909

[193] Wang K, Wu Y, Zhang W, Dawe RK, Jiang J. Maize centromeres expand and adopt a uniform size in the genetic background of oat. Genome Research. 2014;**24**(1):107-116. DOI: 10.1101/gr.160887.113

[194] Henikoff S, Ahmad K, Malik HS. The centromere paradox: Stable inheritance with rapidly evolving DNA. Science. 2001;**293**(5532):1098-1102. DOI: 10.1126/science.1062939

[195] Hasson D et al. Formation of novel CENP-A domains on tandem repetitive DNA and across chromosome breakpoints on human chromosome 8q21 neocentromeres. Chromosoma. 2011;**120**(6):621-632. DOI: 10.1007/s00412-011-0337-6

[196] Gong Z et al. Repeatless and repeat-based centromeres in potato: Implications for centromere evolution. Plant Cell. 2012;**24**(9):3559-3574. DOI: 10.1105/tpc.112.100511

[197] Harrington JJ, Van Bokkelen G, Mays RW, Gustashaw K, Willard HF. Formation of de novo centromeres and construction of first-generation human artificial microchromosomes. Nature Genetics. 1997;**15**(4):345-355. DOI: 10.1038/ng0497-345

[198] Ananiev EV et al. Artificial chromosome formation in maize (*Zea mays* L.). Chromosoma. 2009;**118**(2):157-177. DOI: 10.1007/s00412-008-0191-3

[199] Phan BH et al. Transformation of rice with long DNA-segments consisting of random genomic DNA or centromere-specific DNA. Transgenic Research. 2007;**16**(3):341-351. DOI: 10.1007/s11248-006-9041-3

[200] Lermontova I, Schubert V, Fuchs J, Klatte S, Macas J, Schubert I. Loading of Arabidopsis centromeric histone CENH3 occurs mainly during G2 and requires the presence of the histone fold domain. Plant Cell. 2006;**18**(10):2443-2451. DOI: 10.1105/tpc.106.043174

[201] Sorge E, Demidov D, Lermontova I, Houben A, Conrad U. Engineered degradation of EYFP-tagged CENH3 via the 26S proteasome pathway in plants. PLoS One. 2021;**16**(2):e0247015. DOI: 10.1371/journal.pone.0247015

[202] Hsu PD et al. DNA targeting specificity of RNA-guided Cas9 nucleases. Nature Biotechnology. 2013;**31**(9): 827-832. DOI: 10.1038/nbt.2647

[203] Biggins S. The composition, functions, and regulation of the budding yeast kinetochore. Genetics. 2013;**194**(4):817-846. DOI: 10.1534/genetics.112.145276

[204] Earnshaw WC, Rothfield N. Identification of a family of human centromere proteins using autoimmune sera from patients with scleroderma. Chromosoma. 1985;**91**(3-4):313-321. DOI: 10.1007/BF00328227

[205] Okada M et al. The CENP-H-I complex is required for the efficient incorporation of newly synthesized CENP-A into centromeres. Nature Cell

Biology. 2006;**8**(5):446-457. DOI: 10.1038/ncb1396

[206] Beh TT, Kalitsis P. Centromeres in health, disease and evolution. In: Felekkis K, Voskarides K, editors. Genomic Elements in Health, Disease and Evolution. New York: Springer Science+Business Media; 2015. pp. 221-244 ch. 9

[207] Burger A, Amemiya Y, Kitching R, Seth AK. Novel RING E3 ubiquitin ligases in breast cancer. Neoplasia. 2006;**8**(8):689-695. DOI: 10.1593/neo.06469

[208] Hu X et al. Cul4 E3 ubiquitin ligase regulates ovarian cancer drug resistance by targeting the antiapoptotic protein BIRC3. Cell Death & Disease. 2019;**10**(2):104. DOI: 10.1038/s41419-018-1200-y

[209] Quail DF, Joyce JA. Microenvironmental regulation of tumor progression and metastasis. Nature Medicine. 2013;**19**(11):1423-1437. DOI: 10.1038/nm.3394